SEX, GENES AND CHROMOSOMES

Biological sex is central to our continued existence. Yet the mechanism that determines sex, rather than being rational and conserved, is a battlefield of warring genes, sexual antagonism, unbalanced expression and self-destructing sex chromosomes that come in bewildering varieties. Drawing on 50 years of research in three fields in which the author has played key roles, this unique book explores the organization, function, activity and evolution of sex genes and chromosomes in humans and other animals. Providing eye-witness stories of early discoveries and modern genomic insights, it examines how genes and chromosomes determine sex, delving into key breakthroughs, theories on gene and chromosome function and the enormous scope for variation in body traits and behaviour. Reflecting the huge advances in genetics and genomics that help explain the wonders of human existence and evolution, this book is a fascinating resource for anyone interested in the biology of sex – particularly students and researchers in reproduction, genetics and evolution.

JENNIFER A. MARSHALL GRAVES is Distinguished Professor and Vice Chancellor's Fellow at La Trobe University in Melbourne, Australia. Jenny has spent her working life comparing weird animal genomes to discover how sex works and how it evolved. She was a pioneer of comparative genomics and of epigenetics and was involved in discovering genes important for sex in humans as well as other mammals and reptiles. She has written hundreds of science articles and won many awards, including Australia's top science prize, the Prime Minister's Prize for Science (2017), and the L'Oreal-UNESCO International Prize for Women in Science (2006). Jenny is a Fellow of the Australian Academy of Science and an International Member of the US National Academy of Science. She writes popular science articles for a combined readership of 6 million and recently wrote and performed in an oratorio that celebrates the beautiful science illuminating the *Origins of the Universe, of Life, of Species, of Humanity.*

Sex, Genes and Chromosomes by Jenny Graves is the culmination of a lifelong interest and curiosity in a very special area of developmental biology, the programming of sexual fate. How do organisms become male or female – and how surprising is the diversity of mechanisms by which this decision is made and enacted in different organisms? Jenny is a pioneer in this field and a gifted communicator. She unpacks the dynamic interplay of sex chromosomes, genes, regulatory networks, epigenetic influences and a role for environment. The book is intellectually rigorous but enlivened by a strong personal element. It spans theory, landmark discoveries and the passage to the modern genomic era. A must-have for readers interested in sex determination and its environmental reprogramming – and an outstanding resource for anyone with an interest in developmental biology more generally.

—Arthur Georges, University of Canberra

Sex, Genes and Chromosomes is the master work of Distinguished Professor Jennifer Graves, whose unique perspective stems from her decades of paradigm-shifting research on sex chromosome evolution and function. In it, Professor Graves makes available a vast knowledge to a broad audience. The book is a landmark in the history of the study of sex chromosomes and will be a starting point for coming generations of investigators interested in the genetic basis of sex and sex differences.

—Arthur Arnold, University of California, Los Angeles

Sex might seem straightforward to all of us; however in reality, it is complex and highly nuanced. Graves explores sex in all its guises in her comparative sweep across vertebrates, bringing up-to-date genetic insights and an omnipresent evolutionary perspective. The work is highly accessible, with compelling prose, deft use of asides and aphorisms. This is an important book at a time when simplistic, non-scientific pronouncements on sex abound. It is a valuable 'must read' text for university students, lecturers and research academics interested in the field.

—Andrew Sinclair, Murdoch Children's Research Institute and
University of Melbourne

Sex, Genes and Chromosomes offers a comprehensive journey through vertebrate genomes, blending the organization, evolution and activity of sex chromosomes with the molecular mechanisms that govern sex determination. It integrates historical discoveries with modern genome biology and evolutionary perspectives, providing a clear foundation for understanding sex chromosome biology. Clear and wide-ranging, *Sex, Genes and Chromosomes* will appeal not just to students and researchers but also the broader general public who seek a deep understanding of sex, genes and chromosome biology.

—Paul D. Waters, University of New South Wales

SEX, GENES AND CHROMOSOMES

Their Discovery, Function and Evolution

JENNIFER A. MARSHALL GRAVES

La Trobe University

CAMBRIDGE
UNIVERSITY PRESS

Shaftesbury Road, Cambridge CB2 8EA, United Kingdom

One Liberty Plaza, 20th Floor, New York, NY 10006, USA

477 Williamstown Road, Port Melbourne, VIC 3207, Australia

314–321, 3rd Floor, Plot 3, Splendor Forum, Jasola District Centre,
New Delhi – 110025, India

Cambridge University Press is part of Cambridge University Press & Assessment,
a department of the University of Cambridge.

We share the University's mission to contribute to society through the pursuit of
education, learning and research at the highest international levels of excellence.

www.cambridge.org
Information on this title: www.cambridge.org/9781009688000
DOI: 10.1017/9781009688024

First published 2026

A catalogue record for this publication is available from the British Library

A Cataloging-in-Publication data record for this book is available from the Library of Congress

ISBN 978-1-009-68801-7 Hardback
ISBN 978-1-009-68800-0 Paperback

Contents

Part II Activity of Sex Chromosomes

Part III Sex Determination

Figure Credits

Preface

This book was born in 1993. It has come a long way in the 32 years since then.

I began writing in hospital, after a near-fatal brain bleed. It saved my sanity during the first months of a long convalescence. I couldn't walk and I couldn't see. But I could think and I could touch type. And I was fascinated by sex chromosomes, having done research in all three aspects: their organization and evolution, their unique regulation and their ability to determine sex.

In 1993 it was going to be quite a thin little book, because we didn't know much. Classic microscopic study of sex chromosomes could now be enhanced by gene mapping, but it was a tedious business. Genes could be identified and little bits cloned and sequenced. The male-determining gene on the human Y had just been identified (by a group led by my PhD graduate Andrew Sinclair), and we all assumed that the pathways it controlled would be easily unravelled. X chromosome inactivation that balances gene dosage between men and women was beginning to reveal its epigenetic secrets.

Unfortunately (for the book) I recovered before I finished it, and had to put it aside for years while I restarted my lab, then chaired the department. By the time I picked it up again, my simple-minded faith in simple answers was shaken as I saw new depths in how sex and sex chromosomes worked and how they evolved. Everything was much, much more intricate and complex – and more interesting – than we had expected; sex chromosome structures and sequences were bizarre and their evolution quixotic, sex was not a direct pathway but a pushmi-pullyu network and X inactivation involved layers of failsafe mechanisms. I would have to start again.

Over the next few years I read and rewrote and read and rewrote. It felt like painting the Sydney Harbour Bridge; as soon as I finished one version, I had to start again on the next. I was drowning in new sequence data, new functional data and new epigenetic data. In despair I asked colleagues who were world experts to co-author chapters: Christine Disteche on the latest greatest thing in X chromosome inactivation and Andrew on the latest greatest discoveries on the biochemical pathways of sex determination. These collaborations produced detailed chapters and inspired me to match their degree of detail in the other seven. More frantic painting of the Sydney Harbour Bridge ensued.

The result was a weighty tome with some thousands of references. With a sinking heart I realized that this fat book would be useful for standing on to get things off the top shelf, but nobody would want to read it. Time for another rethink and rewrite. Over the next year, I worked on translating the text into plain language and thinning the trees to see the forest, identifying the major themes and making connections between them. *Sex, Genes and Chromosomes* is the result.

From the beginning I recognized that I was one of a few with an intimate and hands-on knowledge of all three aspects of sex and sex chromosomes that most fascinated me: the structure of these peculiar chromosomes and how they evolved to be so weird, the regulation of their activity to compensate for gene dosage inequity between the sexes, and the action of genes to determine sex.

The book fell naturally into these three parts: Part I, Organization and evolution of sex chromosome; Part II, Activity of sex chromosomes; and Part III, Sex determination. And each part fell naturally into three chapters: background, function and evolution. The first would be telling the story of the basic discoveries and the discoverers, the second would be a slimmed-down account of our knowledge of the system and how we gained it and the third would summarize what we know or surmise about how it evolved. As the book evolved, so did the chapters; the history became longer as time went on, and discussion of the 'latest-greatest' discoveries became simpler and more focused on themes. The third grew with my growing recognition that evolution provides the only sensible explanations of the weirdness of sex and sex chromosomes.

With the widening of the scope and the rewriting for a broader readership (along with my increasing interest in talking and writing for a non-scientific audience), I added an introductory chapter that explains the basic biological processes behind reproduction, the principles behind the tools that geneticists and molecular biologists use and the background of animal phylogeny that you need to interpret comparisons between species. Readers with little (or old) scientific background should find this simple guide useful; biology students and professionals might skip it or pounce on one or a few sections that describe approaches or techniques they have never quite understood, such as meiosis or Southern blotting.

This book is about biological sex. I had not initially intended to discuss gender or mate choice; it wasn't a thing in 1993 and it is not my area of expertise. I never expected that sex and gender could ever be a controversial topic – but it is. So, hoping that a genetic and evolutionary perspective might be useful, I have included a final chapter addressing issues in which biological factors interact with social factors to produce the variation we now recognize.

I have many people to thank for making it possible to complete this project. First, my patient editors (four of them over the years) at Cambridge University Press. Then my colleagues and collaborators, especially Christine Disteche and Andrew Sinclair (to whom I also offer profound apologies for wasting their time and profound gratitude for their insightful reviews born of their chapters, to which I point readers), and my many students and postdocs over the years who shared (and continue to share) my fascination with sex and sex chromosomes. And last, my family, particularly my husband John, who kept me sane (and alive) when I was sick, and then put up with my 30-year preoccupation with sex and sex chromosomes.

One thing sacrificed by the prolonged gestation of this book was the honour that my hero, Dr Susumu Ohno, bestowed on it by agreeing to write this preface. Indeed, he gave me an outline before he died in 2000. Ohno was a giant in classic cytology and contributed many important insights to the development of molecular evolution. He is famous for his bold thinking about sex chromosomes, dosage compensation and sex determination. His inspirational book *Sex Chromosomes and Sex Determination* was my bible as a student, and consciously or unconsciously, I have mirrored his organization in this book. Many of Ohno's ideas turned out to be wrong, or limited, but it didn't matter – we are still testing them 50 years later.

Indeed, it was Susumu Ohno who showed me that it's alright to be wrong; I happened to be visiting my collaborator Art Riggs at the City of Hope Medical Center in Los Angeles, just before our gene mapping of the marsupial X chromosome was to be published. It contradicted 'Ohno's Law' that the mammalian X must be absolutely conserved because of selection for chromosome-wide dosage compensation. I nervously knocked on his door to inform him of our devastating conclusion. 'So?' he responded with a shrug as he invited me to observe what he was working on – using DNA sequences to compose music. Laws are made to be broken, we agreed; that's how science progresses.

1

Introduction to Vertebrate Genomes and Sex

Sex chromosomes are nothing but trouble. The human X and Y don't pair very well, leading to infertility. The single representation of X-borne genes in males leads to some nasty sex-linked diseases in boys. The unequal representation of the X in XX females and XY males requires complex work-arounds to equalize the dosage of genes between the sexes. And the sex-determining gene on the Y is at risk of being moved around or inactivated, leading to atypical sexual development. You have to ask whether sex chromosomes represent an optimal strategy for determining sex or whether they are a terrible evolutionary accident.

Sex chromosomes exist as a consequence, as well as an enactor, of sex determination. Sexual reproduction is almost universal among eukaryotes. Gametes are produced in the gonad (eggs in the ovary of a female, sperm in the testis of a male) and then fuse to form an individual with unique combinations of parental traits.

In mammals, birds, snakes and most other vertebrates, most insects and a few plants, the sex of an individual is determined by a gene on specialized sex chromosomes which trigger either the male or the female developmental pathway.

For instance, in humans and other mammals, females have two copies of a large chromosome called X, whereas males have a single X and a tiny Y that contains the sex-determining gene. At meiosis, the reduction division that produces gametes with a single genome, the sex chromosomes are distributed up into gametes. The XX female makes eggs, all of which carry a single X. The XY male, however, makes sperm, half of which carry the X and half the Y. Fusion of an X egg with an X sperm produces a girl baby; fusion with a Y-bearing sperm produces a boy baby.

Differentiated sex chromosomes are widespread among vertebrates, but they are not universal. Mammals and birds all have strongly differentiated sex chromosomes, but frog sex chromosomes are rather subtle. Strongly differentiated sex chromosoes are common in reptiles and fish, but there are a variety of alternatives.

The pair of specialized sex chromosomes may be very different in size and gene content, as for the human X and Y chromosomes. Or they may differ only in the single gene that specifies male or female development, as in many frogs.

Some animals don't have sex chromosomes at all. For instance, alligators develop as males or females in response to the temperature at which the egg is incubated, and some fish respond to social stimuli.

Sex chromosomes are peculiarly fascinating. First, and most obviously, they are interesting because they bear the genes that determine sex. But also because they possess unique features imposed upon them by their possession of this magic bullet.

To understand the organization, function and evolution of sex chromosomes requires some fundamental knowledge of biology and genetics. In this chapter, I will skim lightly across the critical concepts and some of the methods we use to test them for readers with no background in biology.

1.1 Genes and Genomes

We've known for decades that genes are made of deoxyribose nucleic acid (DNA), and we know, more or less, how they work. Watson and Crick's model of a double-stranded helix, two sugar-phosphate chains held together by pairs of purine and pyrimidine bases, brilliantly offered explanations for the two most profound properties of life. The order of the four bases, adenine (A), thymine (T), guanine (G) and cytosine (C), on the strand provided the capacity to carry huge amounts of genetic information, encoded in the order of bases. The complementary base pairing A-T and G-C gave it the ability to make exact copies of this sequence to multiply cells in a body, or bodies in the world.

The information encoded in DNA is copied ('transcribed') into a single-stranded nucleic acid (ribonucleic acid [RNA]) and shipped out into the cell as messenger RNA. Here the code is 'translated', triplet by triplet of bases, into a string of amino acids that make up proteins that build components of bodies and the enzymes that make other molecules. This is the basis of all life on earth.

The number of human genes was guessed to be about 100,000; however, molecular analysis gives us a total of only about 20,000 protein-coding genes, not even twice as many as the humble fruit fly.

The genome is the name we give to all the genes, and other DNA sequences, that specify an organism. In humans, the genome constitutes 3,234,830,000 base pairs (about 3.2 gigabases, Gb), amounting to about a metre of DNA. This size is rather uniform in mammals.

Birds and reptiles have about one-third the DNA (but pretty much the same number of genes). Some plants and some amphibians have giant genomes. In plants and bony fish, the whole genome has duplicated, providing multiple copies of genes.

What is often not understood is that the genomes of vertebrates are highly conserved; they all have pretty much the same number of genes doing the same jobs. Many genes that code for basic functions can be traced far back beyond the dawn of vertebrates 400+ million years ago (Mya).

A metre is a lot of DNA to squash into a tiny (~10 nm) cell and a lot of DNA to replicate and segregate properly when the cell divides. For ease of handling, the metre or so of DNA is cut up into segments. In humans it is cut into 23 smaller DNA molecules, ranging from about 250,000,000 to 50,000,000 bases (250–50 megabases, Mb). These DNA molecules cannot be seen individually with a microscope, but they bind to protein and become contracted and compact during division, and are seen in the microscope as chromosomes – the name literally means 'coloured body'.

1.2 Chromosomes

Chromosomes each contain a single long DNA molecule that bears genes and sequences that regulate them. But they are much, much more than DNA (so please don't call a long sequence on your computer a 'chromosome'!).

The molecules that bind to DNA contain both structural and regulatory proteins. At their core are the basic histone proteins that assemble as 'nucleosomes', around which DNA is wound in a regular array of little spools. Modifications that are placed on or removed from the tails of these histones have striking effects on the activity of genes, as we will see in Chapter 5. Some act locally, others induce large-scale reorganization of the structure in three dimensions, something we are just beginning to understand.

The earliest microscopic observations of genomes revealed squashed blobs under low power. You can't see much differentiation in a nucleus unless it is dividing. However, it was noted that the stuff of nuclei (called chromatin) was heterogenous; some stained very pale, while there were blobs of darker staining material. Cytologists called these 'euchromatin' and 'heterochromatin' in 1928. We will see that this definition lies at the core of our investigations of how sex chromosomes work and how genes are regulated.

Chromosomes, even at the low resolution visible under the microscope, have some landmarks. The most obvious is a region called the centromere, which appears as a constriction and holds together the two daughters of a chromosome that has duplicated but not yet separated. This structure contains the kinetochore, which is responsible for hitching to fibres that pull the daughter chromosomes ('chromatids') apart when the cell divides. The centromere divides each chromosome into two arms, short (called 'p' for 'petite') and long ('q').

Images of chromosomes may be cut out from a photograph of a cell and arrayed in order of size in a 'karyotype', usually lined up at the centromere with their short arms at the top (look at Figure 2.5). There are two copies of each in all diploids (such as humans and other mammals), one copy from the mother ('maternal') and one from the father ('paternal'). For instance, there are 23 pairs of chromosomes in each cell of a human. The members of these pairs are called homologues.

In different mammals, essentially the same mammal genome may be cut up into a few big chromosomes (the deer mouse has only 3) or many small chromosomes (the dog has 39). The number of chromosomes in a genome is designated n.

The ends of chromosomes, too, are specialist structures ('telomeres') that cap the DNA molecule and ensure it is properly replicated at division. Telomeres comprise highly conserved repeated DNA sequences and are buffered by many subtelomeric repeats.

Another landmark is the position of genes that are transcribed into ribosomal RNA, which is assembled into bodies called ribosomes (RNA-protein structures that translate messenger RNA into protein). This region, called the Nucleolar Organizer (NOR), often appears as a non-staining gap in a chromosome.

Other chromosomal landmarks visible in the microscope are blocks of the funny-staining 'heterochromatin'. These are generally made up of highly repetitive sequences with an odd base composition and atypical protein binding that stains preferentially with some dyes and fluorochromes. Human chromosomes all have a block of heterochromatin at the centromere, and there are also large chunks on chromosomes 1 and 16. The long arm of the Y chromosome is practically all heterochromatin; it literally glows in the dark when stained with fluorescent dyes.

There are exceptional regions of the genome that, although they contain the same DNA and the same genes, adopt a heterochromatic appearance. The X chromosome is the classic case of this 'facultative heterochromatin', which is programmed to become genetically inactive (Chapter 4) by means of modifications to DNA and the histone proteins that bind to it.

Important landmarks on chromosomes are the bands of staining and non-staining material that can be elicited by various treatments and stains. C-banding (centromeric) differentially stains constitutive heterochromatin. G-banding (with Giemsa stain) elicits a complex pattern of dark and light staining according to the base composition (AT-rich stain dark and GC-rich light). The pattern is consistent for a particular chromosome and different between chromosomes, so it enables recognition of individual chromosomes. Bands are given numbers from centromere (proximal) to telomere (distal). This technique was critical for the characterization of human chromosomes, separating members of large groups (e.g., chromosomes 6–12 that included the X).

1.3 The Human Genome under the Microscope

The earliest studies of the human genome were cytological, and the observation of human chromosomes through a microscope is still an important adjunct to molecular studies.

Initially, it was hard to resolve the bunched-up chromosomes of humans or other mammals in fixed sections. For years the chromosome number in humans was thought to be about 44, and even to show natural variation. In 1921 an insect cytologist (Theophilus Painter) counted 48 little blobs, and this number was accepted for decades. Only when techniques were developed in the 1950s to arrest mitosis with an inhibitor and spread chromosomes in a dilute salt solution was it shown to be a constant 46, composed of 23 pairs. When the famous cytologist TC Hsu looked at Painter's microscope drawings later, he pronounced it 'amazing that he even came close'.

Improvements in technique made it possible in the 1950s to group human chromosomes by their size and shape (dictated by the position of the centromere). The discovery of chromosome banding techniques in the 1960s made it possible to identify individual human chromosomes. Indeed, chromosomes came in pairs that had the same banding pattern and were identical in all humans tested. This is so for 22 of the chromosome pairs (called 'autosomes').

But one pair, the sex chromosomes, breaks all these rules. Whereas females had a pair of medium-sized chromosomes (called the X), males had a single X and a pathetic little, oddly staining male-specific chromosome called the Y (Chapter 2). It is the peculiarity of this XY pair and its unique organization, activity and evolution that has inspired this book.

The revolution in cytology came with the development of fluorescent probes that lit up particular sequences under ultraviolet light.

These techniques make it possible to identify variant chromosomes in human patients. Originally it was known that some patients had too many or too few chromosomes or a chromosome(s) with an unusual shape or size. Now we know that babies with Down syndrome (mental retardation and developmental delay) are born with an extra copy of the smallest and most gene-poor chromosome 21 (trisomy 21). Trisomy of two other chromosomes is known to have severe effects; others are lethal at an early stage of embryonic development. Deficiency of a chromosome or even a small region is usually lethal. Rearrangements of chromosomes, for instance, the addition of a region of one chromosome to another, or the reciprocal exchange of two regions of different chromosomes (translocation), are not so severe, but patients may produce children with deletions or duplications of the translocated region. Even small changes to the line-up of chromosomes cause genetic mayhem because some genes are present in too many, or too few, copies.

The sex chromosomes present a striking exception. Abnormal numbers of X chromosomes, or deletions and rearrangements of the X, are extremely benign, and abnormal numbers of the Y are usually not detectable. I will discuss why in Chapter 5. Many cases are known where an X chromosome has been broken in two (fission), internally rearranged by inversion or fused to parts of an autosome as the result of translocation. Other abnormal X chromosomes ('isochromosomes') contain a doubling up of one chromosome arm or the other. Many similar rearrangements have been observed in mice, and many X chromosome rearrangements are available. Humans provide a huge storehouse of X chromosome abnormalities because they are much more likely than mice to seek medical help for their chromosomally challenged offspring.

1.4 DNA Sequence Classes

One of the shocking revelations of molecular analysis is that most of the genome of higher organisms (eukaryotes) is not genes. Protein-coding sequences ('exons') are a tiny minority of the human genome, about 1.5%. Most genes are interrupted by noncoding 'introns' that have to be excised from messenger RNA, some very large, and together constitute 26% of the human genome. Sequences that regulate gene action may comprise another 8%. What comprises the rest of the genome and what does it do?

About half of the mammal genome is made up of highly repetitive sequences. Eight percent consists of simple sequences (two, three or four nucleotides) repeated thousands of times. These blocks of tandem repeats constitute parts of the chromosome ('heterochromatin') that stain oddly because of their biased base constitution, and they can be detected as 'satellite bands' because of their different buoyant density on centrifugation. One class of AT-rich ~171 basepair (bp) tandem repeat satellites is concentrated at human centromeres. These regions are very variable, and often not shared even by closely related species. The Y chromosome is particularly rich in this class of noncoding sequence.

Another very large size class are transposable elements that owe their origin to the insertion of bits and pieces of DNA from ancient viruses (or DNA copies of RNA viruses). They may be transcribed into RNA, and DNA copies can be inserted elsewhere in the genome, so these 'jumping genes' can amplify and move within the genome. These include long interspersed repeat sequences (LINEs) and short interspersed repeats (SINEs). Amplification of these sequences can produce a very high copy number. These mobile elements are scattered throughout the genome, even within genes, constituting 'introns' that are spliced out to form mature messenger RNA. The human genome includes 8% LTR retrotransposons with long terminal repeats, 20% LINEs and 13% SINEs. One of these is a sequence called the Alu repeat (after the restriction enzyme that cuts within it and revealed strong bands on a DNA gel). There are a million Alu copies that make up about 10% of primate genomes but do not occur in other mammals, pointing to an invasion of an ancestral primate genome by a virus 70–80 Mya.

Another class of apparently functionless DNA in the genome consists of inactive copies of genes, so-called pseudogenes. These were born by gene duplication but were inactivated by mutations. They are almost as frequent as genes in the human genome. The Y chromosome, particularly, is full of dead and dying genes.

Variation of these DNA sequences in the population is high, so differences in these sequences make good markers for gene mapping or identification (e.g., in forensics).

There is an ongoing argument into whether noncoding repetitive sequence is 'junk DNA'. I think the answer is that some classes like the highly repeated simple sequence are, indeed, 'hard core junk', and maybe the bulk of retrotransposons sequence also represents 'selfish DNA' that is there because it invaded the genome aeons ago and it is hard to get rid of. However, there are well-documented cases in which sequences within retrotransposed elements have evolved a function in gene regulation.

There are many DNA sequences that do not code for protein but are transcribed into long noncoding RNAs (ncRNA). We know that at least some of these have important regulatory roles, including the control of X chromosome inactivation, which provided the first, and the best, evidence that ncRNAs could regulate gene action (Chapter 6). There is also a class of microRNAs, averaging only 22 bp, which target untranslated parts of the mRNA and control their translation.

1.5 Ploidy – Number of Genomes

Sorry, but you really can't get sex unless you understand ploidy, meiosis and recombination.

The whole point of sexual reproduction is to mix genomes up into new combinations. Genomes from two parents get together at some point in the life cycle. This doubling of the genome is offset by a halving at another point in the life cycle.

In humans and other mammals, indeed all vertebrates, parents contribute specialized cells called gametes that bear a single genome. Fusion of two gametes produces an egg that develops into an animal with two genomes. We call the single genome state haploidy and designate the haploid chromosome number n. The state with two genomes (one from each parent) is referred to as diploidy, and the diploid chromosome number is $2n$.

All vertebrates, indeed most complex life on earth, are normally diploids, maintaining two copies of the genome. The advantages are obvious; a mistake in one gene can be made up for by a good copy of the same gene in the other genome.

Since there are two copies of the genome in a diploid organism, one from each parent, it follows that there are two copies of each chromosome, one from each parent. This is why human chromosomes can be lined up in 23 pairs in the human karyotype. There are two copies of chromosome 1, two copies of chromosome 2, and so on. They are called homologues.

Homologous chromosomes are the same size and bear the same genes. However, there may be differences in DNA sequence between the parents; variants of coding genes (called 'alleles') may manifest as a difference in gene function that may have a major or little effect on the organism, depending on what the product of that gene does. Differences in noncoding and repeat sequences are usually quite benign but are handy for geneticists to use.

1.6 Cell Division in the Body – Mitosis

All life depends on the ability of the cell to divide into two identical daughters bearing the same complement of DNA and genes. This requires that the sequence of bases in the long DNA molecules is exactly replicated, as well as the other components of chromosomes, and that the daughter chromosomes are exactly segregated so each daughter cell gets an identical copy of each.

Faithful replication of the DNA is accomplished by an amazing series of steps called mitosis. At the heart of replication is the AT and GC base pairing discovered by Watson and Crick in 1953. The two strands of the DNA double helix separate, and each replicates by attracting complementary nucleotides, A with T and G with C. However, it is not as simple as they envisaged because the enzymes available will copy only one strand; the second must be physically broken, primed with RNA, replicated and repaired, a complex process that requires many different enzymes. Indeed, DNA replication is a great example of what I call 'Dumb Design', processes which seem functionally crazy, but are easy to explain in terms of evolution; the process probably evolved initially to replicate single-stranded RNA.

DNA replication is only a start. Assembling the other components of the daughter chromosomes and separating them into two daughter cells requires that each contracts into a rod-shaped body (that's why we can see them with a microscope). Chromosomes are arranged in a plane so that their kinetochores can become attached to fibres of a spindle. For each chromosome, the microtubules of the spindle must attach correctly and pull the two daughter chromosomes (called 'chromatids') to the opposite ends of the cell, which then must divide accurately into two.

If any of these steps fail for any chromosome, the daughter cells will die because they are missing bits or have an unbalanced genome. Or even worse, abnormal cells will survive, become misregulated and cause cancer.

Not surprisingly, therefore, mitosis is a highly regulated process involving hundreds of genes. How they work to stick chromatids together, separate centromeres and get pulled apart is a very active research field.

1.7 Making Gametes – Meiosis

At the very centre of sex is how haploid sperm and eggs are made from diploid somatic cells. Making gametes with a single genome (one copy of each chromosome), from diploid cells with two genomes (two copies of each chromosome), requires a specialized division. The stakes are high. A mistake in segregation will produce a gamete – and maybe a child – with missing or duplicated chromosomes.

Meiosis is a complex process, and it is invariably taught as sets of stages with funny old names that obscure the magic of this 'dance of the chromosomes'. But really the logic of the process is simple and beautiful, something that was recognized a century ago. It is the logic, not the names or the molecular details, that we need to recognize in order to understand sex and the peculiarities of sex chromosomes.

Essentially, there is one round of DNA replication and two rounds of chromosome segregation. The special trick of meiosis is to reduce the two sets of chromosomes in a diploid organism to a single set in haploid gametes. This is accomplished by arranging for homologous chromosomes to pair up so they can be sent in an orderly manner to one daughter cell or the other.

As well as making sexual reproduction work, meiosis mixes up the alleles in each individual. During the process of pairing, the two homologues break and rejoin to make new combinations of the variants contributed by each parent. This recombination is why you're unique – and so is everyone else.

The first thing that happens in meiosis is the exact replication of DNA and other chromosome components; this stage is similar to mitosis. But instead of the daughter chromatids separating into daughter cells, they stay glued together by a protein called cohesin.

The highlight of meiosis is the point by point pairing ('synapsis') of homologous chromosomes (already duplicated), during which they exchange regions. This crossing over ('chiasma') occurs by breaking both DNA strands ('double strand breaks') and then repairing them. Repair either rejoins the same ends, or joins ends from the two homologues to form DNA strands with new combinations of alleles ('recombination'). These crossovers (or 'chiasmata') hold the homologues together, and are essential for organized segregation of chromosomes, one into each gamete. This first division therefore separates homologues, thereby reducing the chromosome complement from diploid to haploid (two genomes to one).

But each chromosome has already replicated, though the two sister chromatids remain glued together at the centromere, shielded by another protein. They separate in a second division when this protein is modified and loses its grip. The second stage of meiosis looks much like mitosis, except that segregation is not preceded by DNA synthesis. Thus, each primordial germ cell produces four haploid cells. In the ovary, a diploid oocyte produces four haploid cells, one of which develops into an ovum. In the testis, a spermatocyte produces four haploid sperm.

Meiotic pairing is an extraordinarily accurate process. It is thought to occur in two phases; initial alignment by matching special repeated sequences with the help of chromatin proteins bound to the chromosome axis, then point to point pairing between loops of homologous DNA sequences. Proteins accumulate to form a synaptonemal complex over the paired regions, a conserved structure that 'zippers' chromosomes together. The synaptonemal complex even looks like a zipper, with two lateral and one central element traversed by proteins that hold homologues together. This complex can extend beyond regions of DNA homology.

Within the synaptonemal complex, point by point pairing is achieved by initiating double strand breaks. This exposes single DNA strands that search for homologous sequences. This remarkable process makes use of a bevy of enzymes that break down nucleic acids (nucleases), repair DNA and recombine DNA strands that mediate breakage and reunion of strands at recombination sites. The synaptonemal complex disassembles after recombination, and the homologues are held together only at the chiasmata.

This is a risky process, and not surprisingly there are many molecular backup systems to detect and eliminate errors and fill gaps.

It's easy to see why sex chromosomes run into trouble at meiosis. In mammals, meiosis in XX females works just fine. The two X chromosomes pair along their length, recombine and segregate into egg cells just like autosomes. But in XY male meiosis the extreme differences in size and DNA content of the X and Y make pairing difficult. However, although behaviour of the X and Y chromosomes is a bit bizarre, the X and Y do pair and recombine over a limited region called the 'pseudoautosomal region', abbreviated PAR. This pairing is critical for fertility, at least in human and mouse spermatogenesis, so there is a mandatory single crossover in the PAR of each XY pair.

Male meiosis is what determines sex in humans and other mammals. An XY father will produce two kinds of sperm in equal proportion. X-bearing sperm fertilizing an X-bearing egg will produce XX zygotes that develop as females. Y-bearing sperm fertilizing an X-bearing egg produce XY zygotes that develop into males. Equal segregation of the human X and Y chromosomes at male meiosis produces the equal ratios of boys and girls.

The X chromosome pairs and recombines normally in female meiosis, but not in male meiosis, so its opportunities for swapping alleles are more limited than for the autosomes. The Y chromosome, by definition, is handed down through the male lineage and never has a chance to recombine with another Y. This asymmetry has important consequences for sex chromosome function and evolution.

Recombination provides a means of ordering genes on the same chromosome and estimating relative distances between them (Section 1.9).

1.8 Genome Variants

Our ability to discover the location and function of genes and the sequences that regulate them depends almost entirely on our observation of genetic variation. Variants of genes (alleles) with different characteristics (phenotypes), biochemical characteristics of the protein product, or molecular differences in DNA provide markers to characterize chromosomes, as well as to divine the functions of genes. There may be many normal allelic variants (called polymorphisms) in a population.

Human gene function has traditionally been determined by observing the phenotype (outward manifestation) of mutants. For instance, the bleeding disease haemophilia was found to result from the absence of blood clotting Factor VIII, which acts in a cascade of enzymatic changes to blood components.

A new source of markers which were naturally variable (polymorphic) within the population was provided by recombinant DNA technology. In the 1970s it became possible to clone random fragments of the human genome, and use them to detect differences in the sites cut by various enzymes that cut DNA at defined sequences (restriction enzymes, which revealed restriction fragment length variation, or RFLP). DNA fragments were separated by size on a gel (a slab of jelly like material that DNA molecules are pushed through by an electric current, the small molecules faster than the large). Fragments carrying the sequence of interest were detected by hybridizing with a labelled 'probe' of single-stranded DNA ('Southern blotting'). Even sequences whose functions were unknown, and sequences which appear to have no function, could be used as markers for linkage mapping, as long as they detect variation between people.

The supply of variants was enormously enriched when it became possible to detect differences in DNA sequence directly, often in a single nucleotide (single nucleotide polymorphisms, or 'SNP'). SNPs are common in the human genome, estimated at 1 per kilobase in euchromatic regions.

Another powerful source of variation are variable numbers of tandem repeats (VNTR), usually di, tri or tetranucleotides. These 'microsatellites' can be detected as different sized fragments by polymerase chain reaction (PCR see 1.11), primed from flanking unique sequences.

A particularly common feature of the human genome is the number of segmental duplications, large regions of the genome that may be amplified and moved to new sites within the same, or a different chromosome. Whole regions of the genome may be duplicated or further amplified, and if these regions contain genes, their amplification often has a phenotype. Changes in copy number are frequent at some sites that have duplications because unequal crossing over can amplify or diminish number of copies (see Chapter 6). Screening patients for small deletions and duplications reveals that more than half of known genetic disease is caused by rearrangements and amplifications and deletions.

Human genetic variation is an important area of study for many reasons, including tracking down genetic diseases (medical genetics) and tracking down criminals (forensics) and fathers (paternity testing), as well as charting differences within and between different human populations.

Gene function has been intensively studied by observing mutations in mice. Hundreds of spontaneous mutations, and mutations induced by treatment with radiation or chemical mutagens, have been collected and curated over the years; at least 40 of these are located on sex chromosomes. Because the genomes of all mammals (indeed all vertebrates) are strongly conserved, and the X chromosome especially conserved, these genes usually have counterparts in the human genome, and disease states can be studied in this model mammal.

Genetic variants can also be induced in laboratory mammals, either by mutagenesis (chemical or radiation), or by molecular tricks. A gene can be inserted into the mouse genome ('transgenic'), or deleted or inactivated ('knockout').

A bacterial defence system – the bacterial equivalent of immunity – provides the means to alter base sequences in deliberate ways ('gene editing'). It uses a family of long-known bacterial repeated DNA sequences ('clustered regularly interspaced short palindromic repeats', or CRISPR), which were captured from ancient virus infections. Bacteria use transcripts of these short (~20 base) sequences to seek and destroy DNA from similar viruses during subsequent infections, guiding an enzyme (Cas9) to recognize and cut complementary strands.

Guide RNA sequences that bind to the enzyme can be made to recognize and cut any complementary sequence in the mammal genome. The cell's DNA repair pathways then kick in. Emergency error-prone repair will put any old thing in the gap and probably inactivate the gene, so it's a great way to do knockouts. However, the homologous repair system can be hijacked by providing DNA with homologous sequences on either side of the desired insertion. It is much quicker than transgenesis and knockouts (from weeks to days) and cheaper (from thousands to hundreds of dollars).

The CRISPR-Cas9 system offers a whole new set of techniques for studying gene function by inactivating targeted genes, or modifying the base sequence of a gene in known ways. It is also being explored intensively to create new strains of plants and animals.

1.9 Gene Mapping

A defining feature of any gene is its location in the genome. Many genes have been discovered by pinpointing their location on a gene map, by 'positional cloning'. Indeed, the sex gene *SRY* itself was the one of the first human genes to be discovered by this means, and its target gene *SOX9* was also revealed by its position.

Maps of the sex chromosomes formed the basis of our understanding of their organization, function and evolution, the meat of this book. Strategies for mapping genes have improved out of sight in the last few decades.

The first maps were constructed by comparing the amount of recombination between two genes in human families, or in mouse crosses. Classically, this required variants (alleles) of the two genes that could be detected by phenotype differences (e.g., normal and disease), or biochemical differences in an enzyme (e.g., variants that differ by charge). If two genes are close together, there will be little recombination between them, and they are said to be 'linked'. All markers can be placed on a linear linkage map. Recombination frequencies are additive over short distances, but double recombination over longer distances means they are progressively underestimated and can never get above 50%. Linkage is measured by percent recombination, measured in centimorgans (1cM = 1% recombination).

In this way, gene order within a linkage group can be determined. For markers close together, recombination values are additive, but when they are too far apart, double recombination can restore the parental combination of alleles. Also, because recombination is not uniform over the chromosome, units of recombination do not represent physical distances between markers.

For eutherian mammals, recombination is less in male than female meiosis, and skewed towards the ends of chromosomes. Using thousands of markers of all kinds, very large studies on hundreds of human families show that recombination over the whole human genome is far from uniform. The rate of recombination varies wildly across all chromosomes, presenting recombination 'deserts', especially near centromeres, and 'jungles' at other sites. There is a clear sex difference in numbers and distribution of crossovers; more in females, particularly around centromeres, but fewer in male meiosis, and a peak towards telomeres.

Linkage maps of domestic mammals were constructed in the 1990s, again using molecular markers to order genes and phenotypes of economic interest like milk in cattle, wool quality in sheep and fatness in pigs. Linkage maps were also constructed for carnivores (cats and dogs), horses, rats and other rodents, even marsupials. These have been progressively filled out with markers detected at the DNA level. Linkage mapping, long regarded as a rather fuddy-duddy pursuit of geneticists, is now a high-profile international activity with clear practical goals of locating and isolating disease genes in model mammals, and economically significant genes in domestic mammals. Nothing substitutes for a genetic map based on recombination frequencies within families because linkage map can order and locate any kind of marker, including phenotypic differences (including disease states) whose genetic basis is unknown.

Species differences also provide a huge resource of genetic variation. Biochemical differences (e.g., in the charge of an enzyme) and DNA sequence differences could be used to assign genes to chromosomes by somatic cell genetic mapping. In this technique, pioneered in the 1970s, cells from different species (usually mouse x human) were fused to make cell hybrids, which preferentially lose human chromosomes (we still don't know quite why). Any gene or other sequence that is different between mouse and human can be assigned to a chromosome by correlating its presence and absence in hybrids and revertants. Genes that are retained or lost together are called 'syntenic' (literally 'same thread'). Somatic cell genetic mapping was crucial to establishing the first human gene maps.

Gene maps in other species were also built using somatic cell genetics. This was particularly useful for mapping in primates, since it required only a cell line rather than an expensive (and ethically challenging) monkey or chimpanzee colony. It was the only feasible genetic mapping method in species, such as whales, which cannot easily be captive bred.

However, assignment of genes to a chromosome provided no information on the order or positions of genes within a chromosome. Regional mapping can be accomplished using cell hybrids which retain only a part of a human chromosome, derived from patients with a deletion or rearrangement.

Much greater refinement was possible by deliberately shattering the human genome by pre-irradiating cells before fusion to form 'radiation hybrids' that contained very small fragments of the human genome. Panels of radiation hybrids were constructed and were screened for many DNA markers to put together 'radiation maps' of all human chromosomes.

The terms used for different kinds of mapping have got terribly confused, and it is worth trying to sort them out at this stage. The term 'sex-linked' or 'X-linked' refers to a pattern of inheritance in a family and is not appropriate for a gene which has been shown to lie on the X by some other method, such as somatic cell genetic or genome sequencing. 'Linked' refers to two genes that tend to segregate together in families, and is not appropriate for genes found to be close on the same chromosome by some other method. 'Syntenic' refers to co-segregation of two markers in cell hybrids, which are presumed to lie somewhere on the same chromosome. I will try to use these terms correctly, or to choose some neutral term like 'X-borne' when the mapping method doesn't matter.

1.10 Physical Mapping of the Human Genome

Genetic maps are powerful, but they are just lines on paper and don't align immediately to the position of real genes on real chromosomes. There are now several ways to map genes and other sequences to chromosomes, and align these to linkage maps.

Physical maps of the human genome could be constructed using material from patients with abnormal chromosomes, some from which material has been deleted. Lining up the deletions and detecting a critical interval deleted in all patients can reveal the site of a gene mutated in a particular disease. For instance, patients with a particular disease may have deletions that all overlapped to reveal a critical interval that was missing in all the patients, thus locating the disease gene (Section 2.2). Defining the intervals requires recognition of marker sequences. Initially, patterns of restriction cut sites ('restriction maps') provided this evidence of overlap. More efficient was to develop sets of unique 'sequence tagged sites' (STS) of a hundred or so base pairs which could easily be recognized by PCR (Section 1.11). Clones which share a particular unique STS must therefore overlap.

I used to fantasize that I could see where a particular gene was on a chromosome just by looking down the microscope. My dreams were realized in the 1980s by the development of in situ hybridization.

Key to this technique was the ability to hybridize single DNA strands that had complementary sequences and would pair to reconstitute a stable double-stranded helix. DNA strands can be readily separated by boiling (thermal denaturation), or by some chemicals. So if you add single-stranded DNA from a known gene (the 'probe') to denatured DNA trapped in fixed chromosomes spread on a slide, it will home in and bind at the location of the same sequence.

All that was required was some means to label the DNA probe. The first in situ hybridizations were performed by labelling the probe with radioactive bases (tritiated thymidine), then covering the hybridized chromosomes with photographic film and observing where it activated silver grains (autoradiography). Much more powerful and convenient were discoveries of fluorescent compounds that would bind to a DNA probe and reveal its location under ultraviolet light. The labelled probe binds to the homologous sequence in the chromosome, and emits a signal at that location, a fluorescent dot can be seen at the location of the gene and recorded by a very sensitive camera. This fluorescence label is very efficient, and the location of a sequence can usually be judged by inspection of just a few chromosome spreads.

Fluorescence in situ hybridization (FISH) mapping can resolve markers down to just a few Mb apart, and order three or more markers distinguished by colour. A modification uses interphase cells in which chromosomal DNA is much less contracted ('DNA FISH' or 'fibre-FISH').

Fluorescence in situ hybridization (FISH) came into its own when a source of purified 'probe' DNA became available through DNA cloning or PCR amplification (below). The longer the probe, the better the signal.

RNA-FISH is a modification that detects the chromosomal site of transcription of a particular gene. It has been particularly useful because it can detect expression on individual X chromosomes in single cells. A single-stranded DNA probe that hybridizes to a particular gene is fluorescently labelled, then hybridized to chromosome spreads. It homes in on the chromosome loci that are actively transcribing that gene by hybridizing to primary transcripts at the point of transcription.

Another beautiful cytological technique that has been used to compare genomes is 'chromosome painting'. Here the fluorescent probe is DNA prepared from a whole chromosome. Chromosomes may be flow sorted by size and DNA content using a dual laser flow cytometer that measures DNA content of single chromosomes separated in microdrops and directs them into different containers. This yields many copies of a single chromosome, and DNA can be prepared from them. Alternatively, a particular chromosome can be dissected out from cells under the microscope, and DNA prepared and amplified. DNA from a single chromosome is tagged with a fluorochrome and hybridized in situ to chromosome spreads of the same or a different species. Whole regions of homology are revealed in colour. Several chromosomes may be painted in different colours, providing stunning demonstrations of chromosome rearrangements that distinguish species.

FISH was extremely useful for establishing positions of genes and other sequences in a physical map of the human genome, as well as the genomes of many domestic species. It has been a crucial tool in establishing maps in many exotic species, such as the elephant and the platypus, where other mapping methods are difficult.

There are now many ways to characterize genes and chromosomes, but just looking at the chromosomes remains one of the most direct and complete. A computer-generated 'scaffold' or 'contig' of DNA sequence, no matter how gap-free, can never be a chromosome. The non-DNA components of chromosomes are vital for the regulation of genes, as shown by the many studies of 'epigenetic regulation'. In combination with molecular techniques, it remains a powerful method to integrate molecular data.

1.11 The DNA Revolution

Underpinning many of the advances in genome technology is the ability to isolate and characterize DNA sequences.

The discovery of the structure of DNA double helix in 1953 marked a revolution in our way of thinking about biology and inheritance. But the technological revolution had to wait some decades for the discoveries that underpin analysis and manipulation of DNA.

The problem with DNA was its enormous length, and its physical and chemical uniformity. How could a molecule packed with the code of life present as so chemically bland? Limited information was gained by using physical and chemical properties to cut, measure and fractionate DNA into compartments with different base compositions (therefore density). One property that was critical to Watson and Crick's insight about complementary base pairs holding the helix together by weak hydrogen bonds was that two strands with complementary sequences could be separated by heat, and would snap back together on cooling (thermal denaturation).

Breakthroughs in characterizing and manipulating DNA came, as they usually do, from very fundamental biology – for instance, from characterizing the enzymes that normally replicate DNA and repair it.

An early breakthrough came from studying bacterial defences against virus attack. The discovery of enzymes that cut DNA at specific sequences ('restriction endonucleases') provided the first means of cutting DNA into specific fragments, whose size could be recognized by the rate at which fragments are pushed through a gel by an electric current (electrophoresis).

Some of these cuts had overhangs which could splice different fragments of DNA by complementary base pairing, paving the way for recombinant DNA technologies. This enabled fragments of human DNA to be spliced into a small self-replicating genome ('vector'), such as a virus or a bacterial plasmid (a small circular self-replicating genome), to make 'recombinant DNA'.

Cutting up the DNA of human cells and splicing them into one of these vectors produces a 'library' of sequences that together represent the whole human genome. A single fragment of human DNA can be separated from others by isolating and growing ('cloning') a single recombinant organism and tricking it into making thousands or millions of copies of just one fragment.

Fragments of the human genome may be cloned into a number of different vectors, whose characteristics determine the size of human DNA insert they can carry. Bacterial viruses (usually lambda or a derivative) can accommodate only 15 kilobases, and bacterial plasmids and constructs of plasmid and phage sequences 30–50 kb. Much larger regions (up to hundreds of kilobases) may be inserted into the larger bacterial genomes to make bacterial artificial chromosomes (BACs), and Mb sized pieces into yeast artificial chromosomes (YACs).

Alternatively, coding regions (only 1% of the genome) could be cloned into a vector by isolating messenger RNA sequences from a particular tissue, copying them into complementary DNA (cDNA), and cloning into a vector to make a cDNA library.

In this way, 'libraries' of clones could be generated in which the whole genome, or a specific part of it, was represented many times over. These libraries could be 'probed' for particular sequences using DNA hybridization of denatured strands. Many clever tricks were developed to extend the cloned region by 'chromosome walking' (or jumping), using sequences from the end of one clone to probe the library for adjoining clones.

Other breakthroughs were in ways of separating and detecting specific fragments. Restriction fragments could be separated on a gel, blotted onto a nylon filter, and detected by complementary base pairing with a cloned radioactive or fluorescent probe ('Southern blotting', after its inventor, Ed Southern).

Basic study of replication enzymes from hot springs bacteria also gave us heat-tolerant DNA polymerase, an enzyme that powers the polymerase chain reaction (PCR), an alternative means to replicate particular short regions of a mammal genome. Short DNA sequences (primers) with known sequences are used to begin replication from known sequences on either side. The region between them (up to about a kilobase) is then copied thousands of times. The fantastic sensitivity of this method enabled detection of DNA repeat variation even in single sperm.

A molecular scale map of the human genome was produced as part of the Human Genome Project, the huge effort in the 1990s to clone and sequence the entire human genome. Large and small regions of DNA were cloned, assigned to chromosomes, then put together in order with the idea of constructing a 'golden pathway' of contiguous clones with minimal overlap ('contigs') that could be efficiently sequenced. At yearly meetings of the Human Gene Mapping Consortium, 23 committees put together mapping and sequencing data for human chromosome 1, chromosomes 2, 3, 4 … and the sex chromosomes.

Restriction cut sites and short marker 'sequence tagged sites' (STS) were used to order the cloned fragments by recognizing sequences that overlap ('fingerprinting'). DNA sequences of BAC ends were also used to order the fragments. Overlapping clones identified by shared molecular markers were ordered into contigs extending for many hundreds or thousands of kilobases. By 2001, a PAC and BAC map of the entire human genome had been produced by the International Human Genome Sequencing Consortium. It contained hundreds of thousands of BACs in about 7,000 clusters that contained 93% of markers.

1.12 The DNA Sequencing Revolution

Genome sequencing seemed like an impossible dream 50 years ago when the base sequence could be revealed only by chemical cleavage. Robert Holley won a Nobel Prize sequencing a 77 nucleotide transfer RNA in 1965. But three generations of innovation, using the properties of the four bases for detection and strategies for sequential readout based on fundamental knowledge of DNA structure and replication, have doubled capacity and halved time and cost every five months.

First generation sequencing methods were based on creating fragments that terminated in a particular base, then separating these four families of fragments by size on a polyacrylamide gel, from which the sequence could be read off. One technology for creating the fragments used chemical cleavage at each of the four bases, others created fragments either by limiting one of the four nucleotides, or by synthesizing fragments in the presence of modified ('chain-terminating') nucleotides that could not be added onto. Separation on a monster flat gel was later replaced by detection of a fluorescence-labelled base by capillary electrophoresis, enabling the first crop of automated sequencing. These techniques required a lot of purified or cloned DNA, a great deal of patience and produced fragments of less than a kilobase.

A new 'second generation' of sequencing methods used the discovery that addition of a nucleotide could be coupled to a reaction that produced a flash of light whose strength was determined by the nucleotide added. Light was monitored in real time as each nucleotide was washed over template DNA which was tethered to a solid substrate (slides or beads). Beads with a single molecule were separated, the molecule amplified sequenced. An alternative detection is by changes in pH caused by release of protons as new nucleotides are added. These techniques had the huge advantage of being adapted to sequencing thousands or millions of short DNA molecules in parallel and are highly accurate.

The problem was that these sequencing methods produced only short runs (30–50 base pairs) of sequence. Advances in this technology have produced several methods that produce longer runs and will read off both ends of the molecule ('paired end' sequencing). However, sequence of less than a kilobase would not encompass even a single gene. Assembly of these short sequences was the job of bioinformatics, and huge efforts were put into developing algorithms to compare and assemble sequences *in silico*.

Next generation sequencing has now been developed to sequence long single molecules trapped in nanoscale wells or perforated film with a single molecule of DNA polymerase. Addition of fluorescent nucleotides is detected by a laser focused on a tiny region of synthesis. This long-read sequencing can produce reads of many kilobases, and has the added advantage of eliminating any DNA amplification step that could introduce errors. Initially slow, expensive, and error-prone, many new technologies are being developed that use techniques to drive DNA (or RNA) molecules through nanopores, and to block ion flow proportional to the length of the chain that can be easily measured. The promise is to make long-read sequencing so cheap and technologically non-demanding that it can be done by a tiny device (MinION) that can even be taken to obtain sequences in clinics or in the Amazon.

Many variations of these techniques now extend our intimate knowledge of chromosomes. It is possible to sequence defined regions, such as flow sorted single chromosomes or exomes, or huge assemblages of organisms, such as gut flora or the sea bed (metagenomics), the components of which can be sorted out by a computer.

Chromosome conformation capture (3C) is a new technology that can produce and assemble long-read sequence, but also provides additional information about three-dimensional arrangement of the DNA molecule (Figure 6.4). This family of methods chemically crosslinks DNA sequences that are close in space in an interphase nucleus, and ligates them. Deep sequencing of the two fragments together registers the frequency of all possible pairs of fragments and plots them in two dimensions. Of course, sequences that are adjacent on the same molecule will form a strong diagonal on the graph, and this may be used to assemble sequences over long (chromosome scale) distances. In addition, close interactions between DNA sequences separated by varying distances on the same molecule register as boxes of increasing size on the diagonal, and interactions with sequences on a different chromosome as more distant clusters.

ChIP-seq (chromatin Immunoprecipitation sequencing) identifies DNA sequences that bind to specific proteins. The protein of interest is precipitated by a specific antibody, and the DNA molecules it binds are subjected to massively parallel sequencing.

The almost magical rate of progress in genome sequencing can be gauged from the effort to produce the first genome, which took 13 years and $3 billion, to the cost of doing a novel genome today (a few thousand dollars in a few days).

The rise of second generation sequencing methods has generated an avalanche of new data. All these techniques require massive efforts to sort, store and analyse huge bodies of data on increasingly enormous computers. Indeed, novel software packages have been developed to aid sequence assembly, alignment between sequences, database searches. This enables relationships between genomes, prediction of genes and regulatory sequences, and prediction of the structure and function of the gene product. Most of the data are publicly available, contributing to very many discoveries of genes and comparisons of genomes in many species, enabling many hoary old biological problems to be investigated. Many venerable questions about sex, genes and chromosomes can finally be answered.

1.13 Other 'Omics' Techniques

There are now techniques for the global study of DNA, RNA and protein in a cell or tissue (Box 1.1).

DNA may be the storehouse of genetic information, but the daily business of running cells and bodies falls to RNA. The role of messenger RNA in copying the gene sequence, transporting it to ribosomes in the cytoplasm to be translated into proteins has been long recognized, as have the classes of RNA that are critical for translation; ribosomal RNA (rRNA) and transfer RNA (tRNA) that reads the code. Transcription turns out to be far more complex than initially imagined: the primary transcript must be processed to splice out introns, and alternative splicing may produce several variant proteins from the one gene. Transcription of mRNA is also highly regulated; some genes are transcribed in all tissues, whereas specialized genes are expressed in one or a few tissues, and many are expressed only during certain developmental stages. There are now several other classes of RNA that regulate transcription and translation and are critical for proper gene expression, including long noncoding RNA (lncRNA) and micro RNA.

Box 1.1
The -omes and the -omics

-omics is a suffix given to describe inventorying all sequences in an individual or a tissue or cell.

Genome describes the complete set of genetic material (DNA sequence, including all genes and all non-coding DNA) of an organism. It amounts to a haploid set of chromosomes.

Transcriptome describes the assembly of all RNA transcribed by a given cell or tissue (and **exome** the assembly of all messenger RNA sequences).

Methylome describes the position of methyl groups on all bases in the genome.

Proteome describes the assembly of all proteins in a given tissue.

Many of these techniques depend on **Next-generation DNA sequencing;** several techniques for massively parallel sequencing of short stretches of DNA immobilized on a flat substrate

---

Box 1.2
Methods for qualitative and quantitative assay of RNA transcripts

RNAseq: next-generation sequencing to quantify transcripts of every gene in a particular tissue/time

Expression microarrays: assay for the presence of RNA sequences in a sample by binding sample to cDNAs (DNA copies of mRNA) arrayed on a substrate.

qPCR: quantitative (real time) PCR measures the amount of a specific sequence in a sample by tracking the production of amplification products over PCR cycles.

Allele-specific assays: these methods can be made to distinguish between two alternative forms of a gene by using probes or primers that bind only to one or the other alternative nucleotide sequence.

It was therefore critical to develop methods for detecting and measuring different RNA molecules (Box 1.2). RNA, being single-stranded, is far less robust than DNA and very sensitive to enzymatic breakdown. Nevertheless, techniques have been developed in parallel with techniques to measure, isolate and identify and ultimately sequence RNA molecules.

A particular RNA may be identified on Northern blots (the RNA analogue of Southern blots: there is no Professor Northern), which separate and detect messenger RNA molecules by size on a gel, blotting onto paper and detection by hybridizing with a single-stranded DNA probe that has been labelled with radioactivity or fluorescence. PCR was also used to detect particular transcripts by choosing primer sequences that are specific for a particular gene product and amplifying this sequence in the presence of label. This procedure is very sensitive but not very quantitative, so it is useful for detecting presence or absence of a particular mRNA, but not for assessing its dosage.

The transcription profile of a particular tissue was initially explored using microarrays constructed with denatured DNA fragments from genes to be examined bound to a solid surface. Preparations of RNA from living cells, labelled radioactively or with fluorescent probes, can then be bound to the matrix, and transcription can be detected by the amount of labelling of each spot.

These techniques have now been superseded by whole genome sequencing techniques. The whole transcriptome (all the cytoplasmic mRNA) can be fully sequenced, and transcripts can be assigned to individual genes. This can be made specific to particular alleles if there are polymorphisms that distinguish the two X chromosomes. For instance, a recent study detected transcription of genes from X chromosomes of different mouse strains which differed by 800,000 SNPs (many for most expressed genes).

RNA-seq is a technique for assessing the presence, and the quantity of RNA transcripts from every gene in the genome in a particular tissue at a particular developmental stage. All the RNA molecules are copied into DNA (cDNA), adaptors are ligated to each end, then subjected to short-read sequences from one or both ends. Sequences can then be aligned to genomic sequences, and the numbers of transcripts of all genes computed. It can be modified to look at different RNA populations (e.g., mRNA, noncoding RNA, microRNA). It is so sensitive that it can be done on single embryos, even single-cells. If SNPs are known, the computer can sort out transcripts from different alleles.

We now know that there is a level of epigenetic ('above the gene') control exercised by chemical modifications to DNA, or to the histone proteins bound to them. An important one for our consideration of sex chromosome activity is DNA methylation, the covalent attachment of methyl groups to a carbon atom in the cytosine ring, which is heritable, and reversible. Initially, methylation could be examined only in bulk, or over short stretches using methyl-sensitive restriction enzymes. A big breakthrough was the invention by Marianne Frommer in Sydney of a method to map methylated cytosines at a nucleotide level in a sequence. This was accomplished by chemically converting cytosine to thymidine with bisulphite, sequencing DNA and spotting where unmethylated cytosine had been converted to uracil (non-methylated sites) and where it had remained (methylated sites). There are now several methods to map methylated cytosines over the entire genome ('methylome'), using direct electronic sensing, or detecting a distinct fluorescent flash of individual molecules as they pass through a nanopore.

It is also valuable to be able to identify and quantitate the proteins in a cell. Western blotting separates proteins on a gel and identifies proteins of interest using antibodies. It is also possible to catalogue the complete array of proteins in a population of cells (the 'proteome'). Different proteins could be separated on a gel in two dimensions by their size and charge. Proteins can now be identified by mass spectrometry, using sophisticated programmes to identify proteins by the composition of fragments.

New techniques to separate and identify transcripts and protein products have been particularly useful in investigating the activity of sex chromosomes in dosage compensation and sex determination, as we will see in Parts II and III.

1.14 Phylogeny and Vertebrate Relationships

A great deal of information about mammalian (and human) sex and sex chromosomes has come from comparisons between the three groups of mammals, and wider comparisons with other vertebrates. To interpret these data, we need to know the basics about vertebrate phylogeny. I will briefly describe the strategies for assessing relatedness.

Initially, animals were grouped together by their phenotypic similarities, especially skeletal characters that could be related to extinct fossil lineages. This enabled evolutionary milestones to be established by dating fossil remains or their surrounding sediments. Thus we know that innovations such as a backbone were invented in a chordate ancestor more than 500 million years ago (Mya), jaws were invented in fish 440 Mya, and the decision to go for four limbs was made in an amphibian ancestor about 370 Mya. Other landmarks, for instance, the invention of a waterproof egg in land-based reptiles, could be dated by tracing back amniotes to a common ancestor 350 Mya, and hair in an ancestral mammal about 200 Mya.

This phylogeny gave us the four vertebrate classes; mammals, reptiles (including birds), amphibians and fish. However, bones and teeth could not resolve the endless arguments about the relationships of sharks to other fish, birds to reptiles, and the position of turtles, with their dramatically modified skeleton. Teeth, though providing incredibly detailed information, could also be misleading because of convergent evolution for function – for example, two marsupial carnivores were claimed to be closely related although one came from South America and the other from Australia. Morphological characteristics can also be similar by convergence, for instance, spiny mammals that eat ants would include the distantly related European hedgehogs and egg-laying Australian echidnas.

Phylogeny has utterly changed since the provision of DNA sequence data in the 1990s. First available for mitochondria because these energy-producing cellular bodies have tiny genomes available in many copies, sequence comparisons provided a quantitative estimate of divergence between any two species. Data from sets of nuclear genes followed, and it is now routine to compare whole genomes, or 'exomes' which comprise the protein-coding fraction of the genome.

These can provide a more reliable 'molecular clock' than full sequencing, because they lack repetitive sequences, which can evolve rapidly and quixotically.

We therefore have a good idea of how different vertebrates are related (Figure 1.1). Sequences are now available for many of the more than 60,000 vertebrate species, and though many of the details remain to be resolved, some large-scale corrections are obvious. Firstly, birds are reptiles. And turtles are quite closely related to birds and crocodiles, rather than being the distant reptilian outgroup suggested by their peculiar skeleton. Fish are by no means a monophyletic group, and sharks and rays (cartilaginous fish) diverged from bony fish early in the vertebrate lineage; certain ray-finned fish such as lungfish and coelacanths are probably the fish lineage that gave rise to tetrapods and ultimately us.

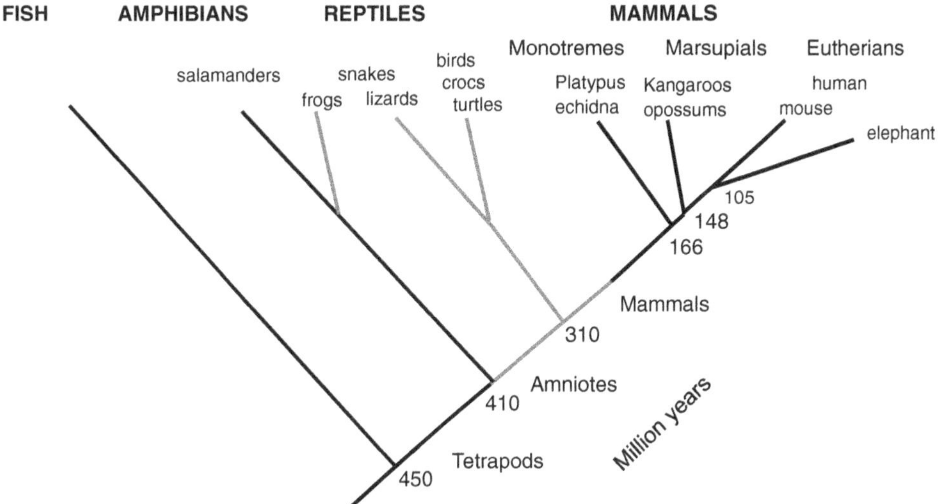

Figure 1.1 Vertebrate relationships. Diagram of the vertebrate family tree showing divergence of fish, amphibians, reptiles and the three mammal clades. Humans last shared a common ancestor with elephants 105 million years ago (Mya), with kangaroos 148 Mya, with platypus 166 Mya, with birds and reptiles 310 Mya, with amphibians 410 Mya and with fish 450 Mya.

Mammals are defined by possessing hair and feeding their young with milk. They evolved from a group of reptiles more than 200 million years ago, and are the only living descendants of this group.

Mammals are grouped into three major extant mammal groups, the familiar placental mammals (Eutheria), marsupials (Metatheria) and monotremes such as the fabled platypus (Prototheria). Eutheria and Metatheria are grouped together as mammalian subclass Theria (which means 'beast'). Their names remain contentious. The Eutheria are traditionally known as placental mammals, but this is quite incorrect because marsupials also have placentae. However, I also resent the term 'Eutheria', which means 'true beasts', as if there is anything less true (or less beastly) about marsupials. I suppose I shall call them eutherians here.

There are also longstanding arguments about the timing of the important nodes of divergence. Though the ultimate datings must be with reference to fossils and sediments, molecular dating tends to go back to more distant ancestors than fossils ('long fuse models'). Classical studies of the fossil record, anatomy and physiology, as well as limited DNA sequences of proteins and genes, agreed that Theria diverged from monotreme about 166 Mya and marsupials diverged from eutherians 120–150 Mya. The order of divergence has been reinforced by many DNA sequence comparisons (although early comparisons of mitochondrial DNA sequences lead to the opposite tree). Whole genome data recently provided dates for divergence of these three mammal groups of 218 Mya for the divergence of monotremes from other mammals (together called subclass Theria), then a split between marsupials and eutherians 169 Mya.

I will not quibble about dates because my evolutionary arguments rest only on the order of nodes, which has wide agreement. And I shall stick to the long-accepted, although perhaps too short, dates of 166 Mya for our divergence from monotremes, 148 Mya for our divergence from marsupials, and 105 Mya for the eutherian radiation.

1.15 The Mammals

Eutherian mammals are by far the most abundant and diverse group. There are about 5,400 species on the planet. Four groups diverged from each other very rapidly in the Cretaceous, sometime between 80 and 100 Mya. Afrotheria (e.g., the elephant) and Xenarthra (including sloths and the armadillo) have long been contenders for the most basal placental mammal groups, diverging from the other lineages 95–105 Mya (at present, Xenarthra seem to be ahead by a nose).

The other two groups diverged about 87 Mya. Laurasiatheria includes bats, carnivores (such as dogs and cats) and cetartiodactyls (including hoofed animals and whales). Euarchontoglires include rodents and primates, which diverged about 78 Mya.

Marsupials are famous for their distinct mode of reproduction, in which immature young, born as small as a pea or a grain of rice, complete development attached to a teat, often (but not always) protected in a pouch.

There are about 350 marsupial species. Most are confined to Australasia (Australia and New Guinea), but two groups of small arboreal possums are found in South America, with a single species in North America.

Although marsupials are popularly considered to be quintessentially Australian, fossil evidence shows that they originated in Asia and migrated through the Americas into the supercontinent Gondwana. Australian fauna were isolated when the land bridge with South America was disrupted by continental drift 55 Mya. Old Australian fossils are scant; we might have to wait till global warming melts the kilometres of ice above Antarctica to fill out the earliest fossil history of Australian marsupials and provide reliable dates for divergence of Australian species.

There have been many efforts over decades to sort out marsupial phylogeny, and again comparisons of DNA sequences (mitochondrial and several nuclear genes) now provide the most detailed and reliable measures of relatedness. These relationships reflect early marsupial migration, with the deepest nodes between two groups of American marsupials and Australasian species.

Two marsupial cohorts are recognized, Australidelphia and Ameridelphia which diverged about 82 Mya. Ameridelphia contains two orders of small arboreal possums. Australian marsupials are much more diverse, containing five orders. One contains the big-footed kangaroos and wallabies that Australia is famous for, called macropodids for obvious reasons. Another contains carnivorous marsupials (dasyurids), including the famous Tasmanian devil and many species of small rodent-like insectivores with no pouches. Oddly, a single species of Australian marsupial returned to the mountains of South America, bamboozling geneticists and cytologists for decades.

The egg-laying monotreme mammals are represented by two extant families, the platypus and echidnas, all confined to Australasia. The platypus (*Ornithorhynchus anatinus*) is still relatively common (though vulnerable) along the waterways of eastern Australia. The echidna (or spiny anteater, *Tachyglossus aculeatus*) is widespread in southern Australia, and there are several (the number is disputed) rare echidna species of *Zaglossus* in New Guinea that are larger.

In contrast to therian mammals, which give birth to live young ('viviparity'), monotremes lay eggs. Platypus lay one or (usually) two leathery little eggs, which the female tends in a deep burrow, then feeds the hatchlings on milk sucked from fur on her abdomen (they have no teats). Monotremes also retain other ancient characteristics from ancestral reptiles, including skeletal features (they walk more like a lizard than a dog) and the ability to make venom.

It is important to recognize that marsupials and monotremes represent independent experiments in mammalian evolution equivalent to that of eutherian mammals. Platypus have been evolving for exactly as long as man and mouse, and they represent just as much divergence, even if they are less abundant and widespread, and are represented by fewer species. Marsupials and monotremes are in no sense 'primitive' or to be regarded as 'intermediate' in the evolution of eutherians. As for eutherians, some of their features may represent the retention of ancient characteristics, whereas others are highly derived.

1.16 Model Mammals

Genetic work on mammals relies heavily on the availability of model species which can be bred in captivity.

There are several eutherian species that have been adapted to the laboratory, but mouse must be the standout for genetic work. There are several strains of laboratory mouse adopted all over the world. Mice have the advantage of being small vegetarians, and have a very rapid life cycle, and well-understood reproduction. There is an amazingly well-curated set of strains with mutant genes or aberrant chromosomes. The laboratory mouse has been studied genetically over decades, and mutant strains and experimental manipulation (knockouts and transgenics) have been vital to our understanding of mammalian sex chromosome organization and function.

However, other mammal groups, particularly artiodactyls of economic importance such as cattle, sheep and pigs, or the carnivores that are companion animals such as cats and dogs, are now the subjects of intense genetic scrutiny. My favourite placental animal is the elephant, which shares many features of sex chromosomes with humans despite being maximally divergent from them; however, I have never managed to convince animal house directors that the elephant is a suitable laboratory animal.

However, there are problems in relying on models. Rodents and primates diverged about 78 Mya, and their very different size, lifespan and physiology mean that we must be cautious about interpreting data from mice. In addition, the mouse genome is unusually rearranged. Indeed, we will see that sex chromosomes provide several cautionary tales.

Marsupials are not easy or cheap to study in the laboratory, but they are not impossible. Three species have been captive bred for some years in attempts to develop a 'laboratory marsupial'. Colonies of the mouse-like dasyurid *Sminthopsis crassicaudata* (the fat-tailed dunnart) and a small member of the kangaroo family *Macropus eugenii* (the tammar wallaby), are used extensively for genetic studies in Australia, and an American marsupial, *Monodelphis domestica* (the Brazilian opossum), has been adopted in the USA and UK for genetic and particularly embryological studies.

Several endangered and vulnerable Australian marsupials (e.g., koala, and the Tasmanian devil) have been studied genetically as a means to aid conservation of wild populations and management of captive breeding. Another Australian species *Trichosurus vulpecula* (the brushtailed possum) receives a lot of attention for the opposite reason; it has become a serious pest in New Zealand since its introduction in the 1920s as a fur animal.

Platypus and echidna are not easy animals to study. They are difficult to breed in captivity, requiring space (platypus like privacy and long tunnels) and time and a great deal of money. Even the sex of an animal is hard to establish in the field.

1.17 Comparative Genomics to Study Sex

A huge source of genetic variation to study aspects of the human genome lies in the genomes of other species; other mammals, other vertebrates. Comparing gene maps and sequences between species can inform us about highly conserved processes like sex determination and dosage compensation.

We can compare different species at the level of their chromosomes, gene maps, or DNA sequence; it is most powerful to have all three levels of information.

Comparisons of sex chromosomes within primates, or within rodents, can inform us of recent changes in sex chromosome organization and function within these lineages. Comparisons between humans and mice inform us of changes that occurred since their divergence 78 Mya, and comparisons between humans and elephants may provide some insight into the organization and function of sex chromosomes in a common eutherian ancestor more than 100 Mya.

At the DNA level, genomes of placental mammals all look much the same, consisting of about 3 billion base pairs (3 gigabases, or Gb) and containing a very conserved suite of about 20,000 genes doing much the same jobs.

However, the eutherian genome is broken up into a wild variety of chromosomes (Section 1.2). In some lineages, such as primates and cats, the karyotype is extremely conserved; whereas others, like rodents, have rearranged quite insanely. Chromosome painting shows that ancestral blocks of chromosomes have been shuffled, with many evolutionary break points. The genome of our model mammal, the mouse, is particularly rearranged.

An even deeper view of mammal sex chromosomes can be provided by comparisons between eutherians and their most distant mammalian relatives, marsupials and monotremes. They are different enough to have evolved distinct genome organization and function, but similar enough to make comparison meaningful. These mammals have genomes of similar size, but their chromosomes are strikingly conserved.

Marsupials have a highly conserved basal karyotype comprising seven pairs of large chromosomes (diploid number $2n = 14$, haploid number $n = 7$), which are almost invariant between 60 species. Chromosome painting and gene mapping revealed that they are all built from the same 19 segments that are arranged and rearranged in all living species. It has been easy to deduce an ancestral karyotype, which has been changed only minimally in groups of Australian marsupials.

Monotremes also have very conserved karyotypes; platypus and echidna chromosomes differ from each other mostly by translocations involving the centromere. Their karyotype looks somewhat reptile-like, with a few large and many very small chromosomes.

Why eutherian chromosomes are so malleable, we don't know, but it may be the consequence of frequent rearrangements between families of repetitive sequences that infected an ancient eutherian mammal more than 100 Mya. This impression is fortified by looking at karyotypes of other vertebrates, where conservation is the rule. For instance, chickens and emus have virtually identical chromosomes, and even turtle and snake karyotypes are very similar.

Comparisons of mammal genomes with those of other vertebrates are particularly enlightening for study of sex because it turns out that sex is determined and expressed very differently in birds, reptiles, amphibians and fish, despite the obvious similarity of their genes and chromosomes.

One oddity of reptile (including bird) chromosomes is that their genome is divided into six large chromosomes and a flotilla of tiny 'microchromosomes'. It was initially thought that microchromosomes were inconsistent between individuals and represented some kind of B-grade DNA. However, gene mapping and sequencing verify that they are real chromosomes with real centromeres, and actually carry a much higher density of genes than macrochromosomes, which are replete with repetitive sequences.

We are accustomed to consider the macrochromosomes as normal and the microchromosomes as weird products of fragmentation. However, it is likely to be completely the other way around, since reptile microchromosomes are almost unchanged from the tiny chromosomes of chordates such as Amphioxus, from which our lineage diverged more than 500 Mya. So microchromosomes are the real animal chromosomes, that are conserved, with much the same gene content they have had since they diverged from a common ancestor half a billion years ago.

To get the most out of comparative genomics, we need good maps to be able to compare genomes meaningfully. Detailed genome maps of non-human eutherian species have been made by linkage and deletion studies, somatic cell genetic and radiation hybrid mapping, and molecular mapping methods. These maps initially contained mostly microsatellite markers, which are useless for cross-species comparisons, but with improvements in detection of single nucleotide polymorphisms in coding genes, there are now many anchor points for comparative mapping in very dense linkage maps.

Sex chromosomes of other mammals and other vertebrates offer us much insight about the organization, function and evolution of our own sex chromosomes. Comparisons can be made at many levels; phenotypic, cytological and molecular.

Closest to home, we observe remarkable conservation of sex chromosomes among all (or almost all) eutherian mammal, despite their greatly differentiated genome arrangements, and there is unanimity among eutherians about how dosage of sex chromosomes is to be managed, and what gene directs sex determination. However, there are differences in gene content and arrangement that provide us with information about how our sex chromosomes got to be so weird.

Differences between the sex chromosomes of eutherians, marsupials and monotremes are more interesting, and these have provided major insights, defining the building blocks of our sex chromosomes and a start date for their differentiation. And some rare exceptions provide us with a larger view of how sex determination works and how it evolved. In addition, differences in the mechanism by which gene dosage is equalized between the sexes provide insights about the epigenetic mechanisms of dosage compensation.

As we widen our comparisons to reptiles, amphibians and fish, our view back in time becomes longer and longer. These wider comparisons, between the sex chromosomes of other vertebrate groups show that our XY system is far from typical. The gene that determines human sex is a recent acquisition and our sex chromosomes are not nearly as old, or as stable, as we believed. The many independently evolved sex chromosomes of other higher vertebrates have led parallel lives; they help establish the ground rules for sex chromosome function and evolution. We can even learn from animals that do away with sex chromosomes entirely in favour of environmental sex determination.

A famous geneticist, Theodosius Dobzhansky, once commented that 'nothing in biology makes sense except in terms of evolution'. Sex is a wonderful illustration of the truth of this observation. The story of sex, genes and chromosomes starts with humans and mice but is enlivened by stories of some really weird systems in really weird animals.

FURTHER READING
Books

There are any number of good up-to-date biology, molecular biology, biochemistry, genetics and zoology texts.

Alberts B, Heald R, Johnson AD, et al., 2024. *Molecular Biology of the Cell*, 7th ed. Wiley Direct (digital or print)
Deakin JD, Waters PD, Graves JAM eds. 2010. *Marsupial Genetics and Genomics*, Springer
Griffiths A, Doebley J, Peichel C, Wassarman DA, 2020. *An Introduction to Genetic Analysis*, 12th ed. W.H. Freeman and Company
Hartl D, Cochrane B, 2017. Genetics: Analysis of Genes and Genomes, 9th edition
Johnson MH, 2018. *Essential Reproduction*, 8th ed. Wiley-Blackwell
Nelson DL, Cox MM, 2021 Principles of Biochemistry, 8th edition
Pough FH, Janis CM, 2022. *Vertebrate Life*, 11th ed. Oxford University Press USA
Strachan T, Lucassen A, 2022. Genetics and Genomics in Medicine, 2nd ed. John Wiley & Sons Ltd

There are also many popular versions, such as *Genetics for Dummies* and so on, even a Manga Comic version.

Reviews and Research Articles

Some interesting history:

Gartler SM, 2006. The chromosome number in humans: a brief history. *Nature Reviews Genetics* 7, 655–660
Paigan K, 2003. One hundred years of mouse genetics: an intellectual history – The classical period (1902–1980). *Genetics* 163: 1–7

The old and new in DNA sequencing:

Heather JM, Chain B, 2016. The sequence of sequencers: the history of sequencing DNA. *Genomics* 107: 1–8

Logsdon G, Vollger MR, Eichler EE, 2020. Long-read human genome sequencing and its applications. *Nature Reviews Genetics* 21: 597–614

Latest contributions to endless debates about mammal relationships and web-based searches of eutherian relationships:

May-Collado LJ, Kilpatrick CW, Agnarsson I, et al., 2015. Mammals from 'down under': a multi-gene species-level phylogeny of marsupial mammals (Mammalia, Metatheria). *PeerJ* 3: e805. https://doi.org/10.7717/peerj.805
Murphy WJ, Foley NM, Bredemeyer KR, 2021. Phylogenomics and the genetic architecture of the placental mammal radiation. *Annual Review of Animal Biosciences* 9(1): 29–53
Zoonomia Consortium. 2020. A comparative genomics multitool for scientific discovery and conservation. *Nature* 587: 240–245. https://doi.org/10.1038/s41586-020-2876-6

Part I

Organization and Evolution of Sex Chromosomes

2

Sex and Sex Chromosomes

2.1 Sex

Sex is the most profound normal difference between human individuals. Not only has sex facilitated the evolution of all complex life, it has helped shape human history, culture and society. It has been called 'the greatest invention of all time'.

Sexual reproduction is almost universal amongst organisms with a nucleus (eukaryotes). Nearly all animals and most plants reproduce by means of gamete production and fusion, alternating between a diploid somatic generation in which cells have two copies of the genome, and a haploid germ cell generation in which cells have a single genome.

In vertebrates haploid male and female gametes (sperm and eggs) are produced by diploid parents at meiosis (Section 1.6). This specialized division, in which the genome divides once but the cell divides twice, produces haploid cells with unique combinations of parental traits.

The whole reason for being male or female is to produce gametes; sperm or eggs. In animals, meiosis occurs in specialized gonads; testes in the male and ovary in the female. Testes produce sperm, often in profligate numbers; these specialized cells comprise a tightly packaged genome and means of swimming to the egg. The ovary makes eggs (ova), large and packed with nutrition and messages for the developing embryo.

But gonads are by no means the most noticeable features that differentiate male and female animals. Male and female phenotypes are specialized for their different roles in the process of mate selection, gamete production and nutrition of the young. This produces many somatic differences between the sexes. Most obviously, genitalia are purpose built for delivering and receiving semen. But there may also be body size differences, dramatic in many vertebrates and many-fold in some insects. Or colour differences as in fish. Many birds have spectacular male plumage; the male and female eclectus parrots were thought for many years to belong to different species based on their completely different colouration. Behavioural differences between male and female animals, particularly in courtship, are legion.

In humans there is another dimension to sexual differentiation; taking in social, psychological, cultural and behavioural aspects. It is important to distinguish biological sex from socially constructed gender, and I will try to honour this distinction throughout. Mostly I will be considering biological sex; how it works and how it evolved, but I will discuss gender in Chapter 11.

Human males and females are not as spectacularly different as for parrots. However, as well as differences in genitalia, there are different skeletal characters that ease childbearing in females, and different distributions of hair (a male display) and fat that can complement breast development. More controversially, there are differences in behaviour. How profound these secondary sexual differences are is underlined by the recent finding that more than a third of our 20,000 genes are expressed differently in at least one organ in males and females (Section 8.6).

The same general rules apply to insects and other invertebrates, and even to plants and many single-celled eukaryotes, in which sex is optional, but differences in gamete size and motility, and contributions to the early development of the embryo, permit us to define a 'male' and 'female' state.

2.2　Sex – Why?

Why is sex so omnipresent among higher organisms? What advantages does sexual reproduction offer?

The question becomes more puzzling when you realize that sexual reproduction should not be favoured by any organism whose aim in life is to maximize its genetic contribution to the next generation. In sexual reproduction, one parent contributes only half of their genes of each offspring (Figure 2.1). Cloning oneself should be a popular option because a parent contributes all their genes to each offspring.

However, in the animal world, such vegetative reproduction is rare, occurring sporadically and usually rather briefly. For instance, there are hybrid lizards whose chromosomes won't pair at meiosis and have no other option but to reproduce by parthenogenesis, in which the female makes eggs out of just her genome. This option is not open to mammals, in which a set of at least 30 developmentally crucial genes must pass through sperm to be active in the embryo (Section 7.14).

The long-accepted answer to the problem of this 'two-fold cost of sexual reproduction' is that sexual reproduction is favoured because genetic recombination offers increased genetic diversity. It was argued by RA Fisher and HJ Muller in the 1930s that mixing alleles from two parents each generation allows new beneficial combinations of characters from two parents to be created on the same

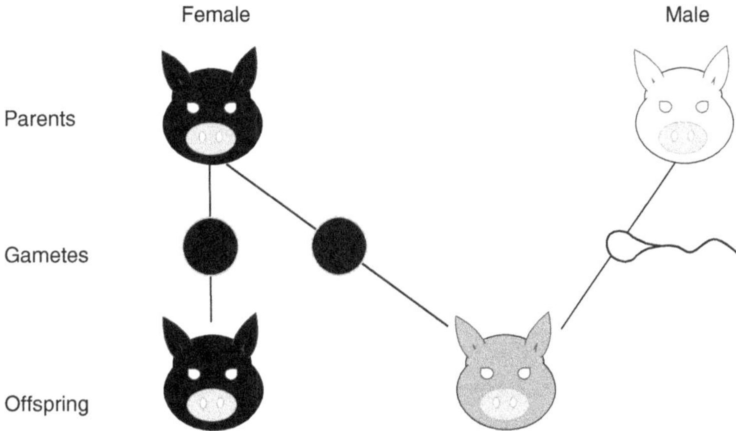

Figure 2.1   Why share your genes? The twofold cost of sex. If a female reproduces asexually, her offspring (essentially a clone) gets her full complement of genes (black). If she contributes an egg that is fertilized by a sperm from a male, the offspring inherit a set of her genes (black) and a set of his genes (white).

chromosome. This is much quicker than if multiple mutations had to occur on the same chromosome in the same lineage.

But why should genetic variability be such a good thing? It has long been argued that increased variability provides raw material for adaptation to new niches. This might be particularly useful to organisms with a long generation time, which might otherwise have little capacity for rapid response to environmental change.

Sex may also provide the variation needed to counter the unwelcome attentions of parasites, and this is thought to be its major role. It is certainly true that small populations of animals with little genetic variation, particularly in the major histo-compatibility locus (MHC), are especially vulnerable to pathogens. For instance, cheetahs have little variation at the MHC locus, and a large captive colony of cheetahs was almost wiped out by a new distemper virus to which all animals were sensitive.

Genetic variability might also ensure that cancer cells cannot be transmitted between animals; witness the transmissible cancers in the genetically depauperate Tasmanian devil that is driving the species toward extinction.

2.3 Sex – How?

Questions of how babies are born male or female are as old as humanity. Greek philosophers 3000 years ago considered that primaeval humans were originally giant bisexuals, who were later cut in half by Zeus, then came together in sexual embrace. The original Adam and Eve story may have been an offshoot if 'rib' was a mistranslation of 'side'.

In ancient Greece, warmth was considered a male characteristic, and male determination was ascribed to warm semen or warm winds. Theories also abounded that credited the right (more perfect) side with maleness and the left with femaleness (conveniently this established a hierarchy of perfection with men at the top). One theory was that the position of the foetus on the right or left side of the womb determined its sex, another was that sperm from the right testicle produced boys, and sperm from the left produced girls.

Discovery of eggs and sperm had to wait for the invention of microscopes in the seventeenth century. Sperm were first seen as tiny eel-like 'animacules' by van Leuwenhoek, and for a time it was supposed that the egg simply nourished their development into embryos. Eggs (or at least follicles of cells nurturing them) were discovered later in female rabbits, and the opposite idea was then put forward that the sperm were nothing but parasites (hence their name 'spermatozoa').

It was not until the nineteenth century that fertilization was observed, first in the sea urchin, in which sperm motion and penetration of a whole microscope field of eggs is a dramatic sight. It was accepted that both eggs and sperm were crucial, and that an embryo was formed by fertilization of a single egg with a single sperm, followed by fusion of the nuclei.

Understanding of how sex was determined in human embryos came from two sources; the discovery of sex chromosomes, and the observation of a special pattern of sex-linked inheritance.

2.4 Discovery of Sex Chromosomes

Sex chromosomes were identified in insects long before they were seen in mammals (Figure 2.2). They were first observed in 1891 as a dark-staining blob in the cells of a male, but not a female, plant bug. Indeed, the origin of their names derives, not from their shapes at metaphase of mitosis, but from the mysterious sex-specific nature of the X (for 'unknown') chromosome in an insect species in which females have two and males have a single unpaired chromosome.

In the early twentieth century it was observed that in some insect species, sperm were of two types; half had a dark-staining blob and the other half lacked it. Later (1905) Netty Stevens compared the chromosomes of male and female mealworms, showing that males possessed a small chromosome that was absent in females. She proposed that this small chromosome was responsible for determining maleness, and called it 'Y', as a follow-on from the bug X. Meiosis produced two kinds of sperm, containing either an X or a Y.

The first sex chromosome pair to be investigated in detail was in the fruit fly *Drosophila melanogaster*. In this species, females have two X chromosomes, and males have a single unpaired X plus a heterochromatic (different staining)

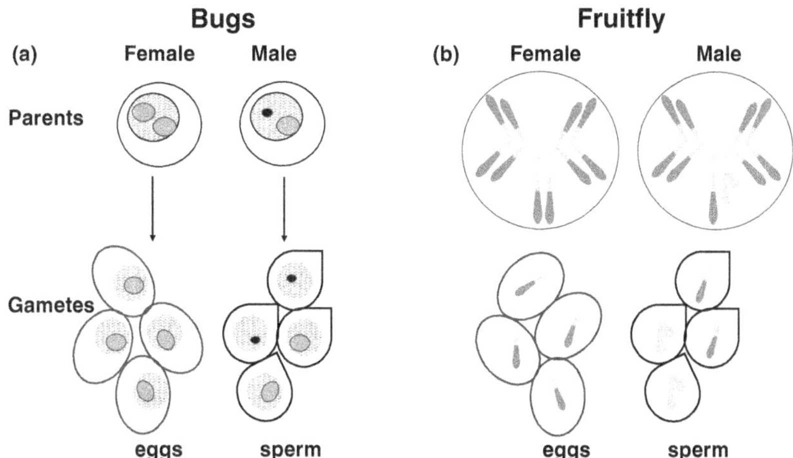

Figure 2.2 Discovery of sex chromosomes. (a) Bugs and mealworms were described in which males and females had different sized heterochromatic blobs in the nuclei of somatic cells and sperm. These blobs represent X and Y chromosomes. Here an X chromosome (large grey blob) is present in two copies in females and one in males. The male has a smaller, darker blob that represents a Y chromosome. All the eggs laid by the female contain one X, but sperm are of two types, half bearing an X and half a Y chromosome. (b) The genome of the fruit fly *Drosophila melanogaster* contains only four pairs of chromosomes. Three pairs (two large and one tiny dot) are autosomes, present in both sexes. Females have two copies, and males have a single copy of a large X chromosome, and males have a large Y that is heterochromatic (pale grey). Eggs all contain a single X, but sperm are of two types, half bearing an X, and half bearing a Y (autosomes not shown for gametes).

element specific to males, the Y chromosome (Figure 2.2). Pioneering geneticist Thomas Hunt Morgan went on to discover other XX female: XY male insects, as well as species with no Y (XO male: XX female, like the bug that provided the 'unknown' blob).

In XX / XY systems, XX females make only one kind of egg, bearing an X chromosome, whereas XY males make two kinds of sperm, carrying either an X or a Y. Females are therefore called the 'homogametic' and males the 'heterogametic' sex.

Morgan described a completely opposite system in birds, in which females rather than males are the heterogametic sex. We call these chromosomes Z and W to avoid confusion: ZZ birds are male and ZW birds are female.

Mammal sex chromosomes were not discovered for some decades. Theophilus Painter (who had erroneously insisted humans had 48 chromosomes) observed a male-specific Y chromosome in human testis preparations. When mammalian chromosomes became tractable, cells from men and women were found to have a complement of 22 pairs of autosomes (non-sex chromosomes), plus the sex

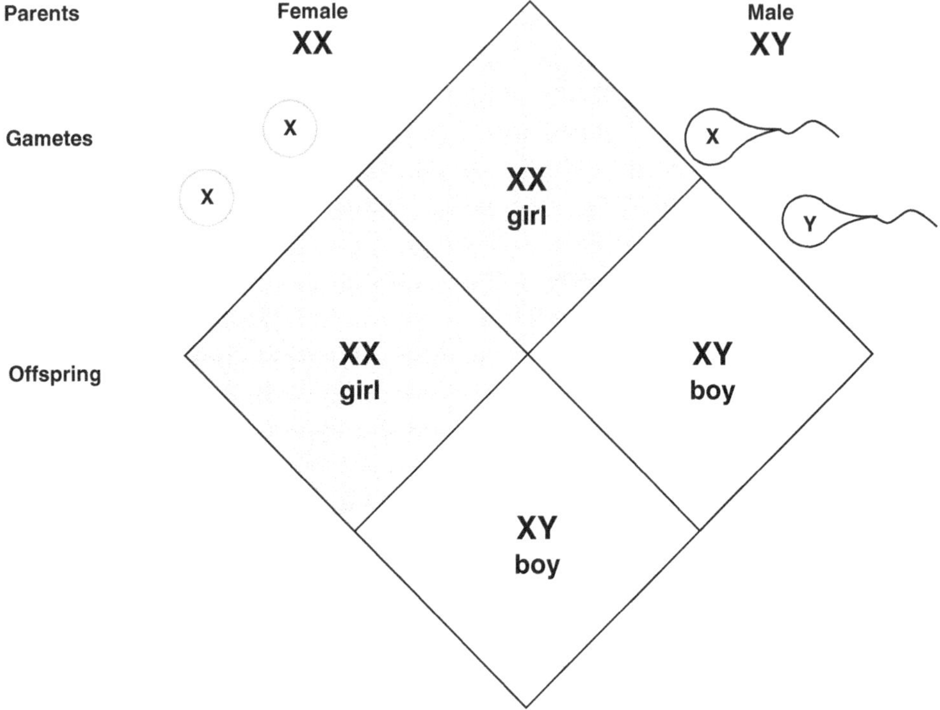

Figure 2.3   How sex chromosomes determine sex in humans. Women have two X chromosomes that segregate one each into eggs. Men have one X chromosome and one Y, so sperm are of two types (hence 'heterogametic'); half carry an X and half carry a Y. X-bearing eggs fertilized by X-bearing sperm are XX, and are born females (shaded). Eggs fertilized with Y-bearing sperm are XY, and are born male. This produces a 1:1 ratio of boys and girls.

chromosome pair. Females had two copies of a medium-sized X chromosome, and males had a single X and a much smaller, and oddly staining (heterochromatic), Y chromosome.

Sex is determined by the sex chromosome borne by the sperm. If it contains an X, the baby will be a girl. If it contains a Y, the baby will be a boy (Figure 2.3).

A limitation to human cytology was getting human material (a cytologist bemoaned that he had to 'wait at the foot of the gallows' to get testicular samples). So many studies were done of chromosomes in rodents and domestic species. They all also proved to have a dimorphic sex chromosome pair, in which the Y is much smaller, and frequently heterochromatic, pairing with the X only over a tiny region at their ends. Most mammal species subscribe to the usual XX female: XY male sex chromosome constitution, although there are variants which have resulted from recent additions of autosomes to either the X or Y. Thus, the male is the heterogametic sex in mammals.

Genetic determination of sex by XY or ZW chromosomes has the big advantage that it ensures a sex ratio of 1:1. This is the optimal ratio, and is expected to be achieved by any system, as famously argued by the great statistical geneticist RA Fisher. If the ratio favours females, for instance, the rarer sex (males) will have a big advantage in mating with many females, and will spread his genes further. This will tip the ratio the other way until males have no further advantage.

Thus sex, and systems to determine it, are crucial concepts for animal growth, reproduction and evolution.

2.5 Sex Linkage and Sex-Linked Diseases

The first sighting of human sex chromosomes was not through the microscope, but was revealed by the odd pattern of inheritance of some human genes.

This followed early experiments in fruit flies that tracked an eye colour variant. Thomas Hunt Morgan in 1910 was baffled to find that for some traits reciprocal crosses produced different classes of progeny. Crosses of homozygous red-eyed females with white-eyed males produced all red-eyed progeny, as you would expect if the red allele is dominant over white. But surprisingly the reciprocal cross of white-eyed females to red-eyed males produced females with red eyes and males with white eyes. The link between this criss-cross pattern of inheritance and the passing on of X and Y chromosomes was suggested by cytologists (including Nettie Stevens), and embraced (though reluctantly) by Morgan.

The pattern of eye colour inheritance in fruit flies was explained if the eye colour gene lay on the X chromosome. Males have a single X which they inherit from the female parent, and a Y chromosome that contains no copy of the eye colour gene. So the male progeny of a white-eyed female will have white eyes whatever the genotype of the male parent. However, the female progeny gets a red allele on an X inherited from the male parent, which is dominant to the white eye allele from the female parent.

The occurrence of only one X in males leads to this criss-cross pattern of inheritance for alleles of genes on the X chromosome in humans as well as flies. This 'sex-linkage' occurs because an XY father can hand on his X chromosome to a daughter, but never to a son, who must (by definition) receive his Y chromosome. An XX mother, on the other hand, hands on her X chromosomes equally to sons and daughters (Figure 2.4).

The special mode of inheritance of genes on the human X chromosome makes it easy to spot sex-linked variants because they are exhibited mainly by boys. For instance, it was known as early as 1876 that the common form of red-green colour blindness was exhibited more commonly by boys (with a frequency of about 1 in 20) than girls (1 in 400). The trait was never passed from father to

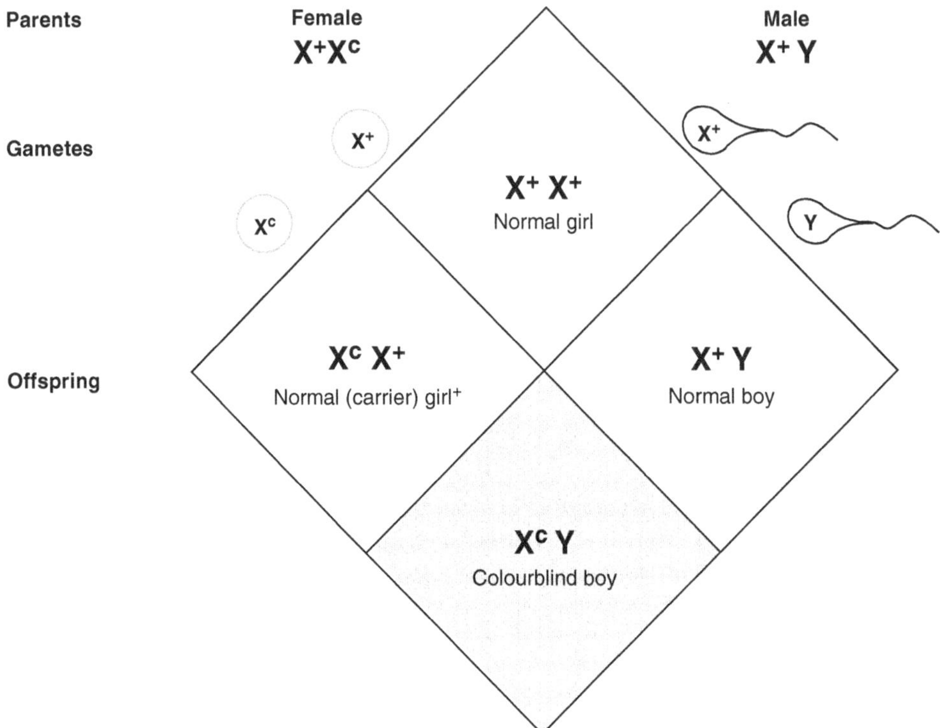

Figure 2.4   Sex-linked inheritance. A gene on the X makes normal red-green visual pigment (wild type, + allele). A mutant version (c) makes no pigment. A woman who has one + and one *c* allele (heterozygous) has normal vision because the + allele makes enough pigment to see colour. Half her eggs contain the X carrying the +allele, and half carry the X with the mutant c allele. If her partner has normal colour vision, half his sperm carry an X with the normal +; so all his daughters will have normal colour vision (but half will carry the c allele). Half his sperm carry the Y chromosome, which lacks a colour pigment gene; thus half his sons will have normal colour vision, and half will be colourblind. This inheritance pattern is called sex linkage.

son (sons only get a Y chromosome from their dad). Colourblind boys usually had mothers with normal vision but who harboured a recessive colourblind allele (often from a colourblind maternal grandfather).

Many sex-linked diseases have been known for decades. For instance, the pattern of inheritance of a blood clotting defect, haemophilia B, was known in the tenth century, and first described in 1802. The mutant allele does not make functional blood clotting Factor IX, rendering affected males deficient and smitten with a severe bleeding disease. The occurrence of haemophilia in the royal lineage is classic. Queen Victoria was a carrier of this disease but was not affected because her other X chromosome carried a normal allele that ensured an adequate supply of Factor IX in her blood. However, one of her sons and some of her grandsons (but not her daughters or aunts) inherited this disease.

A peculiarly horrible, and fortunately rare, sex-linked disease was discovered in 1964 when it was observed that boys, but almost never girls, manifested a familial disease characterized by severe gout and kidney problems, poor muscle control, mental retardation and self-mutilating behaviour. The disease is a manifestation of the lack of an enzyme (hypoxanthine-guanine phosphoribosyl transferase, or HPRT), part of the biochemical pathway that recycles purines and prevents the build-up of uric acid. Symptoms appear in the first year of life, and rapidly become severe and are difficult to manage because of the behavioural and mental impairment.

It was striking, too, that mental institutions were full of boys. Several mental retardation conditions displayed a sex-linked mode of inheritance in families, so were attributed to mutations in one of several X-linked genes involved in brain development and function (termed 'X linked mental retardation' genes, XLMR).

2.6 Sex Chromosomes of Mammals

All eutherian mammals (bar a few weird ones which I will discuss in Section 4.7) have an XX female: XY male chromosomal sex determination mechanism, or some variant of it (Figure 2.5).

Sex chromosomes are unusual from every angle. The X is a medium-sized chromosome, taking up about 5% of the haploid chromosome length in nearly all eutherian mammals, regardless of the number and size of their autosomes. This observation, along with some primitive gene mapping, led Susumu Ohno to predict that the mammal X is invariant in size and gene content ('Ohno's Law').

The X chromosome in males resembles autosomes in its normal appearance, lack of heterochromatin, and location within the interphase nucleus. But in females, one X has these characteristics but the conformation and position of the other are different, symptoms of its genetic inactivity (Chapter 5).

The mammal Y chromosome is downright weird. It is usually small and stains intensively with dyes that bind to repetitive elements in heterochromatin. Probing chromosomes from humans and several other mammals with unique and repeated sequences reveals a concentration of repeated sequences and a dearth of expressed genes on the Y in every species studied.

Unlike the X, the Y is poorly conserved between species. It varies in size, with sheep posting a low of 1% of the haploid genome. The mouse Y is longer, with the centromere near one end. Variation occurs even within groups. For instance, although the X chromosomes of different primates are g-band identical, their Y chromosomes differ in size, repeat content and centromere position. The cat Y is larger than the dog Y and contains more repetitive sequences. Some sex chromosomes have accumulated inordinate amounts of heterochromatin. The vole *Microtus*

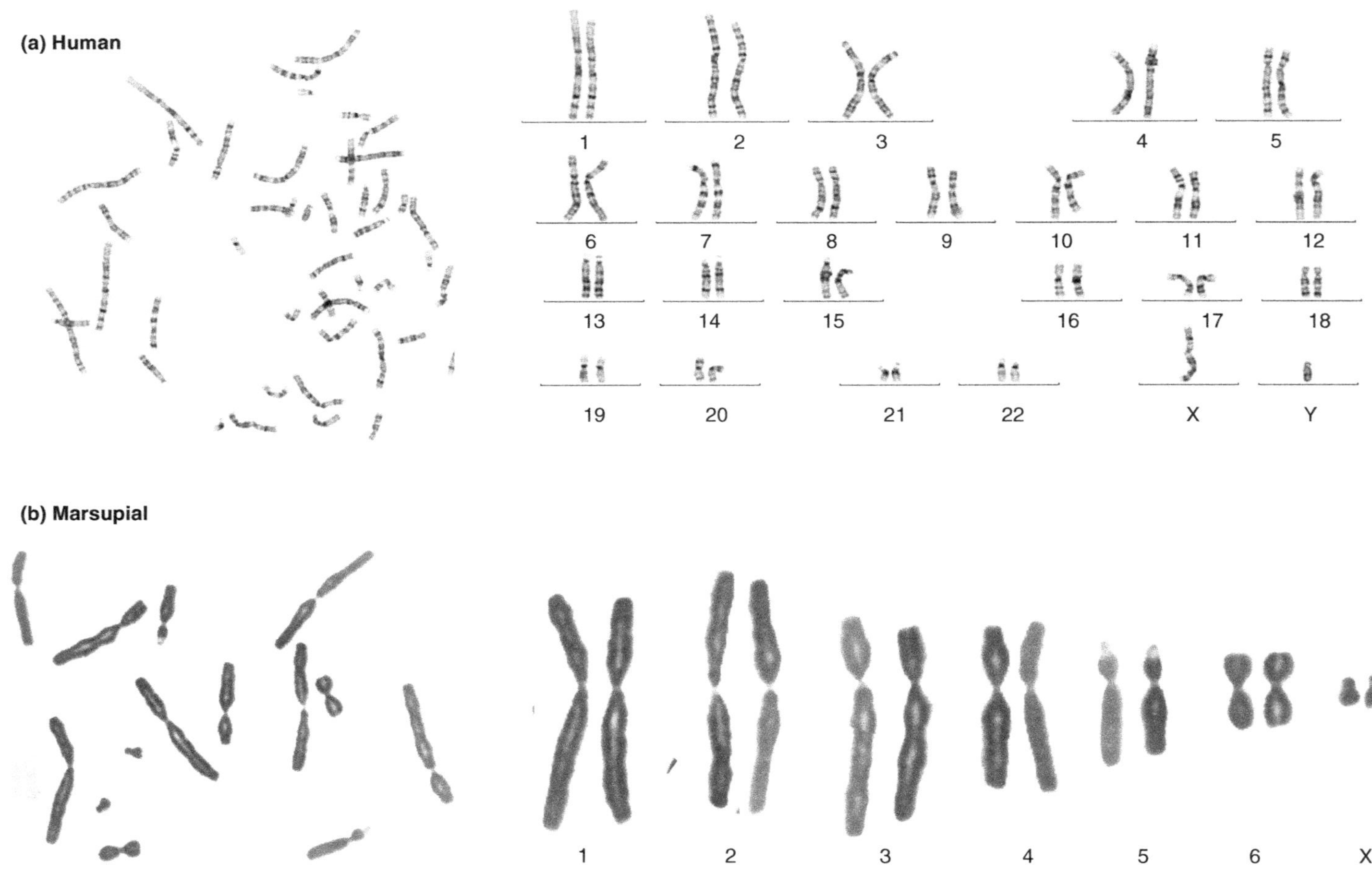

(a) Human
1
2
3
4
5
6
7
8
9
10
11
12
13
14
15
16
17
18
19
20
21
22
X
Y
(b) Marsupial
1
2
3
4
5
6
X

is famous for its enormous sex chromosomes, which stain brightly with dyes that bind to repetitive sequences. Generally, repetitive sequences are species-specific, although there is some cross-hybridization between male-specific sequences of closely related species such as humans and primates, cattle and sheep.

The X and Y chromosomes of eutherian mammals may look entirely different, but at male meiosis, they pair over a small region at one or other (sometimes both) ends. This pairing is crucial for meiosis to proceed, and for the X and Y chromosomes to be segregated properly into different sperm.

Marsupial mammals are exceptional in that the basic marsupial X chromosome is smaller than the X in eutherian mammals, and the basic Y is minute. For instance, in the dunnart (a mouse-like laboratory marsupial, *Sminthopsis crassicaudata*) the X takes up only 3% of haploid length, and the Y much less than 1%. There is no pairing region shared by the marsupial X and Y, which get together at meiosis by a completely different route.

In exceptional species, either the X or the Y has become fused to an autosome, Fusion of the X to an autosome produces an XY_1Y_2 chromosome constitutions, in which Y_1 is the real Y, and Y_2 represents the unpaired autosome. Fusion of the Y with an autosome produces an X_1X_2Y constitution in which X_1 is the real X, and X_2 represents the unpaired autosome. We call these fused chromosomes 'neo-X' or 'neo-Y'. The three sex chromosomes pair at meiosis (forming a 'tri-radial'), then the ex-autosome segregates with the original X or Y chromosome. Sometimes a fusion occurs to both the X and Y; for instance, in the kangaroo family, addition of an autosomal region bearing the nucleolar organizer to both X and Y increased their size and heterochromatin content. The platypus has taken

Figure 2.5 Mammal sex chromosomes. Photographs of the chromosomes of a human male, and a marsupial female. The chromosomes as they are seen in the cell are on the left, and the karyotype derived from this cell by cutting and pairing up chromosomes is shown on the right. (a) Humans have 23 pairs of chromosomes that can be distinguished by their size and shape and banding pattern; 22 autosomes and an XY sex pair. Females have two copies of the X chromosome, and males have a single copy of the X as well as a small, heterochromatic Y chromosome (Photographs of human chromosomes supplied by the Victorian Clinical Genetics Services, Murdoch Children's Research Institute, Melbourne). (b) Marsupials have much the same sized genome as humans, but it is divided into a few large chromosomes, so it is easy to analyse. Nearly all dasyurid marsupials (including New Guinea species *Murexia melanurus*, shown) have seven pairs of chromosomes. Six are autosomes, present in two copies in females and males, and the seventh is the sex chromosome pair. The X chromosome is present in two copies in females (shown) and one in males, and the Y chromosome is a minute element specific to males. (Photograph of marsupial chromosomes supplied by Dr Michael Westerman, La Trobe University, Melbourne).

fusion to an extreme; serial fusions with four autosomes resulted in 5X and 5Y chromosomes whose ends pair in a chain, from which 5X and 5Y chromosomes segregate (Section 4.5).

Of the more than four thousand mammal species, humans and mice are by far the best-known genetically. The mouse has been the favourite mammalian model for classic genetic studies over decades because they are small, cheap, manipulable and have a short generation time and many offspring (1.16). However, the mouse genome has diverged a long way from those of humans and other mammals.

Humans are surprisingly amenable for genetic studies, although they have none of these advantages. This is because patients with aberrant chromosomes and mutant genes are rather frequent. We do not have to search for them by screening the population because they often seek help (or their parents do) from the medical profession for abnormalities and disease states. Secondly, we care more about humans than about other mammals, and are willing to spend billions of dollars exhaustively mapping and sequencing entire human genomes.

Sex chromosomes have been studied in many mammals, but by far the most intensively in humans.

2.7 Human Sex Chromosomes

In humans, the sex chromosome pair is recognized at mitosis in male cells by differences in their size and morphology after different staining regimes (Figure 2.6). The X was first recognized as a member of human chromosome group C (numbers 6–12) of medium-sized submetacentric chromosomes. Originally, it was hard to identify by size and morphology, but it could be readily spotted a decade later when various banding techniques were invented. The X shows little C-banding or heterochromatin distinguished by fluorescence banding – indeed, it looks just like a regular autosome.

The human X is right on the average chromosome size. Its centromere divides it into a long arm (Xp) and a short arm (Xp) half its size.

The human Y is only about a third of the size of the X. It belongs to group G of human chromosomes (which also includes the smallest chromosomes, 21 and 22). It is a cytologist's nightmare; all you can really see with a microscope is a tiny short arm (Yp), and a long arm (Yq) that is just one big blob. There is barely room for one g-band.

The human Y is full of highly repeated DNA material that stains darkly with a treatment that highlights repetitive regions ('C-bands'). The long arm has a large chunk of heterochromatin that stains so intensely with fluorochromes that it literally glows in the dark.

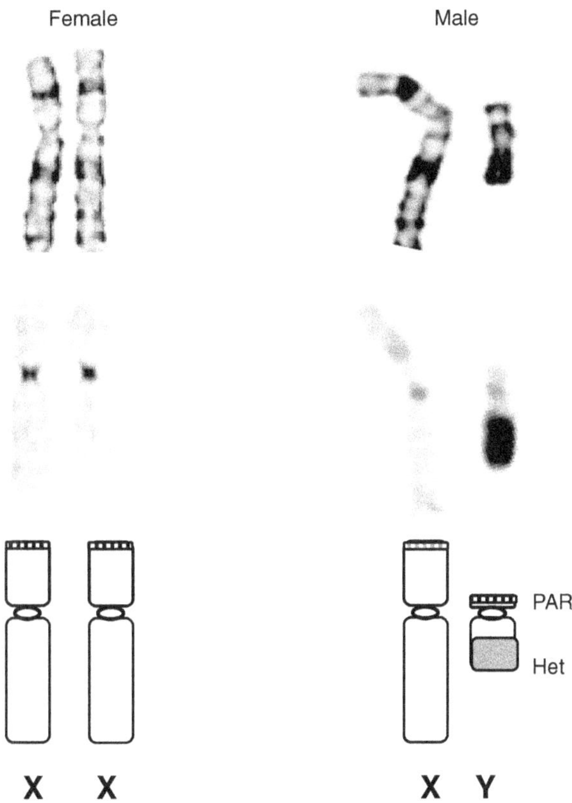

Figure 2.6   Views of human sex chromosomes. Human sex chromosomes, cut out from photographs of dividing blood cells from women (XX) and men (XY). The chromosomes at the top were treated with salts that give normal chromosomes a 'G-band pattern' of darker and lighter bands that distinguish different chromosomes. The chromosomes in the middle row were from cells treated under harsher conditions to reveal highly repetitive sequences ('c-bands') that are concentrated at the centromeres of autosomes and the X, but nearly cover the Y. The diagram at the bottom shows the pseudoautosomal region (PAR, cross-hatched) shared by the tips of the X and Y, and the heterochromatic region of the Y (dark grey) that contains no genes. (Photographs of human chromosomes supplied by the Victorian Clinical Genetics Services, Murdoch Children's Research Institute, Melbourne).

The two copies of the X chromosome pair at human female meiosis, and recombination rates are much the same as those of autosomes. However, at male meiosis the X and Y chromosomes, being largely non-homologous, pair and recombine over only a short pairing region at the tips of their short arms. If this region is deleted or rearranged, meiosis does not go ahead, and no sperm are formed. This short homologous region is called the pseudoautosomal region (PAR), and its strange properties are worth exploring in detail in Section 2.9.

2.8 Meiosis and the Sex Ratio

Fundamental to sexual reproduction is gamete production, during which homologous chromosomes (homologues) from the mother and father must pair at meiosis (synapsis) and segregate to opposite poles, to be incorporated into different gametes (Section 1.6) (Figure 2.7).

In the ovary of a female, the two X chromosomes in an oocyte pair and undergo recombination just like the autosomes, and segregate into haploid cells. The haploid ova all receive a single X chromosome. In the testis of a male, the X and Y chromosomes pair, then segregate into haploid sperm. Half the sperm get an X and half get a Y chromosome.

At fertilization, an X-bearing egg is fertilized by either an X-bearing or a Y-bearing sperm, more or less at random. So half the fertilized eggs are XX and will develop into females, and half are XY and will develop into males. The segregation of X and Y chromosomes into equal numbers of sperm is what ensures that about half of newborn babies are girls and half boys.

But the sex ratio among newborns is not exactly 1:1 as you'd predict. It is more like 105 boys to 100 girls worldwide. And studies of the sex of aborted foetuses suggests that earlier in development there is an even higher proportion of boys. This has been suggested to result from selection to offset the higher mortality of XY embryos (in fact, the higher mortality of males at every life stage).

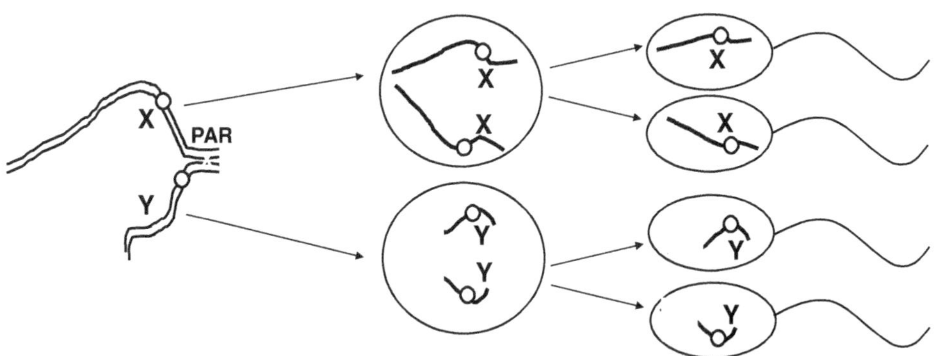

Figure 2.7   Sex chromosome pairing and segregation at male meiosis. 'The human X and Y chromosomes are very different sizes and lack homology over most of their length so they don't pair. However, there is a shared 'pseudoautosomal' region (PAR) of 2.7 Mb at the tips that is homologous between the X and Y. At male meiosis the X and Y pair up and undergo recombination in this region. The X and Y separate from each other (segregate) into different spermatids at the first division of meiosis, then these each divide to produce 2 X-bearing and 2 Y-bearing sperm, each with a haploid complement of autosomes (not shown).

There have been enormous studies of sex ratios worldwide. It is difficult to measure native sex ratio, given the strong preference for boys in many countries that leads to prenatal diagnosis and selective termination of girls (well, this is an advance on infanticide or neglect of girl babies that was practised over many centuries). The sex ratio in China and Pakistan is more like 120 boys: 100 girls, a lopsided distribution that threatens alarming social consequences (a million youths with no hope of marriage and a family; kidnap and rape of girls).

These studies of sex ratio provide some evidence of environmental influence on human sex ratio, in a (rather weak) correlation with latitude. There have been many claims that the sex ratio is more strongly biased toward boys after a war in which many young men were killed.

Such claims may seem quite fanciful, but they receive support from studies of other mammals in which sex ratio is far from 1:1. For instance, sex ratio in a tiny mouse-like marsupial (*Antechinus*) in Australia is strongly biased toward females (nearly 70%), and this bias has its origin before birth. Suggestions that sex is allocated according to nutrients available don't seem to apply because all litters showed this bias, and it bore no relationship to the mother's weight and condition.

A sex ratio that favours males or females may well be a selective advantage to some mammals, but how does it happen? Are Y-bearing sperm lighter and faster than X-bearing sperm? Do they penetrate the egg differentially? Or is there some mechanism that intervenes at meiosis to preference the maturation of X- or Y-bearing sperm, or to degrade it differentially where sperm is stored by the female? Indeed, there are many "meiotic drive" mechanisms that distort sex ratio.

2.9 Meiosis and the Pseudoautosomal Region

Making gametes with just one copy of the genome ('haploid') is the job of meiosis (Section 1.6). In this specialized division chromosomes divide once, but the cell divides twice; this is accomplished by pairing of homologous chromosomes and segregation into different gametes. In males the X chromosome is segregated away from the Y into different gametes. Pairing between homologous chromosomes is essential for correct chromosome segregation into haploid sperm. If pairing of any chromosome fails, meiosis is aborted and no gametes will form.

How does this work with sex chromosomes, which are spectacularly different in size and DNA constitution in males? Pairing requires DNA sequence homology, which is obviously lacking over most of the highly differentiated X and Y of mammals. Yet the X and Y chromosomes segregate into X-bearing and Y-bearing sperm. Many studies of infertility show that pairing of sex chromosomes is essential; pairing failure, for instance, in interspecific mouse hybrids and people with abnormal numbers of sex chromosomes (aneuploids), results in infertility.

Although no cytological homology was apparent between the human X and Y, they were observed to undergo pairing at meiosis, as judged by the presence of synaptonemal complex and visual evidence of crossing over (chiasma). This observation backed up much earlier proposals of synapsis between the rat X and Y in male meiosis. Crossing over was usually limited to the tips of the short arms of the human X and Y, but pairing could involve more, even most of the Y. Later, it was observed that synaptonemal complex is formed at the tip, and may spread into non-homologous regions and involve more than half the Y chromosome.

Visible meiotic pairing predicted that the human X and Y shares a region of genetic homology at the tips of Xp and Yp, which pairs at meiosis and undergoes regular recombination (Figure 2.7).

This region of homology was defined genetically in the mid 1980s with the description of a gene (now called *CD99*) and three noncoding bits of DNA which were mapped to the tips of both Xp and Yp. DNA sequence variants were used in family studies to demonstrate exchange between the X and Y in males. This region was found to comprise about 2.7 Mb of DNA that is genetically homologous between the X and Y.

Because sequences within this region would display an apparently autosomal mode of inheritance, even though they were borne on the sex chromosomes, this putative X-Y homologous region was termed the 'pseudoautosomal region', or PAR for short. The rest of the Y chromosome has been termed the Male-Specific region on the Y (MSY).

To everyone's surprise, it became apparent that there is a second, very tiny PAR at the other ends of the human X and Y chromosomes. DNA segments originating from the X and Y both mapped to terminal Yq as well as Xq, and recombination between them was demonstrated in families. The big and small pseudoautosomal regions are now called PAR1 and PAR2. The 330 kb PAR2 pairs only infrequently and is observed as a circular configuration at male meiosis.

Pseudoautosomal regions also occur on the X and Y chromosomes of other mammals. They are surprisingly diverse in size and position. A small PAR is homologous between the X and Y in mouse, within which recombination is essential for proper segregation of the X and Y at meiosis. In male mice carrying a Y with a deleted PAR, the X and Y fail to pair, chromosome segregation is aberrant, and few normal sperm form.

A PAR is not an absolute necessity. There are several independent examples of PAR-less XY chromosomes. Some exceptional rodent species (pygmy mice, voles, gerbils) have lost the PAR altogether and evolved a completely different way to segregate X and Y chromosomes at meiosis. Marsupial X and Y chromosomes also lack a PAR. In both Australian and American species, X-Y pairing is accomplished by a proteinaceous plate (containing the same protein that accomplishes synapsis in other mammals) that attaches the X and Y side by side during meiosis and ensures the correct segregation.

Pairing between the X and Y chromosomes raises fundamental questions about the control of chromosome pairing and recombination. The restriction of pairing to the tiny region of homology between the X and Y, and the necessity to form at least one chiasma to hold the X and Y together, mean that the frequency of recombination within the PAR must be much higher than the average 1% per Mb for autosomes. It is also higher in male than in female meiosis; recombination between the same two markers at either end of PAR1 is 50% for males, compared with only 3% for females. Similarly in the tiny (1 Mb) mouse PAR, recombination is initiated by double-strand breaks every 0.7 Mb, rather than every 10 Mb for autosomes.

This must mean that recombination within the PAR is regulated by something other than DNA sequence, since this is the same on the X and Y within the PAR. This feeds into long-standing questions of what controls the number and distribution of chiasmata over chromosomes. In the autosomes, recombination occurs at short (1–2 kb) 'hotspots', mediated by a chromatin-modifying protein. Pairing and the double-strand breaks that initiate recombination happen later in PARs. However, once initiated, crossing over occurs faster, possibly because chromatin is held in smaller loops, offering more hotspots. It is likely that hotspots are activated by male-specific factors, one of which has been identified. In the mouse PAR, repetitive sequences have been identified that bind proteins that initiate double-strand breaks in DNA.

The PAR, then, acts very differently from the rest of the X and Y chromosomes.

2.10 Sex Chromosome Variation

Variants of sex chromosome form and function have been of critical importance in characterizing sex chromosomes, examining the control of gene activity and discovering how genes on the mammalian sex chromosomes control sex determination. Much has been learnt about sex chromosomes and sex determination from model mammals with aberrant sex chromosomes, and especially from human patients with abnormal numbers (aneuploid), or with deleted or rearranged sex chromosomes (Figure 2.8).

Infertile patients were described in 1959 with aberrant numbers of X chromosomes. Males with Klinefelter's syndrome had two X chromosomes like a female, plus a Y like a male. These XXY individuals (with a frequency of 1/700) are unequivocally male, but they have small testes, reduced androgen production and no spermatogenesis, and may show breast enlargement.

Females with Turner syndrome have a single X and no Y. The frequency of XO conceptuses is as high as 2%, but most spontaneously abort in pregnancy, so that

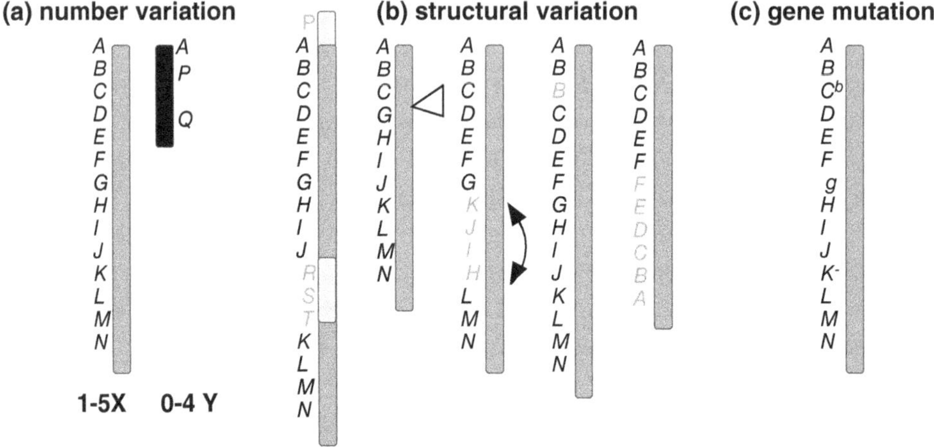

Figure 2.8   Human X chromosome variation. (a) Number variation: X chromosomes may be present singly (as in XY males and XO females with Turner Syndrome), or in multiples up to five. Y chromosomes may be absent, or present in up to four copies. (b) Structural variation: X Chromosomes (shown with genes A-N) may have a region from another chromosome insertion (addition of P from the Y, or of *STR* from an autosome), may have lost a small or large region (deletion symbolized as delta), may have a region switched around (inversion represented by curved arrows) or may have a small or large duplication of one or many genes. (c) Gene mutation: Individual genes on the X may be present in an alternative, but not detrimental form (e.g., polymorphism C^b), may have suffered a debilitating mutation to form a recessive allele that has either lost all function *(g)* or works in an aberrant way *(K-)*.

the frequency among liveborn babies is only 1/2,500. This suggests that surviving XO babies may actually be mosaics (that is, start off with a mixture of XO and XX or XY cells that carry them through early critical stages before being lost). XO women have underdeveloped 'streak' gonads, little breast development and fail to menstruate and produce eggs. They are usually short because they have only one copy of a height gene on the PAR, and often have skeletal abnormalities and other mild physical abnormalities like webbing of the neck. There is some suggestion that XO girls who inherited the X from their mother (X^mO), but not those who inherited the X from their father (X^pO), have difficulty with social skills, suggesting that the X may carry 'social' genes that are imprinted (that is, expressed from the paternal, but not the maternal X, see Section 7.14).

Other X chromosome aneuploidies have been described, including females with three, or more rarely, four or even five X chromosomes. XXX females may be normal and even fertile. Major effects of extra copies of the X are on development of external sexual characteristics, and mild to severe mental retardation depending on the number of extra X chromosomes. Males with XXXY and, rarely, XXXXY constitutions have Klinefelter-like development and mild to severe mental retardation depending on X dosage. Mosaic patients with every possible kind of mixture of normal with aneuploid, or XO/XXX, cells have been described, with a great variability of phenotypes.

The observation that patients with X aneuploidy were male if they had a Y chromosome, and female if they didn't was critical to the conclusion that the human male determining gene lay on the Y chromosome (Section 8.2).

Males with extra copies of the Y chromosome have also been described. XYY males were first found among prison populations, giving rise to the idea that an extra 'killer Y chromosome' produced antisocial and aggressive 'savage supermales'. This notion created a flurry of public debate and was popular with defence lawyers, but it soon became apparent that XYY males were present in the general population (frequency about 1/1,000) and are overrepresented in basketball teams, probably because they tend to be taller, with three copies of the height gene on the PAR.

Sex chromosome aneuploidy cannot be inherited from a parent (because it results in infertility), but stems from an accident of meiosis. It can arise when a mistake in chromosome segregation forms eggs with two, or no X chromosomes, or sperm with XY, YY or no sex chromosome. Alternatively, it can happen if a chromosome is lost or improperly segregated at mitosis in the early embryo, when it produces a baby with a mixture of normal and aneuploid cells (mosaic). The Y chromosome, being small and peculiar, seems particularly prone to loss, and is often lost from cells in culture.

Patients with deletions or rearrangements to the X or Y are also very informative. Some female patients (usually normal but sterile) have gross rearrangements of one X chromosome, such as deletions and duplications. Sometimes either the long arm or short arm is duplicated and the other is missing, producing iso-Xq or iso-Xp constitutions. The X may also participate in translocations with autosomes, which cause sterility because of problems at meiosis.

Also, a number of people whose sex was not concordant with the possession of a Y chromosome proved to have rearrangements in which a part of the Y chromosome had been illegitimately swapped onto the X when recombination occurred outside the PAR. Analysis of the missing bits of the Y was critical in locating the sex-determining gene (Section 8.10).

The effects of sex chromosome aneuploidy, gross deletions and rearrangements on their bearers are remarkably benign, being limited to differences in growth and reproduction. This contrasts with the disastrous effect of duplication of any but the smallest autosome (the small and gene-poor chromosome 21) and the lethality of deletion of even tiny autosomal regions. This tolerance is due to the effects of the very forgiving X inactivation system, which maintains balance between the sexes by the preferential inactivation of the abnormal X (Chapter 5).

X chromosome aneuploidy and rearrangements are also observed in other mammals. XO primates, horses and cattle, like Turner XO women, have undeveloped gonads and are infertile, and the growth differences in XO and XXY primates and horses seem to parallel those of XO and XXY humans. However, XO mice show no embryonic lethality or growth retardation, and are fertile, at least early in life. This suggests that the gene content and activity of sex chromosomes are a bit different in mice and humans, underlining the caution with which we must interpret data from model mammals.

There are many rearrangements of sex chromosomes and autosomes that have been described and preserved in mice. To help distinguish direct effects of sex chromosomes from effects of hormones unleashed by male and female gonads, mice have been manipulated and bred with reversed sex chromosome constitutions. XX males have a sex-determining gene spliced to one X, and XY females are missing this gene from the Y. Thus these 'Four Core Genotypes' comprise XX females and XX males; XY males and XY females. They have been particularly useful for sorting out whether sex differences in behaviours are due to Y genes or male hormones (Section 8.1), as well as discovering major metabolic differences due to X chromosome dosage (Chapter 5).

2.11 Genetic Maps of the Human X Chromosome

Three regions of sex chromosomes – the X-specific region of the X below the PAR, the male-specific region of the Y (sometimes called MSY) and the PAR – have very different characteristics, and it makes sense to discuss them separately. For each, I will describe how genes were discovered and characterized, then recount how complete molecular descriptions of the sex chromosomes have enhanced our understanding of how they work and how they evolved.

We have more and more powerful genetic tools to characterize human sex chromosomes. Here I discuss early work on sex-linked diseases, then genetic mapping of the sex chromosomes, producing maps that were the foundation for their molecular characterization and sequencing (Chapter 3).

The X was historically by far the best-known of any human chromosome. This was because genes on the X can easily be distinguished by their distinctive X-linked mode of inheritance of the traits (phenotypes) that they affect (Figure 2.4).

By the 1970's, there were about 30 genes (mostly diseases and a few enzyme variants) assigned to the X because they showed a sex-linked mode of inheritance. There are now more than 200 disease phenotypes ascribed to the human X chromosome (a complete listing is available through the Online Mendelian Inheritance in Man (OMIM).

Studying sex-linked disease genes was the impetus for constructing the first maps of the human X chromosome. The classic linkage approach (Section 1.8) to ordering genes on the X was taken by scoring recombination among the male offspring of a mother heterozygous for two X-linked genes, colour blindness and haemophilia A (caused by mutations in the X-borne *F8* gene and deficiency in the blood clotting Factor VIII). She can make four types of gamete. Two are parental; that is the combinations of alleles she inherited directly from her mother or father. Two recombinant types are formed when crossing over occurs between the two genes at meiosis. For instance, if she received both mutant alleles from her carrier mother, recombinant gametes are those which carry either the haemophilia or the colour blindness allele, but not both. Since her sons receive no other X chromosome (they receive a Y from their father), the mutants will be expressed, and it is easy to score sons according to whether they received a parental or a recombined X.

If the blood clotting gene *F8* and the visual pigment gene *RCP* were so far apart that they would always recombine, you would expect 50% recombinants. However, over several families, only about 1/8 of boys received a recombinant X. The low recombination means that these genes must be close together on the X ('linked'). Since the frequency of recombination increases as the distance between the genes, the amount of recombination provides a measure of their distance apart, which is therefore 12.5%, or 12.5 centimorgans (cM). A linkage group was established that included colour blindness, haemophilia A and deficiency for the enzyme G6PD. A second linkage group on the short arm encompassed a blood group and three genetic diseases.

This approach was limited by the few families segregating more than one visible X-linked condition. However, phenotypic markers were supplemented by biochemical markers and DNA variants, whose segregation could be followed in families. First RFLP variants and minisatellites (Section 1.3), now mostly single nucleotide polymorphisms (SNPs), have provided recombination values low enough to be additive. These markers filled the gaps and allowed linkage to be established over wider distances, for instance, the genetic diseases ichthyosis (*STS*) and Duchenne muscular dystrophy (*DMD*) (Figure 2.9).

Extremely dense linkage maps of the X have been constructed over decades. Very large studies on hundreds of families using thousands of markers add up the total length of the X to about 180 cM, and include markers from the tip of Xp to the tip of Xq.

Constructing a linkage map of the X chromosome was much easier for laboratory and domestic mammals than it was for humans because crosses can be deliberately set up between parents differing in two or more traits, and their segregation monitored among many progeny.

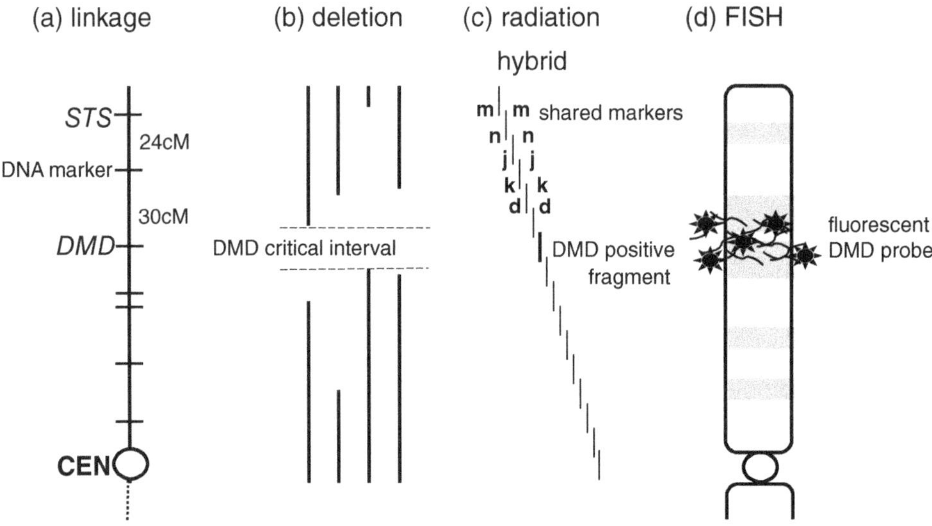

Figure 2.9 Four types of maps are illustrated for the short arm of the human X chromosome. (a) Linkage (genetic) maps order genes, phenotypes and DNA markers by assessing the frequency of recombination between them. The diagram shows that the disease ichthyosis (caused by mutation in the steroid sulfatase gene, *STS*) is linked to a DNA marker because there are only 24% recombinants rather than the 50% expected if they assorted independently. Duchenne muscular dystrophy (mutation in the *DMD* gene) is also linked to this marker, with 30% recombinants. The two disease genes are too far apart to show linkage, but both are linked to the DNA marker between them; thus, linkage values are additive. (b) Deletion maps are constructed by lining up the markers on X chromosomes from *DMD* patients with different bits of X deleted, all of which include the *DMD* gene (gaps in lines). The diagram shows how these gaps align and share a minimal 'critical interval' that must contain the *DMD* gene. (c) Radiation hybrids are constructed by irradiating human cells, then fusing them to rodent cells. Cell hybrids contain fragments of human chromosomes. Diagram shows how they can be arranged in order by observing overlapping markers (e.g., m, n, j …) to obtain a detailed map around *DMD* and isolate a small fragment containing the *DMD* gene. (d) Fluorescent in situ hybridization (FISH): The cloned DNA fragment bearing *DMD* is tagged with a fluorescent dye and hybridized to human chromosomes. The fluorescent dot seen under the microscope (represented by starbursts) physically localizes *DMD* on the short arm of the X chromosome.

Indeed, detailed linkage maps of mice preceded human linkage maps by decades, in attempts to develop model systems to study genetic disease. Many natural and induced mutant phenotypes were mapped to the X with reference to many molecular markers. Sex linkage was observed for several coat colour mutants (one which led to the discovery of X chromosome inactivation, Section 5.2), as well as biochemical and behavioural abnormalities. The availability of different mouse strains and subspecies increased the supply of markers many-fold. Hundreds of phenotypic and biochemical loci were mapped to the mouse X by the 1970s. A mouse X linkage map containing 118 cloned genes (most with human homologues), 18 X-linked traits and 411 anonymous sequences in an unambiguous linear order was accomplished by 1997.

A rat X map followed closely, as well as X maps for many domestic species that were developed in order to locate traits of economic importance. Even the marsupial X was mapped by crossing wallaby subspecies. As first noted by Susumu Ohno, these X maps showed remarkable conservation of the gene content of the X between species, starting with the classic enzyme markers *G6PD, PGKA, GLA* and *HPRT*.

2.12 Physical Maps of the Human X

Several different methods for physically mapping the human X complemented this linkage map; deletion mapping, somatic cell genetic mapping, radiation hybrid mapping and cytogenetic mapping by fluorescence in situ hybridization (FISH) (Figure 2.9).

Deletion mapping of the X chromosome provided some of the first information about the physical location of disease genes on the X chromosome. Deletion mapping is particularly useful for mapping the X chromosome because deleted regions are missing from the genome in males, who have only one X. If two markers were usually deleted together, it could be inferred that they lay close together. If a third marker was always deleted along with two outside markers, it could be inferred to lie between them. Thus, the chromosome can be divided into deletion intervals.

This method was used to find the gene mutated in Duchenne muscular dystrophy, a sex-linked muscle-wasting disease. Many affected boys had deletions of part of the short arm of the X which overlapped in a critical region at Xp21, which was therefore assumed to contain the *DMD* gene. Boys in some families, evidently with larger deletions, also had other abnormalities such as ichthyosis (a skin condition due to absence of steroid sulphatase activity) retinitis pigmentosum (eye cancer resulting in blindness). This contiguous gene syndrome is understandable if the *STS, RP3* and *DMD* genes lie close together on Xp.

Association with a cytological abnormality also allowed the identification of a gene mutated in the most common of the X-linked mental retardation syndromes. This syndrome, exhibiting characteristic facial features (large ears) and large testicles, was observed to be more common in boys than girls. It seemed to be associated with the appearance of an extended 'fragile' site in the distal region of the X, leading to the identification of the Fragile X Mental Retardation gene *FMR1*.

Somatic cell hybridization gave a huge impetus to human gene mapping, and enabled many genes, particularly enzyme loci, to be assigned to the human X in the

1970s. The X chromosome was particularly well served by this approach, because the X-borne selected marker *HPRT* (the gene that is mutated in Lesch-Nyhan disease), and therefore the human X, could be selected for or against (Section 1.8). Rodent-human hybrids retaining a human X were found to express the human, as well as the rodent, forms of housekeeping enzymes *PGK, G6PD, GLA* and *HPRT*. Hybrids lacking a human X expressed none of these human enzymes, but only the rodent form. These four genes were therefore on the same physical chromosome (syntenic), so they could all be assigned to the human X.

Regional assignments were also made using hybrids between human cells with X deletions or rearrangements. The human X was divided into 40 intervals by the break points in a panel of such somatic cell hybrids, allowing sequences to be assigned to each region.

Much greater refinement was possible by deliberately shattering the human genome by pre-irradiating cells before fusion to form 'radiation hybrids'. Some of these contained very small fragments of the X. Panels of radiation hybrids were constructed which retained overlapping regions of the human X, then these were screened for many markers, allowing large-scale ordering of genes and DNA markers.

Large-scale information about the human X chromosome was afforded by cytogenetic mapping using FISH. Many cloned gene sequences were positioned on the X, and the resolving power was sufficient to localize the probe sequence to a single chromosome band. Gene and marker order could also be determined by tagging probes with fluorochromosomes of different colours. Detailed order within a region of the X could be established by in situ hybridization to DNA stretched out from interphase nuclei onto a slide (fibre-FISH). Some very gene-rich regions (e.g., Xq28) were put together early because there were already many genes cloned from the region, so it was well mapped. A large-scale cytogenetic map of the X was built up in the 1980s and 1990s.

The X chromosomes of many other species were mapped with the same combinations of methods; for instance, rat, cat, dog, bovine.

Thus, several mapping methods relate gene and marker positions to cytological views of X chromosomes. The order of genes and DNA markers on the X by genetic and physical mapping are in excellent agreement, although the relative distances of the two techniques, as expected, differ because recombination is not uniform over the chromosome.

2.13 Cloning X-borne Disease Genes

The availability of a detailed linkage map made it possible to pinpoint a disease gene on the X with respect to polymorphic DNA markers (Section 1.7), allowing the gene to be assigned to a region represented by a large insert clone. It was then possible to clone the gene and characterize mutant alleles.

Positional cloning was successful even before the advent of large insert clones. For instance, the X linked *DMD* gene that was located by deletion analysis (2.12) was isolated and characterized, immediately explaining the mystery of mild and severe muscular dystrophy. *DMD* was found to be a huge gene (at 2.2 Mb, the largest in the human genome) and to contain many functional repeating units. It is translated into a large protein called dystrophin, whose function is to link the muscle fibre through other proteins to the extracellular matrix and cell membrane. Out of frame deletions that messed up the genetic code and produced nonsense protein caused severe Duchenne's muscular dystrophy, whereas in frame deletions that resulted in a functional, though shorter, protein caused the less severe Becker muscular dystrophy. This was one of the first successful 'reverse genetics' characterization of a normal protein and a pathogenic mutation.

Fragile X Mental Retardation syndrome was mapped because mutation is associated with an extended 'fragile' site near the tip of Xq, making it possible to positionally clone the *FMR1* gene. The gene codes for a protein that is expressed in neurones and has a role in synaptic plasticity. The disease-causing mutation has some extraordinary features. Amplification of a repeated three-base sequence (triplet) upstream of the gene affects its function, and the degree of amplification increases (along with the severity of the syndrome) as it is inherited through generations. This was the first 'triplet repeat' amplification to be discovered, and greatly increased the understanding of other diseases (like the classic Huntington's chorea) that also showed amplification of a 'premutation'.

Genes on the X could also be detected systematically by screening cDNA libraries with X-specific clones. Transcripts of X-borne genes were selected, for instance, from heart and placenta cDNA libraries, by hybridizing to X-specific genomic clones. This missed many genes expressed only transiently in a specific tissue, and overestimated gene numbers by counting many alternative transcripts that are produced from many or most genes by differences in splicing. However, it has the advantage that cDNA sequence contained no introns and could be directly translated to deduce the protein product.

Some genes were discovered on the human X in a roundabout way. For instance, *UBE1X* (ubiquitin binding enzyme) was first detected because it rescued a temperature-sensitive cell cycle defect in cell hybrids, and it mapped to human Xp.

Thus mapping the genes on the X chromosome was the first step in the molecular characterization of the X, and answered many medical questions. It also raised new questions about the unusual features of the X, which, we will see, is special in many ways.

2.14 The Conserved and Specialized Mammal X Chromosome

Cloning and mapping genes to the X chromosome in mice and many other eutherian mammals (including cats, cows and other domestic animals, even porpoises and bears) reinforced Ohno's observation that the gene content of the mammal X chromosome is remarkably conserved. For instance, the classic enzyme markers *G6PD*, *PGK* and *GLA* were present on the X in all eutherians – and even marsupials. So were homologues of genes cloned from the human X, such as *F8* (haemophilia) and *RCP* (red colour pigment). Cloning homologous sequences and mapping them by recombination, somatic cell genetics or FISH showed that the X is almost perfectly conserved among eutherian mammals. This is spectacularly different from the autosomes, which have been broken and fused and rearranged many times in mammal evolution.

An exception to this conservation was marsupials; the basic marsupial X is much smaller than the eutherian X (and the Y is minute). Somatic cell genetic mapping and FISH using human cDNA probes or marsupial BACs showed that genes on the human Xq had homologues on the kangaroo and dunnart X, but genes further from the centromere (distal) on Xp were autosomal in marsupials. We will see that this difference is a key to understanding how the human X evolved (Section 4.1).

The conservation of the mammal X chromosome provided another route to discovering human X genes by homology to the mouse X. For instance, the sequence of mouse X-borne gene *Amel* (amelogenin), mutated in mice with incomplete amelogenin in their teeth, pulled out a human homologue, which mapped to the human X. An important gene, *KDM5*, that codes for a lysine demethylase was first cloned in mice from its initial discovery of a homologue on the mouse Y; it proved to have homologues on the X (and Y) of all therian mammals. *RBMX* was discovered by its homology to a Y-borne gene in kangaroos!

This homology is important in practice, since many mouse mutant phenotypes reflect human diseases, and they are a lot easier to study in a model mammal than in humans. For instance, a classic X-linked coat mutation (mottled) in mice turned out to be caused by an alteration in a copper transport gene, which has a human equivalent that is mutated in Menkes disease.

So does the mammal X chromosome harbour quite an ordinary collection of genes? Or do their functions promote female development or repress male development? Or are they special in some other way by virtue of their single dosage in males?

The functions of many genes mapped to the X by classic studies of X-linked diseases were inferred from the mutant phenotype, as we have seen. But how do we go the opposite way – from gene to function? Clues come from the tissues in which they are active, by the amino acid sequence and inferred conformation of their protein products, and by comparison with the functions of genes with similar sequences. Genetic ontology terms have been applied to known genes, and we can now just look down the list.

Extraordinarily, half of the 200+ X-linked diseases include mental retardation or brain/behaviour and/or eye abnormalities. These categories make up more than 50% of catalogued X-linked mutants, but fewer than 20% of autosomal mutants. X chromosome variants are linked to differences in brain development and anatomy. Genes on the X therefore seem to affect preferentially the brain and nervous system and the higher sense organs (eye and inner ear). Defects in reproduction (abnormal gonad development, infertility) are also disproportionately sex-linked.

Most extraordinary is that many of the mental retardation syndromes are accompanied by gonadal abnormalities; these genes are expressed in the brain and in the testis, and have been called 'brains-and-balls' genes.

In addition, the X contains sets of genes that are more strongly expressed in one sex or the other (sex-biased genes). There is evidence that the X is enriched in genes that are expressed more strongly in male somatic tissues, particularly in the brain and muscle, and reproductive tissues.

We must conclude that the X chromosome is no ordinary chromosome. Competing models of the human X chromosome are represented in the cartoon (Figure 2.10).

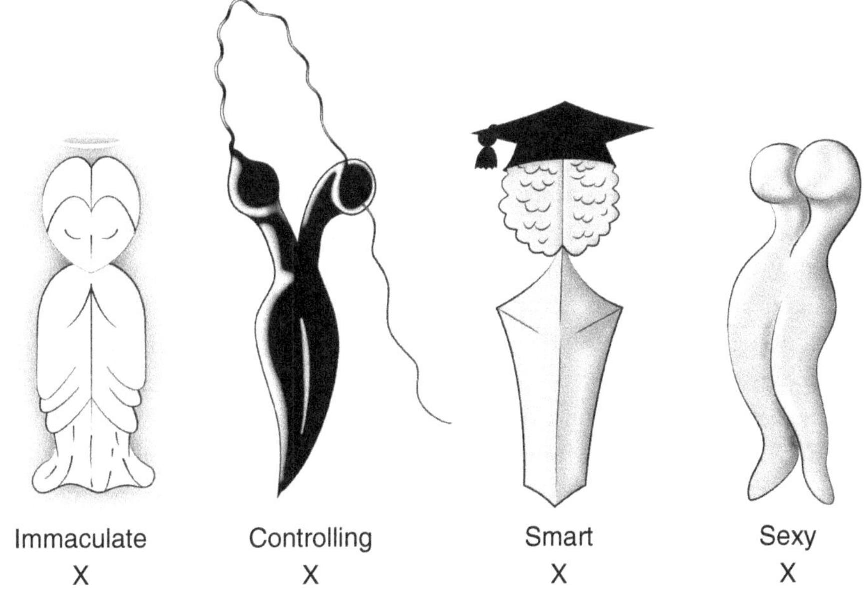

Figure 2.10　My models of the human X chromosome. The human X is no ordinary chromosome, and many models have been proposed to capture its specialness. Here are four of mine. 'Immaculate X' refers to the original proto-X (an autosome), and the other manifestations to observations that the X carries a disproportionate number of genes with functions in regulation, on intelligence and on sex and fertility. Original drawings by Jenny Graves, rendition by Bronwyn Knight, Knight's Design Cavern, knightscavern.myportfolio.com.

In Chapter 4 I will discuss the evolutionary forces that endowed it with its unusual structure and biased gene content.

2.15 Genetic Characterization of the Pseudoautosomal Regions

The pseudoautosomal regions shared by the human X and Y chromosomes act more like autosomes than sex chromosomes. PAR1 on the tips of the short arms and the tiny PAR2 on the tips of the long arms undergo recombination, so PAR genes do not exhibit typical X-linked or Y-linked inheritance patterns in families. Yet they are physically a part of the X and Y, so they get tugged into sex chromosome behaviour.

Since the male-specific region of the Y does not recombine with the X, the entire differential region of the Y, sex determination included, acts as a single locus. Thus genes in the Y PAR are partially sex-linked, depending on how far they are from the boundary of the PAR. A gene within the PAR will show an 'autosomal' pattern (hence its name) if it lies sufficiently far from the boundary of the differential region to undergo recombination in all cells, producing 50% recombinants. Indeed, PAR markers showed a gradient of sex linkage from 50% (independent assortment) for a marker near the telomere to 2% for a marker near the PAR boundary.

Traits have been assigned to PAR1 by observing correlations with disturbances of pairing or PAR deletions. Of these, male sterility is the most obvious. However, it is not specific to a gene because it could result from any chromosome deletion or rearrangement that affects pairing and leads to failure of meiosis.

The only other phenotype assigned to PAR1 was short stature. There was long-standing suspicion that the short stature of XO girls with Turner syndrome was caused by a single dose ('haploinsufficiency') of a gene normally present on both the X and Y chromosomes. Very strong evidence for a growth control gene in the PAR1 was gathered from analysis of terminal deletions of Yp or Xp in patients with inherited short stature in some families. Deletion and mutation analysis pinpointed a gene, *SHOX* (short stature homeobox-containing gene), at the tip of the X. *SHOX* was transcribed from both X and Y alleles in embryonic bone cells. An extra dose of *SHOX* would also explain why XO females are shorter and XXY and XYY males are taller than average.

Other genes were discovered in PAR1, including *MIC2* (now called *CD99*), which makes a surface antigen that is tightly linked to sex, implying that it is near the PAR boundary. Several other genes were discovered between *MIC2* and *SHOX* that included growth factor receptors, enzymes involved in DNA replication, energy metabolism and immune function, splicing regulation and transposition.

Another gene even nearer to the boundary with X-specific sequence is a gene whose variants are detected as blood group *XGA*.

Since crossing over between the X and Y occurs within the PARs, it is possible to construct recombination maps of the two pseudoautosomal regions. A detailed multipoint linkage map of PAR1 was constructed using microsatellite markers within and outside the PAR, and using PCR to detect different size alleles in single sperm. A linkage map of the tiny PAR2 was also constructed using microsatellite markers derived from a large DNA clone containing the blood clotting Factor VIII (F8) gene at the end of Xq.

All told, the two pseudoautosomal regions contain a total of 28 active genes. They have a range of housekeeping and tissue-specific functions. The gene density contrasts with the paucity of genes on the differential region of the Y, and the atypical Y gene functions, which we will discuss in Section 2.16.

2.16 Traits Determined by the Y Chromosome

Genetic characterization traditionally starts by identifying visible characters controlled by genes on the chromosome. This approach drew a blank for the human Y chromosome.

The effects of Y genes, being male-limited, should be obvious from strict father-to-son inheritance. Obviously, male development itself fits this pattern, and the Testis Determining Factor on the Y (TDF) that initiates male development was identified on the human Y (Section 8.7). The search for TDF uncovered a male-specific HY antigen (HYA), raised by inoculating female mice with cells from males of the same strain. This antigen cross-reacts with cells from human males, but not females, and turns out to be a reaction to a ubiquitously expressed gene(s) on the Y (Section 8.9).

Several male-specific traits, such as baldness, were initially thought to be Y-linked. In particular, there were early claims to have identified a Y-linked genetic trait for hairy ears, which showed male-to-male transmission in pedigrees. Behavioural traits like schizophrenia have also been attributed to the Y chromosome. However, most of these traits, rather than being Y-linked, are expressed preferentially in males (male-limited) because of interaction with hormone pathways; male-pattern baldness is actually X-linked.

Fertility is the only well-defined trait, other than sex, that can be assigned to the human Y. It was known since the 1970s that many men who made little, or no, sperm (azoospermic) had deletions in the middle of the long arm of the Y, suggesting that this region housed one or more azoospermia (AZF) genes (Figure 3.4). There is also evidence for an oncogene *GBY* on the Y, which is responsible for gonadoblastoma, a tumour that often occurs in the undeveloped gonads of patients who possess at least some parts of the Y, but lack a functional testis.

There are also normal, fertile men who are found to have Y chromosomes with either a particularly short, or a particularly long Y long arm. The absence of phenotype for these Yq variants suggests that the heterochromatic long arm of the Y comprises repetitive sequences that contribute little to the phenotype.

Genome-wide association studies, which have revealed many loci on the X and autosomes that are involved in various disease states and thousands of complex conditions, come up with a completely blank Y. Thus, apart from sex determination itself, it is difficult to pinpoint phenotypes that can be assigned to the differential region of the human Y chromosome.

In other mammals, too, the Y chromosome seems to specialize in sex determination and spermatogenesis. The function of the Y chromosome as a male determiner is virtually universal among eutherian mammals (Section 10.1). Functions in spermatogenesis were also ascribed to the mouse Y by the observations of infertile males with Y deletions or rearrangements. For instance, a small deletion in the short arm produced sperm with morphological defects, and larger deletions caused meiotic arrest that blocked sperm proliferation. Deletions of different sizes within the long arm produced a gradation of sperm abnormalities, suggesting multiple copies of a spermatogenesis gene that is required later in the process.

The Y of other mammals may also harbour genes involved in growth, revealed by the stature of XO and XXY animals. Also, mice, as well as primates and cattle, seem to share a Y gene that promotes growth of the pre-implantation embryo, first discovered in cattle by observing that when embryos were sorted by size, the bigger ones were mostly XY.

One of the few other phenotypes that have been assigned to the Y in any mammal species is a locus for hypertension mapped to the Y in a particular strain of rats, although, like baldness, this character is androgen-dependent.

There have been many reports, and many dismissals, of evidence that genes on the Y chromosome are associated with behavioural conditions. Certainly men and women have different frequencies and ages of onset for conditions such as schizophrenia, autism and ADHD. It is always difficult to establish that these are direct influences, rather than hormone-induced and therefore sheeted home to sex determination. However, there is some evidence that variants in *PCDH11Y* (on the Y only in humans) are associated with common behavioural conditions. *PCDH11Y* encodes a protocadherin which guides nerve cell connections, and could be involved in brain lateralization. There are reports, too, that a gene on the mouse Y chromosome contributes variants that give mouse strains different levels of aggression.

Thus, the Y chromosome of humans and other mammals seems to bear few genes that influence visible characteristics other than sex and spermatogenesis.

However, as we will see, this search for phenotypes does not capture the several genes on the human Y that have partners on the X chromosome and whose dosage may be critical.

2.17 Mapping the Y Chromosome

The Y chromosome is not amenable to recombination mapping because the entire male-specific portion does not recombine with the X at meiosis. Physical maps are required.

Constructing a physical map of the human Y chromosome using the same techniques as for the X proved to be a challenge. Even the euchromatic region of the Y is bedevilled by a high content of repetitive sequences, making it difficult to find probe sequences that don't hybridize all over the Y. FISH using Y-derived BACs usually showed hybridization to several regions of the Y.

The development of a battery of unique DNA sequences from the Y was vital to provide markers to order deletions or fragments. These DNA markers could also be localized using FISH, although the resolution was too poor to be much use with such a small chromosome.

Deletion mapping provided the first physical map of the Y chromosome. Natural deletions of the Y are frequently detected in infertile men. DNA from many patients with partial Y chromosomes was examined using the unique Y DNA markers. The first physical map of the Y divided it into initially seven, then 43, deletion intervals.

Radiation hybrid mapping of the human Y chromosome greatly refined the deletion map. A radiation hybrid panel was tested for hundreds of unique Y DNA markers, and the fragments could be ordered with respect to the deletion intervals.

In mice, also, extensive deletion analysis enabled regional localization of genes responsible for spermatogenic blocks on the mouse Y. Extraordinary mouse genome engineering, adding different pieces of the Y chromosome back to an XO genotype, also identified regions of the Y responsible for early and later blocks to meiosis.

Thus, genetic and physical mapping techniques were of some use to characterize the Y chromosome. But it took a molecular approach to reveal the peculiar structure and gene content of the human Y, as I will recount in Chapter 3 (Section 3.7).

2.18 Conclusions

Humans and other mammals have an XX female XY male system of sex determination. Females (the homogametic sex) make eggs that all have an X chromosome. Males (the heterogametic sex) make sperm, half of which bear an X and half a

Y chromosome. Random fertilization reconstitutes half XX females and half XY males in the next generation, more-or-less. The human sex ratio worldwide is 105 boys:100 girls at birth, and even greater preponderance of boys early in embryogenesis, selected, perhaps, to make up for higher mortality of XY embryos.

Discovery and mapping of genes on the X and Y in humans and other mammals showed that both sex chromosomes are extremely specialized.

Many techniques are available to discover genes and order them along the chromosome. Genetic maps (based on recombination between markers on the same chromosome) have been useful for the X and PAR, but not the non-recombining Y. Physical techniques for locating genes, such as somatic cell hybrid analysis and radiation hybrid mapping, deletion mapping and in situ hybridization, have been brought to bear on characterizing both the X and Y chromosomes.

The X chromosome seems fairly normal; an average size and euchromatic. It is by no means female-biased, but bears a range of genes with many different functions. Mutations in some of these genes give rise to X-linked diseases that afflict males more than females because they have a single X. However, it bears a disproportionate number of genes involved in neural function and reproduction – including many 'brains and balls' genes that are involved in both. It is conserved to a remarkable degree between species (Ohno's Law). The X is definitely not a typical chromosome.

The Y chromosome is exactly the opposite; small and heterochromatic, with a very high content of repetitive sequences. The few phenotypes that can be ascribed to the Y include sex determination itself, and infertility. Because it does not recombine (it has no pair, and is not homologous with the X), linkage mapping cannot be done, and we have relied on physical methods: deletion mapping and radiation hybrids to characterize it.

Although the human X and Y chromosomes are spectacularly different in size, gene content and sequence make-up, they do share a small region at one end. This PAR pairs and recombines at meiosis, and is crucial for the correct distribution of sex chromosomes into gametes; half receive an X and half receive a Y, leading to a more-or-less 50:50 sex ratio.

These X and Y maps were the basis of molecular-scale maps, which set the scene for complete sequencing of human sex chromosomes (Chapter 3).

FURTHER READING
Book

Ohno S. 1967. *Sex Chromosomes and Sex Linked Genes*. Springer Verlag Berlin

This is a classic. Although many of Ohno's conclusions turned out to be wrong, we are still testing his imaginative hypotheses 60 years later.

Classic Papers

Davies KE, 1985. Molecular genetics of the human X chromosome. *Journal of Medical Genetics* 22: 243–249

Mandel J-L, Monaco AP, Nelson DL, et al., 1992. Genome analysis and the human X chromosome. *Science* 258: 103–109

D Vollrath, S Foote, A Hilton, LG Brown, P Beer-Romero, JS Bogan, DC Page, 1992. The human Y chromosome: A 43-interval map based on naturally occurring deletions *Science* 258: 52–59

Reviews and Research Articles

Boyd Y, Blair HJ, Cunliffe P et al., 2000. A phenotype map of the mouse X chromosome: Models for human X-linked disease. *Genome Research* 10: 277–292

Chelly J, Mandel JL, 2001. Monogenic causes of X-linked mental retardation. *Nature Reviews Genetics* 2: 669–680

Davison MJ, Ward SJ, 1998. Prenatal bias in sex ratios in a marsupial, *Antechinus agilis*. *Proceedings of the Royal Society of London. Series B: Biological Sciences* 265: 2095–2099

Graves JAM, Gécz J, Hameister H, 2002. Evolution of the human X – a smart and sexy chromosome that controls speciation and development. *Cytogenetic and Genome Research* 99: 141–145

Hinch AG, Altemose N, Noor N et al., 2014. Recombination in the human pseudoautosomal region PAR1. *PLoS Genetics* 10(7): e1004503

Mittwoch U, 2013. Sex determination in mythology and history. *EMBO Reports.* 14(7): 588–592. doi: 10.1038/embor.2013.84

Saifi GM, Chandra HS, 1999. An apparent excess of sex- and reproduction-related genes on the human X chromosome. *Proceedings of the Royal Society of London. Series B: Biological Sciences* 266: 203–209

Vogt PH, Edelmann A, Kirsch S et al., 1996. Human Y chromosome azoospermia factors (AZF) mapped to different subregions in Yq11. *Human Molecular Genetics* 5: 933–943

3

Molecular Biology of Mammalian Sex Chromosomes

In the last 50 years we have progressed from recognizing sex chromosomes as peculiar unpaired bodies under the microscope to possessing a full molecular description of their DNA sequence and gene contents. In this chapter I will describe how these techniques have been applied, and the remarkable insights molecular techniques have provided into the structure, sequence and gene content of the X and Y in humans and other mammals.

To zero in on sex chromosomes, the first job was to construct molecular scale maps of the X and Y chromosomes of humans and other mammals, which built on the genetic and physical maps described in Chapter 2. This paved the way to applying the rapidly developing sequencing methods (Section 1.11) to reveal the most intimate molecular details of sex chromosomes.

3.1 Human Sex Chromosomes – Vital Statistics

The molecular make-up of human sex chromosomes has long been of special interest. Even rough quantitative data showed that the human X and Y chromosomes are atypical.

The DNA content of human X and Y chromosomes was measured in several ways. Traditionally, DNA is assessed chemically in a population of cells, the amount per cell calculated, then the amount per chromosome estimated from the relative lengths of chromosomes seen under the microscope. The amount of DNA per chromosome can also be estimated by flow cytometry, in which isolated chromosomes are forced past a laser beam that measures relative fluorescence of a dye bound to the DNA. By these measures, the human X chromosome comprises about 5% of the genome, and its DNA would measure about 53 mm end to end. The Y chromosome is barely 2% of the genome, its DNA measuring about 20 mm.

The X and Y chromosomes are spectacularly different in their base composition and the distribution of sequence classes. Whereas the X looks much like an ordinary autosome, the Y is largely heterochromatic.

Although the X lacks large blocks of heterochromatin, it is highly enriched in interspersed repeats (Section 1.3) (56%, compared to a genome average of 45%). Of these, most are short interspersed repeats (SINEs), including a somewhat lower than average frequency of Alu elements, which are ubiquitous in the human genome. Long terminal repeats are frequent, and long interspersed elements (LINEs) of the L1 family are highly enriched (29%, compared to a genome average of 17%).

The X is particularly rich in segmental duplications (Section 1.7). Regions of the X with more than 90% identity account for about 2.6% of the X. They are hotspots for deletions observed in patients with X-linked diseases. For instance, an inversion at the end of the long arm disrupts the F8 gene that codes for a blood clotting protein, causing haemophilia A. Segmental duplications in the short arm are associated with X-linked mental retardation, presumably because of the altered dosage or disruption to one or more XLMR genes (Section 2.13).

The human Y chromosome is quite bizarre in its sequence make-up. It is atypical of the human genome in every way.

The Y contains three huge blocks of heterochromatin, one on each arm and one at the centromere. At the terminus of the short arm (Yp) subtelomeric interspersed repeats are concentrated. The centromere is buried in repeated sequence, including centromeric alpha satellite repeats and Alu repeats. These are present in all human centromeres, but the sequence and repeat structure of the Y centromere are different from those on the autosomes or the X. The heterochromatic bottom half of the long arm (Yq) is composed of highly repeated simple sequences. A few are shared with the X and/or autosome, but most are Y-specific and do not cross-hybridize, explaining why the Y stains uniquely.

The euchromatic portion of the human Y, too, has a high content of repetitive DNA, nearly all of it unique to the Y, although Y-chromosome LINE repeats cannot be distinguished from LINEs on autosomes or the X. The most extraordinary repeats on the Y, however, turned out to be enormous palindromes – loops in which sequence is read the same in both directions. These loops contain multicopy genes that are expressed only in the testis (3.6).

The sequence make-up of human sex chromosomes, especially the Y, is therefore quite atypical of the genome as a whole, so it was of special interest to map and sequence the entire X and Y.

3.2 Molecular-Scale Maps of Mammal X Chromosomes

The human X and Y chromosomes are huge in molecular terms, so the approach was to break them down into bite-sized chunks that can be analysed minutely.

To construct a molecular scale map of the human X chromosome, DNA from an XX woman was broken into large pieces, which were inserted into the bacterial

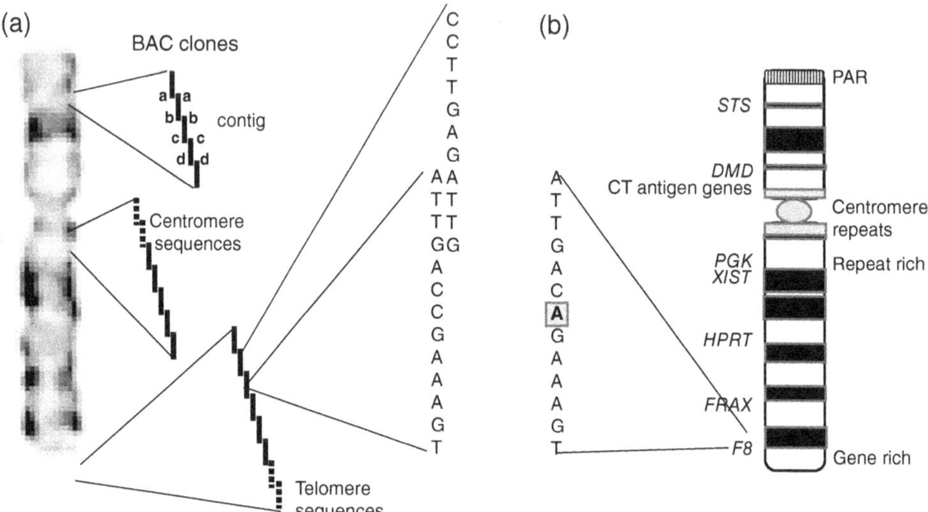

Figure 3.1   Clones, contigs and sequences of the human X chromosome. (a) The human genome was broken into fragments, and these were inserted into a bacterial chromosome to make a library of Bacterial Artificial Chromosomes (BACs). BACs were arranged into contigs by observing overlap of unique markers a, b, c, and so on. Contigs were assigned to the X by comparison with deletion and radiation hybrid maps. BACS were each sequenced, then put back together into contigs, then assembled into scaffolds that describe the whole X chromosome. X chromosome photograph supplied by Victorian Clinical Genetics Services, Murdoch Children's Research Institute, Melbourne. (b) Sequence is aligned with physical and genetic maps. Genes are annotated. Sequence can be compared in patients with conditions like haemophilia A to determine pathological mutations (e.g., in the *F8* gene).

or yeast genome. Single bacterial or yeast cells were grown into visible colonies ('clones'), which then made huge numbers of copies of a piece of DNA, such that single regions of the X were 'cloned' (Figure 3.1).

The whole human X was represented by a 'library' of thousands of these Bacterial or Yeast Artificial Chromosomes (BACs or YACs). They could be ordered with the help of short unique sequences ('sequence tagged sites') which detected over-lapping clones. Then they were sorted into contiguous segments (contigs) by lining up other markers, such as sequences detected by restriction enzymes, and inte-grated with existing maps.

The contigs spanned practically all of the X chromosome, from terminal clones that contained repeated sequences typical of the ends of chromosomes (telomeres), to sequences near the centromere. Thousands of BACs were used to construct a 'golden pathway' with minimal overlap along the X (Figure 3.1b). The physical location and order of BACs on the X could be checked directly by tagging them with fluorescence and mapping them by FISH.

Radiation hybrid and YAC-based maps of the whole mouse genome were produced in 1999, and a detailed physical map using thousands of DNA markers to order hundreds of contigs of large insert clones was released a few years later. The maps could be easily compared with human maps, revealing many conserved segments, rearranged over the whole genome. Each mouse autosome had blocks of homology shared with multiple human chromosomes.

The same combinations of methods were used to map the genomes of many other species, including rat, cat, dog and cow. Comparing these maps revealed strong conservation of the gene content on the X chromosomes of humans, mice and cats. Radiation hybrid maps of cat and human revealed complete conservation of markers and even marker order between the X of humans and cats, and indeed elephants share this order. The mouse X was strikingly different; eight orthologous blocks were rearranged on the mouse X. Since rodents are more closely related to humans than are carnivores or afrotherians, this means the mouse X is unusually rearranged.

A physical map of the human pseudoautosomal region PAR1 was also constructed as part of the YAC map of the X; in fact, it predated the discovery that this region is pseudoautosomal. Long-range restriction mapping ordered the large fragments, which were present on both the X and Y. This confirmed that PAR1 is about 2.6 MB long from a telomeric sequence to the proximal *CD99* gene, a size consistent with the length of the minimal cytological pairing region between the X and Y. A more detailed BAC contig across PAR1 was constructed as part of the X BAC mapping and sequencing.

3.3 Molecular-Scale Maps of Mammal Y Chromosomes

A 'golden pathway' of overlapping large-insert clones was prepared from the human Y chromosome in 1992. The starting material was cells from a man who had four copies of the Y chromosome, which ensured good representation of Y fragments. DNA was broken up and combined with the genomes of yeast cells to make a library of YAC clones, containing large (~600 Kb) stretches of the human genome. The library was screened to identify YACs containing any sequence tagged sites known to be male-specific. Overlapping YACs that shared the same site or sites were identified. Gaps were closed by reference to the Y deletion map that used the same marker sequences. This produced a map of 207 Y-borne sequences in 127 intervals, about 220 Kb apart on average. The end YAC contained highly repetitive DNA that belonged to the heterochromatin on the distal half of the long arm, and was impossible to map because it was so repetitive. The map covered most of the euchromatic portion of the Y, and contained all of the then-known genes on the Y (Figure 3.2).

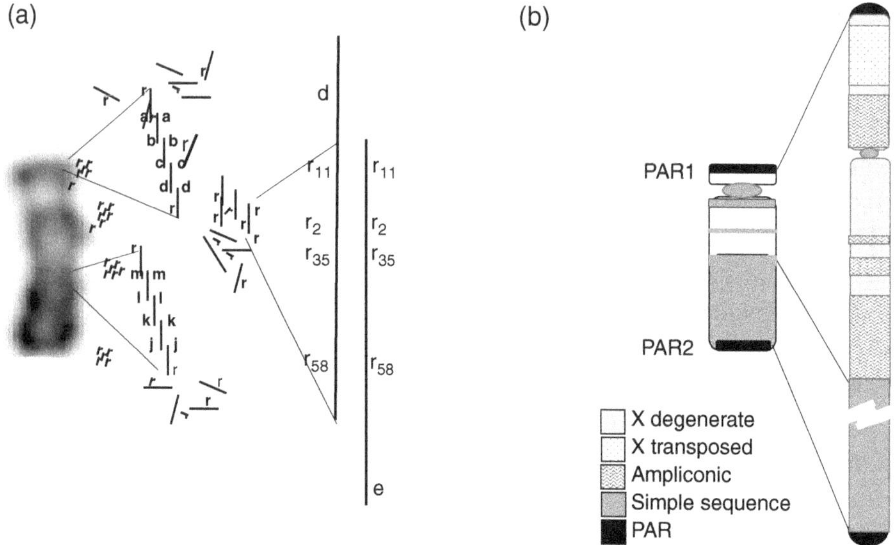

Figure 3.2   Cloning and sequencing the human Y chromosome. (a) The human genome was broken into fragments, which were inserted into a bacterial chromosome to make a BAC library. BACs were arranged into contigs by observing overlap of unique markers a, b, c, and so on. However, repetitive sequences (r) can lead to false detection of overlaps. Each BAC must therefore be scanned for unique copies of the repeat that denote a true overlap (variants r_{11}, etc.). Y chromosome photograph supplied by Victorian Clinical Genetics Services, Murdoch Children's Research Institute, Melbourne. (b) Four types of sequences in the human Y chromosome. X degenerate sequences with copies on the X represent the remains of ancient homology. X transposed sequences were recently recruited from the X. Ampliconic sequences were amplified in giant palindromes. Simple repetitive sequence constitutes thousands of copies and stains atypically (heterochromatin). PAR sequences are homologous with the X.

A more detailed map of the non-recombining male-specific region (MSY) was built by assembling BAC clones within these intervals. BACs from the Y chromosome were obtained by screening a BAC library with the same set of short unique marker sequences known to be on the Y that were used to construct the radiation hybrid map. Using the radiation hybrid map as a scaffold for assembly, more than 1000 BACs were assembled into contigs that covered nearly the whole non-recombining region with only four small gaps.

In situ hybridization was only marginally useful for characterizing the human Y chromosome. Although the resolution of FISH is too poor to correctly locate or order genes, at least their assignment to the Y could be verified, as well as their position on the short or long arm. Multiple copies could be detected. DNA fibre-FISH using stretched out chromatin fibres (Section 1.9), was useful to correctly locate Y genes, and count their number and arrangement in multigene arrays.

Y Maps obtained by different methods were cross-checked. For instance, BACs from the genome physical map were located by FISH, deletion maps and YAC maps of the Y were compared.

These molecular-scale maps set the scene for complete sequencing of the human sex chromosomes – indeed, the whole human genome.

3.4 The Human Genome Project

Sequencing the entire human genome seemed preposterous when it was first proposed in the 1980s when DNA sequencing was slow and expensive. The project was also not universally sanctioned; a prominent Australian biologist labelled it 'intellectually offensive, just a fishing expedition, a waste of time and money'. Forty years later, we can reply, 'Yes, but just look at the fish!'.

The aims of the publicly funded Human Genome Project were to determine the complete sequence of nucleotides in a reference human genome, to identify and map all human genes and to determine their function. It started in 1990, a draft sequence was published ten years later, and it was declared 'finished' (more or less) in 2003. It remains the largest biological collaboration ever, led by huge publicly funded groups in the USA and involving laboratories in several countries. It cost about USD $3 billion dollars, and has been called 'the greatest feat of exploration in history'.

DNA sequencing was to be accomplished by using the molecular scale maps to identify a golden pathway of large DNA sequences (generally BACs) on each chromosome. These were to be sequenced by first-generation (largely chain-terminating) technologies (Section 1.11), then assembled by matching up identical sequences using the physical and molecular scale maps as a guide. This method was considered to be the gold standard.

At the same time, the alternative method for breaking up the genome randomly ('shotgun'), and sequencing small fragments of DNA at enormous depth was also being trialled by a private company. It worked for *Drosophila* – why not for humans? The approach left more gaps and more errors – but was accomplished at 1/10 the cost. The race to publish the human sequence ended in a dead heat in 2001.

The original human sequence was reported to cover 99% of the human genome. This was limited to the euchromatic regions and excluded the repetitive regions at centromeres and telomeres. Sequencing over most of this region was claimed to be 99.99% accurate, but many gaps existed.

The advent of several methods for sequencing much longer stretches of DNA ('longread') from a single DNA molecule has completely changed the sequencing picture (Section 1.12). The ability to sequence kilobase- or even megabase-long molecules makes it very much easier to detect overlap and assemble long contigs. Even repetitive DNA can be sorted out if these stretches are long enough, making it now possible to piece together heterochromatic regions like centromeres – and the Y chromosome. Telomere-to-telomere 'T2T') sequencing of one after another human chromosome has been reported – the X was first and the Y was last chromosome to succumb.

3.5 Fruits of the Human Genome Project

The availability of the human genome sequence has utterly changed the way we do human genetics, discover the causes of disease and devise treatments, as well as assess human variation in space and time and compare our species with others.

To everyone's surprise, the total number of genes that could be recognized in the human sequence ('annotated') was only 19,042 – far fewer than the 100,000 or so that had been predicted. This was not much more than the 13,600 genes in *Drosophila*, leaving us to ponder whether humans are only 50% more complicated than fruit flies.

The hunt for individual genes, mutations in which caused particular genetic diseases, accompanied DNA sequencing from the beginning. Then as sequencing became cheaper, sequencing the entire genome of individuals became practicable, and is now almost routine. The genomes of many patients with unknown diseases or conditions have been entirely sequenced, revealing the genes underpinning several diseases.

Many organizations worldwide now run projects to catalogue and analyse particular facets of human DNA sequence. Encyclopedia of DNA elements (ENCODE) is focused on identifying functional elements in the human genome. Several organizations maintain centralized resources for DNA sequence, including the Broad Institute, UCSC and Ensemble, as well as the US National Institutes of Health Genome Research Institute. Several international projects collect data relevant to, for instance, cancer research. The HapMap project aims to find markers associated with human disease by virtue of their proximity in a founder.

As technology advances, the cost of sequencing has fallen dramatically, so that there have also now been many whole genome sequencing studies of multiple individuals to assess variability. The 1,000 genomes project aims to chart human genetic diversity over global populations.

The availability of human genome sequence has also enabled a whole new basis of comparison with the genomes of other mammals, even other vertebrates. Of great value are sequences of other mammals that can be compared with human sequence. Radiation hybrid and molecular-scale maps were constructed for many species, as a prelude to sequencing. Initially it was possible only to sequence and assemble small regions, and sequencing depended on constructing good maps. New techniques have greatly expanded the availability of sequences from the X – and Y – of a range of mammal species.

Sequencing other mammal genomes began in the 2000s with the mouse, also a clone-based assembly. This revealed 20,210 genes, as well as 439 long noncoding genes, mostly shared with the human genome. Most lineage-specific (mouse only or human only) genes were in multigene families involved with reproduction. Genome arrangement, however, was very different in humans and mice, as had been predicted long ago by comparing gene maps. Lining up orthologous segments showed that every mouse chromosome except the X contained bits orthologous to several human chromosomes.

Genomes of the species most closely related to humans; chimpanzees, then later gorillas, bonobos and orangutans, were also sequenced, then dogs and cats, cattle, pigs and horses. But early projects were also launched to sequence the mammals most distantly related to humans, a marsupial (the opossum) and monotreme (the platypus). Aligning sequence with human sequence showed how conserved the mammal genome is in most groups, particularly primates and cats, as had been predicted from gene maps. Mice are a spectacular exception.

Now that full genome sequencing no longer costs billions, or even millions (we are down to maybe a thousand dollars), full genomes have been mapped and sequenced in many eutherian species, including other domestic and companion animals, pandas, dolphins and elephants. Marsupials are being added at a fearsome rate, and both monotreme mammals now have good sequences. The sequencing revolution has spread to other vertebrates and invertebrates, plants, fungi and protists. Indeed, the Earth Biogenome Project aims to sequence all complex life on earth over the next ten years.

The discoveries about sex chromosomes well illustrate the power of these new techniques and the value of the information they provide.

3.6 Sequencing the Human X Chromosome

A 'finished sequence' of the human X chromosome was published five years after the human genome draft sequence was released. The project started by mapping large-insert clones of X chromosome DNA, which were assembled into 16 contigs using restriction fragment fingerprinting. Clones with minimal overlap were chosen to construct a golden pathway that extended from telomeric sequences (repeats of TTAGGG) at the termini of both arms, and included both pseudoautosomal regions (Figure 3.1).

Sequencing these clones with first-generation techniques was laborious. A problem with the first BAC-based sequencing was that DNA came from five women, so there was considerable allelic variations between the ten X chromosomes represented. This made lining up overlapping clones difficult. However, this sequencing produced 151 million base pairs (151 Mb) of the X. The sequence was checked for the inclusion, and the order of known genes and markers, and some small inconsistencies with the molecular scale X chromosome maps were sorted out.

There were 14 gaps in this X sequence. The centromere was completely missing (as centromeres always were), and other gaps occurred, especially in PAR1 and around regions containing amplified genes. Several projects, using a variety of sequencing strategies, aimed to fill gaps and iron out inconsistencies in particular regions.

The advent of long-read sequencing enabled a telomere-to-telomere assembly of the X in 2019 that resolved all gaps. Even the centromere could be assembled, and found to consist of 3 Mb made up of several levels of repetitive DNA. The segmental duplications were documented with almost perfect accuracy.

Sequencing the DNA of the X chromosome enabled more accurate size measures. The X contains 156 Mb of DNA, consisting of a single DNA molecule that measures about 51 mm end to end (close to the predicted 5% of the total genome length). It is divided by the centromere into a short arm Xp of 61 Mb and a long arm Xq of 95 Mb.

The X chromosome, especially a region amounting to nearly 2% of the X just above the centromere on the short arm, bears a disproportionate number of large inverted repeats (24 of a total of 96 in the genome). These are up to 100 kb long, and each is composed of two near-identical sequences inverted and complementary to each other in a 'palindrome' arrangement of identical forward and reverse sequences. Their percent identity is much higher than for inverted repeats on autosomes. Those with simple inverted repeats could form cross-shaped ('cruciform') secondary structures; loops that might be especially susceptible to recombination, gene conversion and rearrangement.

When genome sequences became available for other primates, it was found that their X chromosomes were practically identical to the human X, including the palindromic loops housing inverted repeats. The mouse X chromosome too, had a gene complement virtually identical to the X of other mammals (single copy genes had about 95% identity), though greatly rearranged by many internal inversions, and with a small terminal deletion.

Undoubtedly, the human X was the best-assembled chromosome of humans or any mammal. Having an accurate readout of bases along the X has enabled studies of human disease, as well as of the evolutionary processes shaping sex chromosomes.

3.7 Using Human X Sequence to Discover Genes and Characterize Variants

Given that mapping and cloning human genes had turned up a disproportionate number of genes on the X, it was surprising that the X chromosome turned out to be rather gene poor. A total of 831 protein-coding genes were detected among

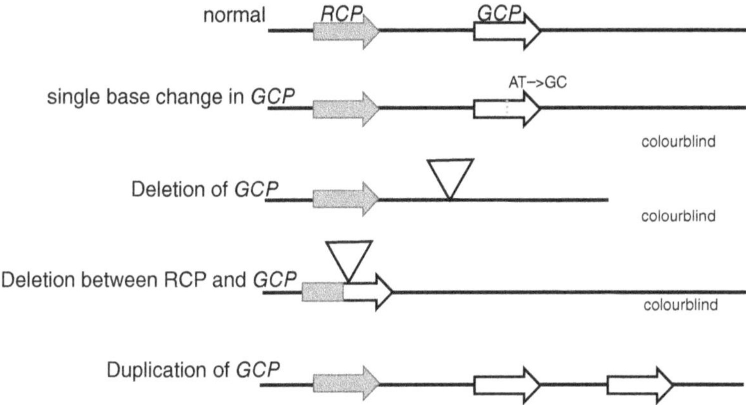

Figure 3.3   DNA sequencing to discover genes and detect mutations. DNA sequencing confirmed that two colour vision genes *RCP* (dark grey, encoding red pigment) and *GCP* (white, encoding green pigment), lie in tandem on the X chromosome. Red-green colour blindness can be caused by mutations in either gene, deletion of one or other gene or fusion between the genes. Duplications of GCP do not affect colour vision.

the sequence, and checked against available cDNA sequences. Of these, 699 were known from other studies and 132 were novel.

Although it might lack the thrill of the chase, the brute force sequencing of a DNA region is the most certain method for obtaining a complete catalogue of genes. Initially this method was employed over short regions; for instance, exhaustive sequencing of 360 kb within the gene-rich region at the end of Xq identified at least thirteen novel genes, as well as the classical genes mutated in colour blindness, haemophilia A (the Factor 8 gene that makes a protein that is a part of the cascade of blood clotting reactions) and the housekeeping enzyme G6PD.

For instance, the colour blindness locus was revealed to be a tandem pair of red-green pigment genes. Sequencing beautifully confirmed the conclusions from genetics that different classes of colour vision were caused by deletion of one or the other, fusion or rearrangement (Figure 3.3).

The complete sequence of the human genome was scanned for genes, both novel and previously mapped or cloned. Annotation of genes is not always straightforward. Recognition of novel coding sequences on the human X depended on the correct assembly of X sequence, and the identification of sequences that signal transcription initiation and mark the boundaries of introns and exons.

The most surprising finding was of 99 genes that are expressed normally only in the testis, but are re-expressed in many cancers. These genes were grouped along the palindromic loops. Cancer-testis (CT) genes are detected by an antigen that recognizes a common (MAGE) domain. Several families of CT genes were known to be located at various sites on several human chromosomes, but their concentration on the X was unexpected; they constitute 10% of the genes on the X. Their number is somewhat variable between individuals, leading to some variation in the X gene count. Detailed analyses of X-borne CT genes showed that their expression is restricted to germ cells, and to brain, in contrast to autosomal family members.

Also detected in the X sequence, as well as in expressed gene libraries, were 173 sequences that were transcribed into RNA but not translated into protein. Two genes are transcribed into transfer RNA, and 13 into micro RNAs that are thought to control gene activity by binding to messenger RNA. The rest produce longer noncoding RNA transcripts, at least one of which (*XIST*) plays a vital function in X chromosome inactivation (see Part II).

More detailed work on sequence in particular regions, and recent telomere-to-telomere long-read assembly refined gene annotation and resulted in some changes to gene numbers. The number of genes reported on different databases range from 900 to 1400, depending on gene definition, and 400-600 noncoding transcriptional units.

A complete listing of X-borne genes and associated phenotypes is available at several sites including Ensemble and the National Centre for Biology Information (NCBI), MedlinePlus, and GeneCards. Gencode lists 1144 gene on the human X; I will go for 'about 1000'.

The availability of sequences has provided explanations for many disease-causing changes, including deletions and rearrangements, especially of tandemly repeated genes. It also provides single nucleotide polymorphisms (SNPs) or other sequence variants for diagnoses. The availability of cheaper and quicker sequencing methods means that it is now practical to sequence the entire coding regions (exome) of patients with suspected X linked disease such as X-linked mental retardation – and patients with atypical sexual development (Section 9.6).

Sequencing the X chromosome of other mammals has also delivered a catalogue of genes (940 in the mouse, and 690 in the pig) which is strikingly aligned to those on the human X, confirming and extending Ohno's observations and guesses.

3.8 Sequencing the Human Y Chromosome

The human Y chromosome was by far the most challenging part of the human genome to sequence. Everything conspired to make it difficult. Its high content of repetitive sequences meant that it was hard to map because marker sequences for identifying overlapping contigs frequently detected more than one non-contiguous fragment. Even sequencing itself was difficult because most methods cannot handle runs of a single base, which are a feature of many repeats.

However, the region that was identified by deletion analysis as containing fertility factors was ultimately cloned and sequenced.

Sequencing the entire euchromatic portion of the male-specific region of the Y was undertaken in David Page's lab in the 2000s. It took a herculean effort of mapping and ordering contigs. Because of difficulties with repeats, constructing this pathway partly depended on sequencing and detecting minor variants that distinguished different repeat copies. Sequencing used a golden pathway of 220 overlapping BAC clones derived from a single Y chromosome. Sequences derived from BACs were used to detect the sometimes trivial base sequence differences between repeated sequences, which avoided collapsing repeats by detecting variants that denoted different copies. Sequencing, finding variants and ordering repeats was an iterative process, extremely accurate, though expensive and tedious.

Sequence of the entire euchromatic part of the Y (excepting the PAR) produced 23 Mb of DNA, representing most of this region except for two gaps, and a large repetitive region on the short arm (Figure 3.4). This is less than half of the expected total size of 57.2 Mb. The centromere is near one end of the Y, so it has a tiny short arm (8 Mb).

Surprisingly, segments on both arms of the Y contained genes and pseudogenes that have homologues on the X chromosome. This was first discovered when the candidate sex gene *ZFY* was cloned from the human X, and discovered to have a copy *ZFX* on the X. Several other genes cloned from the Y before sequence was available were also shown to have X homologues; the ribosomal protein *RPS4Y* also recognized an X gene *RPS4X*, and *SMCY* also recognized *SMCX*. Even multicopy genes with functions in spermatogenesis had copies on the X; the spermatogenesis factor *RBMY* was discovered (in a roundabout way through its marsupial homologue) to recognize a gene *RBMX* on human Xq, and *TSPY* recognized a gene *TSPX* on Xp. In fact, sequencing shows that most genes on the human Y (at least 20 of the 27) have X homologues. These XY homologous regions (called 'X-degenerate') appear to be relics of ancient X-Y homology, as discussed in Chapter 4.

There is another class of human Y sequences present on the X, but it has a different evolutionary origin. On the short arm of the Y lies a 3.4 Mb block of sequences that are nearly identical (99%) to sequences from the middle of Xq. The near-homology of these sequences implied that they had been transposed from the X relatively recently (3–4 Mya), and thus defined an 'X-transposed' region on Yp. Their recent arrival on the Y was confirmed by their absence from the Y chromosome in great apes (Figure 3.5). The block, on both the X and Y, contains few genes and is rich in transposed elements. It is inverted on the Y, so the X and Y blocks do not line up and do not recombine at meiosis.

Spectacularly weird features of the human Y were regions, one on the proximal short arm and others scattered throughout the long arm, that contained massive amplified regions ('amplicons') and practically no unique sequences. Restriction enzymes and marker sequences detected multiple copies that could not be distinguished except by very subtle sequence differences. Since the large insert clones all came from a single male, and therefore represented a single Y chromosome, such variants could not be different versions of the same sequence (alleles), but must be different copies. Some of these copies were tandem repeats, others inverted repeats, with head-to-tail arrangements of complementary sequences that could pair within giant loops ('palindromes').

These palindromes contained several families of genes that were arranged within loops of different sizes. Within each, sequence was highly symmetrical, with identical genes on both arms around a spacer. Half of this sequence was contained in seven large – and one massive – palindromes. A huge (2.9 Mb) palindrome contained subsidiary 24 kb palindromes. As well as genes, it contained inverted repeats and long tandem arrays, some of which were transcribed.

Thus, sequence of the euchromatic regions of the human Y fell into three classes, X-degenerate, X-transposed and ampliconic (Figure 3.4).

Several long-read sequencing projects have attacked the human Y, including PacBio sequencing of racially diverse Y chromosomes as part of the 1,000 genomes project. An 8-fold improvement came with the single-molecule sequencing of flow-sorted Y chromosomes of African origin using a long-read technique (MinION) that lacks an amplification step, so might be expected to introduce less artefact. This delivered an almost uninterrupted sequence; 35 contigs with an average length (N50) of 1.46Mb – and at 1/10 of the cost. The sequence was largely homologous to the reference sequence and confirmed the major features; the single copy X degenerate sequence, and an ampliconic region of almost 10 Mb that contained eight massive palindromes, the largest nearly 3 Mb.

A complete 62,460,029 base pair sequence of a human Y chromosome has recently been assembled that represents the telomere-to-telomere sequence of a single Y chromosome. It reveals the complete ampliconic structures of the *TSPY* cluster on Yp, as well as the palindromes containing *DAZ*, and *RBMY*. It details the structure of some heterochromatin, and adds more genes, all but one of which are additional *TSPY* copies.

Under the microscope, a striking cytological feature of the human Y is the differential staining of large blocks of heterochromatin. One spans the centromere, and another forms most of the distal long arm. Sequencing centromeres of any human chromosome has been next to impossible because they are composed of thousands of near-identical tandem repeats. However, ironically the 300,000 base centromere of the Y chromosome is the smallest of any human chromosome, and the most amenable to long-read sequencing. It has now been sequenced, using a strategy to insert unique markers to break up the array, then do Nanopore sequencing of very long BAC clones that span the region. Not surprisingly, most of the centromeric locus is defined by a boring array of 301 kb head-to-tail repeats.

The distal half of Yq is composed of massively amplified simple sequence repeats of at least six families. Two predominate; one is a repeat of just 5 nucleotides, and the other is a fragment of an Alu sequence. A coding function is unlikely, and I regard these repeated sequences as 'hard core' junk DNA.

Many other human Y chromosomes have been sequenced since this first break-through. There is particular interest in comparing Y chromosome sequences between humans of different ethnic backgrounds, and DNA from fossil hominids, in order to construct a human family tree and document origins and migrations of people (Section 3.13).

3.9 Genes on the Human Y Chromosome

Most genes on the human Y were discovered in the decade before methods were available to sequence the Y chromosome. Some were found by screening large insert clones within deletion intervals for expressed sequences (positional cloning). The sex-determining gene *SRY* (and *ZFY*) was cloned from the critical sex-determining region (Section 8.11, 8.12). *RBMY* and *DAZ* (Deleted in Azoospermia), two genes with functions in RNA binding and processing, were cloned from azoospermia deletion intervals (Figure 3.4a). *TSPY* (testis-specific protein Y), detected via a

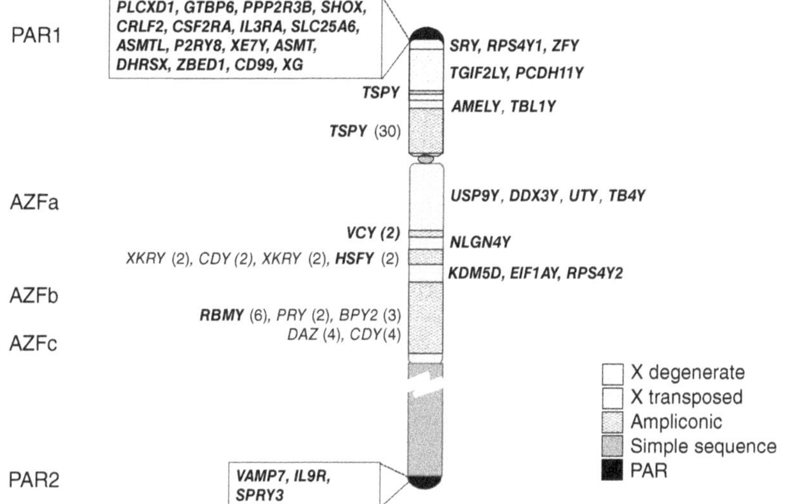

Figure 3.4 Characterized genes on the human Y chromosome (transcribed pseudogenes, long ncRNAs and microRNAs not shown). Genes in pseudoautosomal regions PAR1 (16 shown) and PAR2 (3) are boxed. Amplified genes are on the left (copy numbers in brackets), and single copy genes on right. Not counting amplified copies, the grand total of unique protein-coding genes in the male-specific region is 27. There are none in the heterochromatic region. All the single copy genes and four of the amplified genes (shown in bold) have partners on the X from which they diverged. Deletion intervals (Azoospermia Factor) AZFa–c on the long arm are marked.

testis-specific RNA, was found in the region of Yp that contains the 'gonadoblastoma gene' that promotes tumour growth in women with deleted or mutated Y chromosomes that lack the testis determining gene.

Several Y genes were discovered via their homology with genes on the X. A mouse X-linked tooth enamel gene *Amelx*, identified by a mutation that caused incomplete tooth mineralization, recognized active copies on the human Y as well as the X. Human *AMELY*, though it is largely homologous, is expressed at only 10% of the level of *AMELX*, and men with deletions of the Y that include *AMELY* have no tooth abnormality, suggesting that *AMELY* is largely functionless. This approach also detected several pseudogenes on the Y by their homology with X-borne genes. For instance, neighbouring genes near the tip of Xp *STS* (steroid sulfatase, mutations within which cause the X-linked skin disease ichthyosis) and *KAL* (Kallmann's syndrome, a neurological condition) both detect non-functional homologues on the Y which are missing exons and have frameshifts that knock out function.

Whole BAC-based Y chromosome sequencing allowed systematic cataloguing of genes on the male-specific region of the human Y. Y sequence was searched for matches with previously reported genes, and for the hallmarks of genes, as well as for expressed sequence tags. All told there were at least 156 transcription units, half of which encoded proteins. This total included 27 genes coding for unique proteins. Eighteen of these were single copy genes, and 9 were present as gene families (with 60 members overall). The most recent T2T sequence annotated 693 genes, of which only106 encoded protein. Not counting multiple copies, this amounts to 27 genes that make unique proteins in the male-specific region of the human Y chromosome. With the 28 pseudoautosomal genes, the Y tally is 51.

Recognizing genes and pseudogenes on the Y and determining their role and activity is an ongoing process. Even the names of genes have changed confusingly as their function has been revealed, for instance, *SMCY* began life with a name derived from investigator first names, then was recognised as part of the *JARID* gene family and now is usually known as *KDM5D* to express its function as a lysine demethylase. A list of human Y genes, their aliases and proposed functions is presented as Table 3.1.

All of the 18 single copy genes have homologues on the X: 16 lie in the X-degenerate region and 2 in the X-transposed region. Of the 16 single copy genes in the X-degenerate region, most (13) are ubiquitously or widely expressed, one (*AMELY*) is expressed only in teeth, and the other two are in brain and reproductive tissue. In addition, 13 pseudogenes were detected, which, like *STS* and *KAL*, have functional copies on Xp, near the boundary of the pseudoautosomal region.

In contrast, all but one of the multigene families reside in ampliconic regions, mostly symmetrically arranged within one of the eight palindromes on Yq. For instance, six copies of *RBMY* (plus several pseudogenes) occupy one palindrome, and four *DAZ* copies lie in another. The highly repeated *TSPY* is an exception,

Table 3.1   *Selected active protein-coding genes on the human Y chromosome.*

Gene symbol	Gene name	Tissue expression	Function	X copy
PAR1				
SHOX *PHOG*	Short stature homeobox	Multiple	Transcription	PAR1
CSF2RA	Colony stimulating factor 2 receptor subunit alpha	Wide	Growth and differentiation of cells in bone marrow	PAR1
IL3RA	Interleukin 3 receptor subunit alpha	Ubiquitous	Receptors for interleukin 3, which stimulates cell proliferation	PAR1
AKAP17A *XE7*	A-kinase anchoring protein 17A	Wide	Alternative splicing regulator	PAR1
ASMT *HIOMT*	Acetylserotonin O-methyltransferase	Wide	Catalyzes the final reaction in the synthesis of melatonin	PAR1
CD99 *MIC2*	CD99 molecule	Haemato-poietic cells	Cell surface molecule involved in T-cell adhesion processes	PAR1
XG	Xg glycoprotein (Xg blood group)	Skin, epidermis	Generates a cell-surface antigen	PAR 1 boundary
Yp				
SRY	Sex-determining region on Y	Embryonic gonad, testis	Testis determination	*SOX3*
RPS4Y1	Ribosomal protein S4, Y1	Ubiquitous, strong	Ribosomal protein subunit required for mRNA-ribosome binding	*RPS4X*
ZFY	Zinc finger protein, Y linked	Ubiquitous	Fertility?? meiosis	*ZFX*
PCDH11Y	Protocadherin 11, Y linked	Ubiquitous, esp brain	Cell-cell recognition in brain development	*PCDH11X*
AMELY	Amelogenen, Y linked	Low expression	Tooth mineralisation	*AMELX*

Table 3.1   (cont.)

Gene symbol	Gene name	Tissue expression	Function	X copy
TBL1Y	Transducin Beta-like 1Y	Testis and a few tissues	Corepressor/activator of transcription factors	*TBL1X*
TSPY (66)	Testis-specific protein Y-linked	Testis	Cell cycle regulator candidate gonadoblastoma	*TSPX*
Yq				
USP9Y DFFRY (AZFa)	Ubiquitin specific peptidase 9, Y linked	Ubiquitous	Regulates protein turnover. HYA candidate	*USP9X*
DBY DDX3Y	DEAD [asp-glu-ala-asp] box3 Y-linked	Ubiquitous	Cell cycle regulation, early germ cells. Sertoli Cells only syndrome	*DBX*
KDM6C UTY (AZFa)	Ubiquitously transcribed tetratrico-peptide repeat, Y linked	Ubiquitous, two transcripts	Male-specific histone demethylase, demthylates H3K27me3. HYA candidate	*UTX*
TB4Y TMSB4Y	Thymosin beta 4 Y linked	Wide, especially intestine	Activates natural killer cell cytotoxicity, HYA candidate	*TB4X*
VCY (2)	Variable charge Y-linked	Testis	Intracellular vesicle traffic	*VCX*
NLGN4Y	Neuroligin 4 Y-linked	Ubiquitous	Membrane protein, cell adhesion	*NLGN4X*
CDY (26) *(AZFc)*	Chromodomain Protein Y-linked	Testis	Chromatin remodelling, regulates gene expression	–
HSFY (8)	Heat shock transcription factor, Y linked	Testis (spermatids) brain	Regulation of spermatogenesis	*HSFX1 HSFX2*
KDM5D SMCY JARID1D (AZFb)	Lysine Demethylase 5D	Ubiquitous	Meiotic chromosome condensation, demethylation of H3K4m2, m3	*KDM5C*

EIF1AY	Eukaryotic translation Initiation Factor 1A, Y linked	Ubiquitous	Translation initiation; binds RNA, dissociates ribosome subunits	*EIFAX*
RPS4Y2	Ribosomal protein S4, Y linked 2	Testis, prostate	Regulation of spermatogenesis	*RPS4X*
PRY (8)	PTPN13-like Y linked	Testis	Major spermatogenesis gene, regulates apoptosis	–
RBMY (34) *HnRNPG* *(in AZFb)*	Ribonucleic Acid [RNA]-Binding Motif, Y linked	Testis, kidney	Spermatogenesis, regulates alternate splicing	*RBMX*
DAZ (4) *(in AZFc)*	Deleted in Azoospermia	Testis	Spermatogenesis, regulates RNA translation	*DAZL* on chr 3
BPY2 (2)	Basic Protein Y linked, 2	Testis, kidney	Might bind DNA and regulate cytoskeleton in spermatogenesis	–
PAR2				
IL9R	Interleukin 9 receptor	Several tissues	Receptor for IL9, which stimulates proliferation and inhibits apoptosis	PAR2
VAMP7 *SYBL1*	Vesicle associated membrane protein	Ubiquitous	Membrane fusion	PAR2

Genes are shown in the order of their position on the Y chromosome. Alternate gene names are given where these are still used. Ampliconic genes are in bold (number of copies in brackets).

30 copies existing in a huge tandem array on Yp. These multicopy genes are all testis-specific. Four of these genes have X copies (Table 3.1).

There were almost the same number of transcribed noncoding sequences in the male-specific region of the Y; 13 unique sequences in the X-degenerate region, and 15 repeated units in the ampliconic region (making a total of 65 copies). At least one of these seems to influence the expression of other genes, though, curiously, not in the testis.

Long-read sequencing later confirmed that genes on the Y fell into two classes; single copy genes, most of which are broadly expressed in many somatic tissues, and ampliconic genes present in multiple copies that were generally testis-specific (3.8).

Thus, the human Y chromosome is highly unusual in the content and distribution of genes. Firstly, the gene density within the male-specific region (MSY) is extremely low; there are almost as many genes in the tiny (2.8 Mb) PAR1 as in the rest of the Y. The male-specific region of the human Y bears a grand total of 27 unique protein-coding genes, though many have two or more copies. Secondly, it bears several genes involved in male reproduction, including the sex-determining gene *SRY*, and multiple copies of genes with apparent function in spermatogenesis. Thirdly, it is particularly variable between individuals, showing a high rate of mutation.

The human Y chromosome has been called 'a genetic wasteland', a 'junkyard' full of repetitive sequences and pseudogenes, and responsible for few phenotypes. Now we have a readout of genes on the human Y, we can judge how fair this assessment is. The gene density on the MSY is less than 1/10 that of the autosomes and X (the human Y is a bit bigger than chromosome 22, which accommodates about 450 genes). A grand total of 27 unique protein-coding genes in the human MSY seems rather pathetic. So although the human Y doesn't entirely deserve its categorization as a 'genetic wasteland' or 'junkyard' (because, after all, it has important roles in sex and spermatogenesis), it does look like neglected real estate.

3.10 Sequencing Y Chromosomes of Other Mammals

The picture of genomic mayhem of mammal Y chromosomes has been greatly enlarged by sequencing the male-specific region of the Y in many other mammals.

Because sequencing the Y chromosome is so tough, most genome projects initially used female DNA to avoid problems with assembly. But special projects were undertaken to sequence the Y chromosomes of several therian mammals, using the careful iterative process developed in Page's laboratory. As for the human Y, a golden pathway was constructed from overlapping large-insert clones that could be individually sequenced, and copies distinguished by iterative sequencing. Cheaper and quicker options have been explored, including flow sorting to enrich the starting material by manyfold, or subtracting male-specific

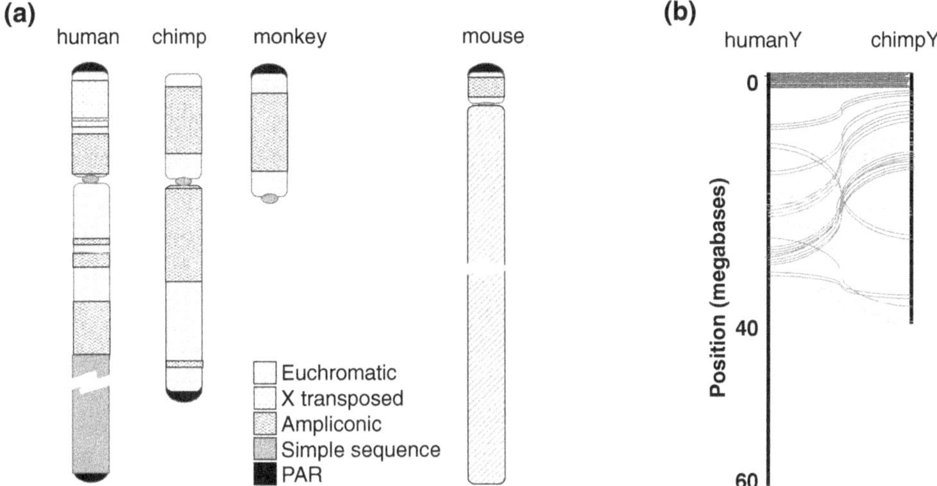

Figure 3.5   Y chromosomes of humans, other primates and mice. (a) Comparison of the structures of Y chromosomes of humans, chimpanzees and rhesus macaques (monkey). X degenerate (ancestral) and ampliconic regions with similar gene contents are scrambled on the rearranged chimp Y, and the human Y contains a unique region that was transposed recently from the X, as well as a large heterochromatic region of simple sequence repeats. The mouse Y retains only part of the ancestral region, arranged with ampliconic genes on the tiny short arm; the long arm is composed of giant repeats of other genes (cross-hatched). (b) Sequence alignments of the human and chimp Y chromosome euchromatin. Lines connect homologous sequences on the human and chimp Y. Dark grey lines represent direct sequence alignments, light grey lines represent inverted sequences. Sequences within PAR1 are perfectly aligned, but there are complex rearrangements, inversions and differential amplifications on the Y, contrasting with near perfect 1:1 alignment of human and chimp X chromosome sequences.

DNA sequences or RNA transcripts by comparing sequences between males and females. The introduction of long-read and single molecule sequencing has at last surmounted the problems posed by repetitive regions, and we are beginning to see Y chromosome assemblies of several mammals.

Eutherian Y chromosomes generally share the same four regions; pseudoautosomal, X-degenerate, ampliconic and heterochromatic. However, the proportions and arrangements of these regions are very variable, even between Y chromosomes of closely related species.

The first, and still one of the most surprising examples, is a comparison between the Y chromosomes of humans and our closest relative, the chimpanzee. The chimp Y lacks the enormous heterochromatic region on human Yq, but the euchromatic regions are of similar size. Like the human Y, the euchromatic region of the chimp Y has discrete regions that are X-degenerate and ampliconic, but there is no sign of the X-transposed regions of the human Y (Figure 3.5).

Comparison of the ampliconic regions of great ape Y also reveals a lot of variation in the number and size of amplicons. The gorilla Y contains the same eight palindromes as humans, containing different numbers of the same genes, but there were an extra 13 short gorilla-specific palindromes. These structures are evidently extremely labile, as expected because their repetitive and complementary nature facilitates duplications and deletions.

Thus, the sequence composition and arrangement of the Y is exceptionally variable even between closely related species, and its erratic changes do not follow the expected relationship with divergence time.

These conclusions have been reinforced by sequencing the Y chromosomes from more distantly related mammals. The mouse Y was first. The small, gene-dense mouse Yp was completely sequenced in 2009. The euchromatic long arm is at last characterized.

Sequencing the Y chromosomes of other mammal groups, carnivores (cats and dogs) and domestic species (cattle, horses and pigs) have confirmed the quixotic variation of the eutherian Y chromosome. Limited genome sequencing and isolation of male-specific transcripts have extended this analysis to Afrotheria (elephant). In each of these species, the Y contains X-degenerate and ampliconic regions, though their arrangement, and the numbers of genes contained in them varies considerably. The tiny marsupial Y chromosome, which lacks large heterochromatic blocks in many species, would seem to be an excellent target for complete sequencing, but so far, no good Y assemblies have been reported.

Sequencing also confirms that heterochromatin also varies wildly in amount and sequence between species. For instance, the human Y chromosome contains a very large block of heterochromatin consisting of simple sequence repeats that is entirely missing from the gorilla and chimp Y (Figure 3.5). The different sequence make-up was obvious even from the grossest cytological comparisons, for probes that light up repetitive sequence on the human Y do not bind to heterochromatin of the Y in other species, even chimps and gorillas. This reflects different origins and sequence of repetitive units: on the human Y the distal block of heterochromatin is made up of degenerate Alu sequences, in cattle it is degenerate *TSPY* sequences – and the heterochromatin of Bennett's wallaby is made up of degenerate telomere sequences. It seems that almost anything can be amplified to supply the Y with heterochromatin.

Thus, the Y chromosome of all mammals is small and gene poor with a high content of repetitive sequences. The few genes it bears form overlapping sets. Most Y genes have copies on the X, some of which were recently transposed.

3.11 Genes on the Y of Other Mammals

Homologues of human Y genes had been discovered in other mammals, initially from Noah's Ark blots included rather casually in papers on human Y-linked sequences. *SRY* and *ZFY* were cloned and sequenced in dozens of mammals, including primates, rodents and domestic mammals, and mapped to the Y in each. It turns out that these two genes are part of a small core of genes present and active across all eutherian mammals.

X and Y homologues of other human or mouse X-Y shared genes, such as *KDM5D*, and *RPS4Y*, were detected at least on Noah's Ark blots in primates, carnivores and artiodactyls. Sequencing confirmed the presence, numbers and locations, coding potential and activity of these genes and several others in a wide variety of mammals.

The picture is therefore of wild variation, in the presence/absence of genes with X partners, their presence or absence within the PAR (Figure 3.6), and additional genes imported from autosomes.

There is surprising variation even within great apes. For instance, although the gorilla Y shares all 16 X-degenerate protein-coding genes on the human Y, the more closely related chimp Y is missing three. Also, the alignable sequences are very jumbled on the chimp Y. Plotting the positions of homologous sequence between humans and chimps shows many short regions of homology, some inverted and rearranged (Figure 3.5b). This contrasts with the dead straight diagonal you get when you compare chromosome 21 sequence between humans and chimps.

Sequence of the Y chromosomes of carnivores (cats and dogs) and domestic species (cattle, pigs, horses) confirmed that most eutherian mammals share single-copy genes from the X-degenerate region of the human Y. However, the presence/absence of these genes is very inconsistent between species (Figure 3.6). For instance, *UBE1Y* is present and active in rodents, carnivores and the elephant, but is a pseudogene in cattle and missing entirely in primates. *KDM5D* is present on the Y in primates, carnivores and elephants, but is a pseudogene in pigs and missing in cattle. Only five genes are present in all eutherian mammals (*ZFY, UTY, USP9Y, DDX3Y* and *SRY*); others are pseudogenes or absent altogether in at least one species. Some genes, like *RBMY*, are single copy in some species and multicopy in others. Such variability is quite atypical of the mammal genome.

Many Y chromosomes also contain genes or regions that are specific to a species, or a lineage. The large X-transposed region that is specific to the human Y contains the human-specific Y gene *PCDH11Y*, which may contribute to male-specific behaviour. Several species have X-Y shared genes that are missing from the human Y. For example, the cat has several feline-specific genes, including a Y-borne copy of *CUL4BY*, and cattle and pigs have *OFD1Y*. The horse has 29 X-Y shared genes, including several specific to horses. Overall, there are 36 X-Y shared genes in the Y of one species or another.

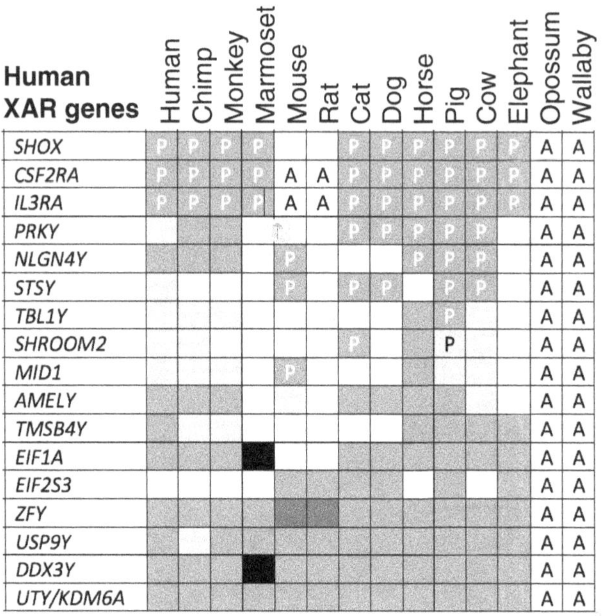

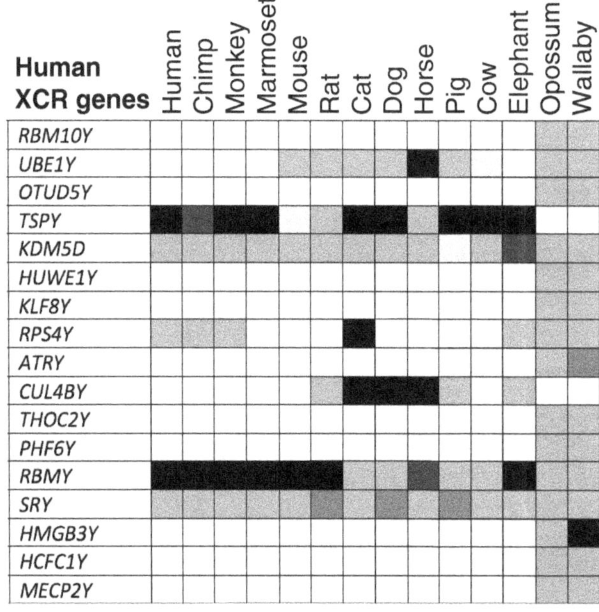

Figure 3.6   Y genes and pseudogenes with X homologues in different mammals. Twelve eutherian and 2 marsupial mammals are presented in the order of their relationship with humans. Genes are listed in the order in which they lie on the human X. Human XAR genes lie in the recently added region and are autosomal in marsupials. Human XCR genes lie in the ancient X conserved region that is shared by eutherians and marsupials. Pale grey – pseudogene, mid grey single active copy, dark grey multicopy, black highly amplified. P = pseudoautosomal, A = autosomal. It's clear that there is a lot of variation in activity and amplification between species, as well as the borders of the PAR; also that there are few genes left on the ancient conserved part of the eutherian X.

This amazing variability in the number and activity of Y-borne genes makes no functional sense, and is explicable only in terms of sex chromosome evolution (Chapter 4).

Noncoding RNA transcription units have also been identified in sequences from all these Y chromosomes, but again their number and identity vary. For instance, the bull Y was found to contain 375 noncoding RNAs transcribed in testis, many more than in primates or mice.

As ever, mice prove to be exceptional. Only nine of the genes on the human Y are present, squeezed into the tiny short arm of the mouse Y. The first gene with a biochemically defined function to be located on the mouse Y chromosome (in fact, the first Y-borne gene to be identified in any mammal) was an active copy of steroid sulphatase *Sts*, which is present only as a pseudogene on the human Y. Its Y location was deduced in the 1980s by the low enzyme activity of its product in XO compared to XY mice. Several genes on the mouse Y were isolated in the 1980s by screening large insert Y clones for transcribed sequences, and screening libraries for homologues of human Y sequences. Exhaustive cloning of the tiny short arm yielded several novel genes that proved to have homologues on the human Y, *Ube1y*, *Smcy* (now *Kdm5d*) and *Uty*. *Sry* and two testis-specific *Zfy* genes were discovered, as well as a cluster of 20–25 *Rbmy* genes.

In some species the Y has received a boost from sundry autosomal sequences. *DAZ*, which is present in multiple copies on the Y in primates and is required for spermatogenesis, is absent from the Y in all other mammal groups and appears to be a copy of a locus on chromosome 6. The euchromatic long arm of the mouse Y remained a mystery for decades until it was shown to consist of huge repeats, each containing copies of three genes with X and autosomal homologues in other species, that are involved in meiosis. Evidently this region was translocated to the Y in the rodent lineage, where it amplified enormously. The pig Y, too, has unique ampliconic regions that appear to contain multiple copies of autosomal olfactory genes (can boars smell better than sows?).

A grand total of 134 protein-coding genes have been detected on the Y over many eutherian species, as well as 214 pseudogenes and noncoding RNA. The suite of single copy genes overlaps, but there are many exceptions, species-specific deletions and additions. Even closely related species (such as humans and chimps, cats and tigers) show many structural rearrangements.

The marsupial Y chromosome turned out to be particularly informative. Early work using human or mouse Y genes as probes established that the marsupial Y contained *SRY*, as well as *UBE1Y, RBMY* and *KDM5D*, all genes with copies on the X in eutherians. But Y homologues of genes on human Xp (such as *ZFY, UTY, EIF1AY and AMELY*) were missing from the marsupial Y.

More surprising was the accidental finding of a gene *ATRY* on the marsupial Y. Its homologue *ATRX* is present on the X in all mammals, but there is no Y homologue in any eutherian. Mutations in the human *ATRX* gene cause sex reversal, so this gene was investigated for a sex-determining function in marsupials. Probing DNA from male and female kangaroos showed clear male-specific bands, and a copy of the huge *ATRY* gene was isolated.

Microdissecting the marsupial Y, and using Y-derived DNA to screen a BAC library yielded several more XY shared genes that are present on the X but absent from the Y in eutherians, including *HUWE1Y*. Sequencing a male opossum genome later confirmed that, as well as five genes also usually on the eutherian Y (*HSFY*, *SRY*, *RBMY*, *KDM5D* and *UBE1Y*), there were another 11 genes that are present on the Y only in marsupials. All had X homologues in eutherians as well as marsupials.

Thus, a set of X-Y shared genes appeared to be shared – more or less – between different mammals, but few genes are consistently present, even in eutherians. The tiny marsupial Y lacks 18 genes that are present on the eutherian Y and have homologues on human Xp, but bears an extra eight genes not present on the eutherian Y.

3.12 Gene Amplification on the Y

Many genes on the Y chromosome are present in multiple copies. For instance, the *RBMY* genes on the human Y are present in at least 23 copies (only 6 of which are active), borne on at least 2 giant palindromes, and *DAZ* is present in at least 6 copies in a single palindrome.

The greatest source of variability of the Y is the degree to which genes and other sequences have been amplified (Figure 3.7). Genes that are multicopy in one species may be present in single copy – or even absent or inactive – in others.

Genes that are present in a single copy on the human Y may be present in multiple copies in other species. For instance, *ZFY* is single copy in primates but multicopy in dogs, duplicated in rodents, and enormously amplified in some old world mouse species. Even the testis determining *SRY* gene is single copy in primates but duplicated in pigs and rabbits – and has amplified 100x in some weird rat species (Section 4.12).

The opposite is also true; genes that are multicopy in humans may be single copy in other species, or even absent. For instance, *RPS4Y* is duplicated in primates, single copy in elephants, multicopy in the cat, but is missing altogether in rodents. *HSFY* is amplified in cats and bulls as well as primates, but is single copy in dogs and elephants, and absent in rodents. *RBMY* is hugely amplified in bulls, amplified in primates, but single copy in dogs and elephants. *TSPY* is

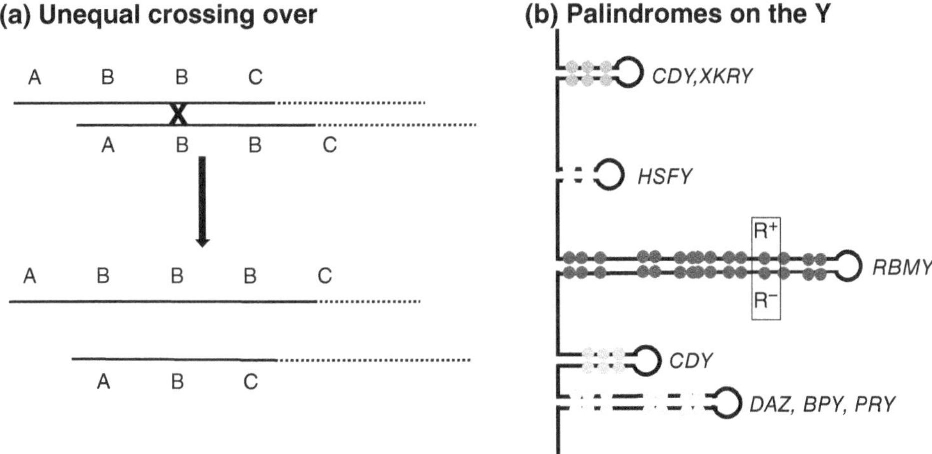

Figure 3.7   Amplification of genes on the Y chromosome. (a) If homologues with a duplicated gene B misalign at meiosis, crossing over generates one recombinant with three copies of B and one with a single copy. (b) Palindromes on the human Y chromosome are giant loops within which multiple copies of genes (represented by dots of different shades) are arranged with complementary sequences along the two arms. Recombination can occur within these loops, as well as gene conversion; for instance, conversion between two forms of *RBMY* (R⁺ and R⁻) homogenizes the sequence so both are R⁺ or both R⁻.

amplified in carnivores, but is present only as a pseudogene in mice. The bull Y is largely composed of 80 copies of a huge repeat (450 kb) that includes *HFSY* and *TSPY*.

Such duplications and amplifications frequently occur within mammal families. For instance, the gorilla Y shares all eight palindromes with humans, containing the same genes (though with fewer *TSPY* and more *RMBY* copies). But although the amplicon region is much larger on the chimp Y and contains twice as many, and more elaborate palindromes, they contain fewer genes (25 rather than 60), and three gene families are missing altogether. *DAZ* is present in four copies in humans and chimps (some of which may be inactive), but only two in gorillas. Even the more distantly related rhesus macaque monkey shares much the same repertoire of Y genes with humans. It seems, then, that the chimp Y is unusually unstable, and indeed there appears to be great variation in the numbers of repeats in amplicons among four subspecies of chimpanzee, and the bonobo.

The 34 species of the cat family have virtually indistinguishable autosomes and X chromosomes. They have a highly conserved repertoire of multicopy genes, but difference in the copy numbers of different repetitive genes make their Y chromosomes vary greatly in length and structure. The horse family (donkeys, zebras and extinct species) show amplifications of several genes in some species but not others, as well as species-specific deletions.

Why so variable? Tandem repeats are intrinsically unstable because mistakes can be made during meiosis (Figure 3.7). For autosomal and X-borne genes, tandem duplicates can line up incorrectly when homologues pair at meiosis. Crossing over can then produce extra copies or deletions. Although the Y chromosome is unpaired during meiosis, unequal crossing over can occur between palindrome arms to duplicate or delete genes. HiC showed that these ampliconic regions have increased contact within and between loops that may facilitate further amplification.

The other interesting capacity of homologous sequences on two palindrome arms is gene conversion, which can replace one variant with another. Conversion occurs when homologous DNA strands align, one breaks and is repaired by copying the other strand. This process tends to homogenize tandemly repeated sequences. Importantly, it is equally likely to replace the normal or the mutant sequence.

Variation within the amplicon regions is frequent because the palindromic sequences on either side of the loop are complementary and can pair and exchange, adding or deleting copies. There is good evidence that gene conversion takes place frequently within the loops and homogenizes sequence of the members of multigene families.

The structure of the Y can influence the propensity to rearrange and cause problems, as is evident in the frequent deletions from the ampliconic 'azoospermia' regions (Figure 3.5), and the frequent deletions of the sex-determining gene that lies between two heterochromatic blocks in the horse.

3.13 Expression and Functions of Genes on the Y

Beginning with the only phenotypes attributed to the Y – sex determination and fertility – the Y chromosome has long been thought to be dedicated to maleness. Indeed, many of the genes on the Y in humans and other mammals are copies of genes that are expressed in the testis and thought to have functions in male meiosis. However, as more genes were discovered, it was found that they may have many basic functions; indeed, several are highly transcribed in somatic tissues or transcribed specifically in differentiating stem cells (Table 3.1).

The expression of Y genes is very atypical. In mice at least, most genes (11/15) are expressed only, or largely, in the testis, and only five have significant expression in other tissues. In humans, several Y genes are barely expressed at all; for instance, *AMELY* is expressed at about 10% the level of its X-borne homologue. However, some genes are highly expressed in non-gonadal tissues; for instance, the translation factor *EIF1AY* (which has an X-borne copy) is more highly expressed in the heart from the Y than the X because it has lost a regulatory microRNA. Generally, Y copies are generally less strongly expressed than their X counterparts (Figure 3.8).

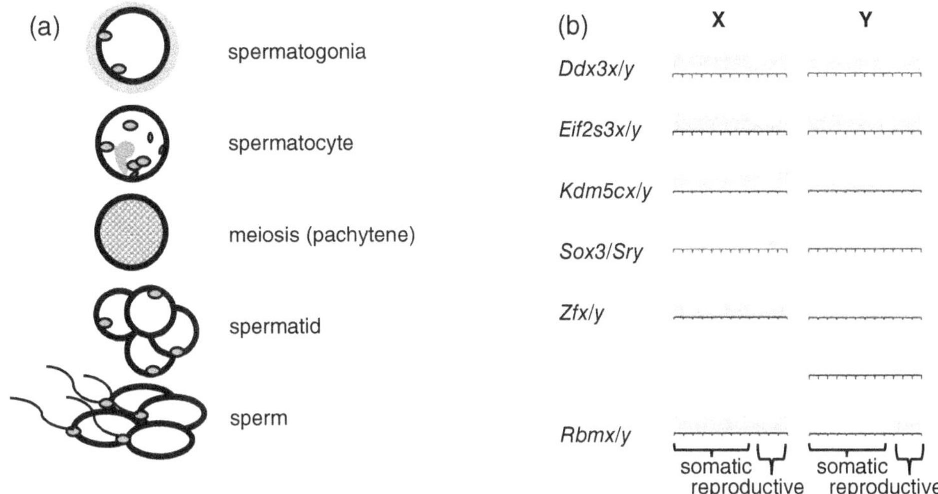

Figure 3.8   Expression of genes on mammal Y chromosomes. (a) Expression pattern of the human spermatogenesis factor RBMY, with functions in RNA splicing. Cells were stained with an antibody to the RBMY protein (grey). Initially localized mainly in the nucleolus, RBMY is activated in spermatocytes and associates with chromosomes that are pairing and segregating at meiosis, then becomes localized in the midpiece of mature sperm. (b) Transcription of gene homologues on mouse X and Y chromosomes in different tissues (including testis with and without germ cells). Y genes are generally less transcribed than their X homologues, and many Y genes (e.g., *Sry*, *Rbmy*) are expressed specifically in the testis. Mouse has two Y copies of gene *Zfy*. (After Godfrey et al., 2020, Genet Res 30: 860).

Making gametes is a complex and exacting process at the very core of successful sexual reproduction. There are some hundreds of genes known to be required at the various stages of meiosis, and most of these are on autosomes. However, several genes on the Y are critical for stages of spermatogenesis.

RBMY, a candidate spermatogenesis gene cloned from an azoospermia deletion interval, contains an RNA recognition motif that interacts with RNA to form RNA-protein complexes in spermatogonia, spermatocytes and round spermatids. Deletion of some *RBMY* genes in human males decreases sperm count; however, *Rbmy* knockout in mice doesn't affect sperm production, although it leads to abnormal sperm. *RBMY* is a member of a gene family that includes its X-borne partner and several autosomal copies, all with functions in RNA binding and alternative RNA splicing. The distribution of RBMY over cells in the premeiotic stage (largely nucleolar), during meiosis (associated with pairing chromosomes) and in spermatids and sperm (localized in the midpiece) suggests functions in splicing during meiosis and sperm differentiation (Figure 3.8).

DAZ, cloned from another azoospermia deletion interval, also codes for an RNA-binding protein important in spermatogenesis. It has no X-borne homologue, but is related to an autosomal gene with an ancient function (even in *Drosophila!*) in spermatogenesis. *TSPY*, present in up to 64 copies on human Yp, has a role in spermatogenesis confirmed by the effects on fertility of partial deletions, but the gene, and its X-borne homologue, are also cell cycle regulators and proto-oncogenes, consistent with a role for *TSPY* in gonadoblastoma. Other multigene families on the Y also seem to have diverse functions, including apoptosis, protein turnover, chromatin remodelling, cell cycle control, that are expressed in spermatogenesis.

Genes in XY shared regions are mostly ubiquitously expressed and seem to have important basic functions. Many are involved in chromatin modification (e.g., *KDM5D*), transcription (e.g., *ZFY*), RNA splicing (*DDX3Y*), translation (*EIF1AYI*) and ubiquination (*UBE1Y*). *AMEL* is an exception, being expressed in tooth buds during tooth mineralization.

It is interesting to consider whether any or all of the transcribed non-coding sequences on the Y have functions. Several novel long ncRNAs were found to be widely expressed in a cell culture, and knockdown of one of them (lnc-KDM5D) produced changes in expression of several other genes, some that affected lipid metabolism, suggesting that this lncRNA gene might have a role in atherosclerosis. Six ncRNAs were described with conserved expression in the central nervous system in developing chimpanzees.

There has been much debate about the possible role of repetitive DNA on the Y, or even whether it has one. Although the simple sequence of most of this material is not compatible with a coding role, geneticists have been hesitant to dismiss most of the Y as useless junk. Possible roles suggested for Y heterochromatin include mechanical functions in pairing, or genetic functions in control of sex. For a long time, repetitive DNA was credited with a controlling role in sex determination (Chapter 8), until deletion mapping ruled out this possibility. Nor is it likely that repetitive DNA regions protect the sex-determining gene(s) from crossing over to the X since they are different ends of the Y. One suggestion is that it acts as ballast, protecting the otherwise tiny Y against loss at division (there is ample evidence that the Y is particularly prone to loss). The strongest evidence for a lack of function of the highly repeated heterochromatic blocks is that the region varies in length between individuals; from undetectable to half the size of the Y. This variation is accompanied by no recognizable phenotypic effect, suggesting that these tandem repeats are truly 'hard core junk' DNA.

3.14 Sequencing the Pseudoautosomal Region

Human sex chromosomes pair regularly only over a shared pseudoautosomal region (PAR1) at the tips of their short arms. The human PAR1 was sequenced as part of the X sequencing project, and the corresponding sequence on the Y was checked against it. It is 2.7 Mb long, within which there is 96–100% homology with the Y.

The sequence structure of PAR1 differs from that of the differentiated regions of both the Y and the X. It has a higher GC content than the X-specific region.

Homology ends at the pseudoautosomal boundary (PAB), below which X and Y sequences diverge. The boundary was first identified by chromosome walking and sequencing from the *CD99* gene towards the centromere, where differences in X and Y sequences went from near-identity to less than 80% identity, then to 45%. An *Alu* sequence was found proximal to *CD99* on the Y, but not on the X. This boundary was located right in the middle of the *XGA* gene!

The second pseudoautosomal region, PAR2 at the tips of the X and Y long arms is only 320 kb long. Pairing and recombination within this small region of homology are seen only occasionally, and, like PAR1, recombination is more frequent in male than female meiosis. Again, there appears to be a sharp transition between homology and non-homology. The region of PAR2 lies on the X in great apes, but not the Y, suggesting that it was transposed to the Y in humans in the last few million years.

Pairing regions were also observed at the tips (top or bottom) of sex chromosome pairs in many animals. From their variety of size and location, it was not clear whether the pairing region was homologous between sex chromosomes of different mammal species. Sequencing PARs of a variety of mammals showed that they share sequence even between distantly related animals such as humans and elephants, as might be expected from a portion of the highly conserved X. However, the position of the PAB varies considerably, such that sequences, and genes, that lie within the PAR in one species are in the male-specific region in others (3.12). For instance, *STS* is pseudoautosomal in cows but X-specific in humans (Figure 4.5).

The mouse is exceptional again, having a much shorter (700 kb) and GC-rich PAR that includes *Sts* but lacks several genes conserved across other eutherians.

Sequencings PARs of humans and other mammals therefore confirms genetic evidence that these regions have special properties. They are GC-rich and gene-dense. They also must have particularly high rates of recombination in males (but not females) to ensure at least one crossover within a few megabases, enough to hold the X and Y together during meiosis. In mice, recombination within the PAR is a hundred times more frequent than in autosomal regions. There is some evidence that the mouse PAR contains repetitive sequences that form 'blobs' of factors that promote double-strand breaks at male meiosis, the prerequisite of crossing over.

3.15 Genes in the Pseudoautosomal Regions

Sequencing the human PAR1 revealed 16 protein-coding genes, 2 pseudogenes and 9 noncoding RNA loci, a much higher gene density than the rest of the Y

(Figure 3.4). Discovery of other PAR genes brings the total to 28 (although some genes are duplicates). Among these genes is *SHOX1*, a bone growth gene, mutations in which produce idiopathic short stature. Haploinsufficiency for this gene is probably the reason that XO women are shorter and XXY and XYY men are taller than average. Other genes within PAR1 include a growth factor, cytokine receptor, methyltransferase, immune cell homeostasis, enzymes required for DNA replication and RNA splicing.

At the boundary of the human PAR1 lies the *XGA* blood group gene. In fact, sequencing showed that the boundary goes right through it; the 5' region is identical on the X and Y, but the 3' ends have diverged.

The PAR2 at the other end of the human X and Y chromosomes is a much shorter region, and contains at last count only three protein-coding genes, five pseudogenes and four sequences that are transcribed into long noncoding RNAs. One PAR2 gene (*VAMP7*) lies on the X in all mammals, suggesting an X-Y transposition, but the other three have chequered histories suggesting addition at other times to both X and Y.

The characteristics and gene content of the PAR amongst even distantly related mammals show deep conservation of gene content and gene order of human PAR1. What does vary is the extent of the homology with the Y chromosome (Figure 4.5).

Sequencing the tip of the Y and Y in great apes and monkeys confirms that the gene content of this terminal region of the X and Y chromosomes is highly conserved. The PAR in chimps, gorillas and rhesus macaques contains the same protein-coding genes including the bone growth gene *SHOX1*, and 9 noncoding loci. An exception is one gene that has been deleted from the very top of this region on the chimp Y. The PAB lies in the middle of the gene that encodes blood group *XGA* in all primates, but only in humans and chimps is the boundary marked by an Alu sequence on the Y. But in prosimians (more distantly related primates), the PAR apparently extends down the X to include two other genes on the conserved X. One of these is steroid sulfatase *STS*, which is represented by a pseudogene on the Y of humans and chimps and is missing from the gorilla Y.

The extent of X-Y homology is also greater in horses. The horse PAR at the tip of Yq is physically smaller (only 1.8 Mb) than the human Y because it contains fewer repetitive sequences. The horse X is virtually identical to the human X in the content and order of 18 genes. The horse Y shows homology to the X past the human XGA gene and the human PAB, to take in two more genes (including *STS*) that have no active Y copy in humans.

The large (>6 Mb) PAR in cattle and pigs, cats and dogs, occupies still more of the terminal region of the X and contains about 40 genes. Again, the gene content and order are conserved on the X, and homology with the Y extends along the X taking in another 13 genes (including STS) and terminates in the conserved gene

SHROOM. The PAR in camels and alpacas take in even a few more genes on the conserved X (including *STS* and *SHROOM*), with a boundary near the *MID1* gene. Thus, PAR seems to comprise simply the terminal portion of the highly conserved eutherian X, the only difference being the point at which XY homology ceases.

Rodents are a glaring exception to this picture of PAR conservation and stepwise reduction. The 700 kb mouse PAR contains only 5 protein-coding genes, none of which are shared with the human PAR. However, an active *Sts* gene is present on the mouse Y, and the PAB interrupts the *Mid1* gene that is conserved on the human Y. These two genes lie on the human X just below the boundary, and are included in the PAR in other species (Figure 4.5). The mouse PAR appears to have lost genes from the sex chromosomes; five human PAR genes have no homologues in the mouse genome, and another three have been transposed to autosomes. Other mouse subspecies retain a larger PAR.

The different extents of the PAR in different eutherian groups means that they terminate at different places, so we would expect the boundaries to have different sequences. The primate boundary lies in the middle of the *XGA* gene, the pig PAR in the middle of *SHROOM* and the mouse PAR in the middle of *MID1* gene; all are intact on the X but functionless on the Y. However, different boundaries do share some molecular signatures, notably the insertion of a transposable element.

The pretty much random functions of PAR genes, as well as variation between species, suggest that the gene content of PAR1 is immaterial to its critical function in XY pairing and fertility. The PAR is necessary only for correctly segregating the X and Y chromosomes at meiosis.

3.16 The Y Chromosome as a Population Marker

Humans vary considerably in their genome sequence. About 1/1,000 base pairs vary from person to person, making it possible to recognize the genetic signatures of different people (or, forensically, different dead bodies). Specific Y variants are inherited, so they are ideal for sexual assault testing, paternity testing or discovering long-lost (male) cousins through various matching services. Such tests can even be retroactive – for instance, typing the Y chromosome from locks of hair souvenired from Ludwig von Beethoven suggested that the great composer did not possess a von Beethoven Y chromosome.

Most importantly, different variants may be present at higher or lower frequencies in different populations. The origin and movements of human populations can be traced by following these variants in space and time.

The Y chromosome is particularly useful for following human populations because it does not recombine. The direct father-to-son transmission makes it very easy to follow particular sequence variants.

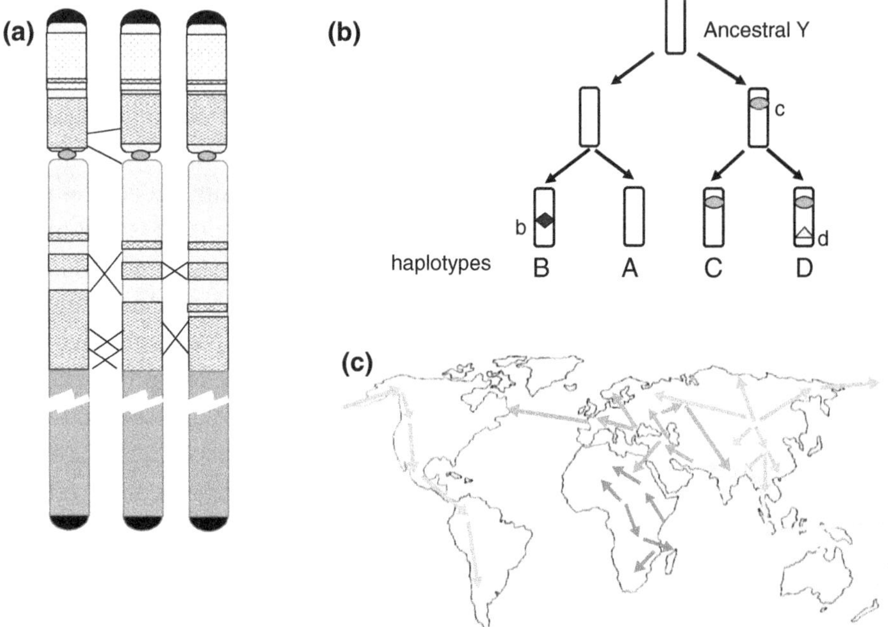

Figure 3.9   Y chromosome variants. (a) Most variation is within the ampliconic regions, where amplifications and inversions are common. (b) Variants define Y haplotypes by specific mutations to the ancestral Y: mutation B (diamond) defines haplotype B, mutation C (circle) defines haplotype C and mutation D (triangle) haplotype D. (c) Y chromosome variants can track human population movements in space and time. Different Y haplotypes helped trace the major migrations of early humans; darkest arrows within Africa, mid-grey west route out of Africa, light grey north route out of Africa and into the Americas, pale grey south route out of Africa to Australia and the Pacific.

Y sequence varies more than the X or autosomes, so there are plenty of changes that can be traced back and followed (Figure 3.9). Sequencing large chunks of different human Y chromosomes (most of the X-degenerate region and half of the ampliconic) from men coming from different backgrounds revealed a lot of variation. The assemblies varied by hundreds of small deletions and insertions, and also in the numbers of genes within five of the nine amplicons. Indeed, a survey of the male-specific region of the Y of 62 men revealed nearly 11,000 single nucleotide variants and 3,000 indels (insertions or deletions). The ability to compare sequences between fathers and their sons (who must inherit the same Y) showed that the rate of new mutation is very high for the Y chromosome. There are now some 15,000 established Y chromosome markers.

Testing thousands of men all over the world for particular Y variants provided markers that defined particular groups of variants, called haplotypes (because the Y is present in only a single 'haploid' copy) (Figure 3.9). It is possible to build a family tree of Y variants and use the rate of variation to figure out where in the world the last common ancestor lived. Such analyses can reveal the disproportionate contributions some men have made; for instance, it is estimated that 8% of men living in a swathe of Asia carry the Y chromosome of Genghis Khan.

Even more excitingly, Y haplotypes can be typed from ancient DNA recovered from fossil hominids, so we can go back in time to reconstruct migrations, and mixing, of our hominid ancestors. The ancestral Y with the constellation of variants that all haplotypes share pointed to an ancestral Y that was present in Africa more than 200,000 years ago in a Y chromosome 'Adam'. African men had by far the greatest variety of Y haplotypes, most of which exist only in Africa. Only a few of the original haplotypes made it out of Africa, and these can be traced to major migrations; north through the Arabian peninsula, then west to Europe (with a substrain returning to Africa), further north to Asia; one haplotype made it across the Bering Straits and into America 15,000 years or more ago. Another haplotype took a southern route around the coast of India and into Australia, New Guinea and Oceania. There are now more than 300 Y haplogroups, which accrued new mutants that enable subdivision of these populations (Figure 3.9).

With strategies to extract Y chromosome sequences from ancient DNA, it has even been possible to track Y chromosomes as far back as the divergence of Denisovans 700,000 years ago, and Neanderthals 370,000 years ago, and to speculate that hybridization with modern humans gradually replaced the Neanderthal Y chromosome.

Of course, this is only half the story, the story of the movement of men. The other half, the movement of women, can be traced through changes in the tiny genome of mitochondria, which are inherited only through females. Largely, they complement, with a mitochondrial Eve in Africa at least 100,000 years ago, and migrations across the globe that match the Y chromosome. But sometimes the trajectories are different, attesting to social patterns of kinship and marriage (or rape and subjugation).

In the same way, Y chromosome markers can be used to distinguish sex of animals (dead or alive, or even extinct), to differentiate populations and examine the history of domestication or patterns of migration of wild populations.

For instance, the paucity of Y haplogroups implies that very few stallions founded domestic horse breeds, and breeding cattle using sperm from a few prize bulls introduces problems of inbreeding in some cattle breeds.

In wild populations of endangered animals, too, Y chromosome markers can yield valuable information on sex ratios and population dynamics and history. For instance, the dingo (the wild dog that accompanied human immigrants into Australia at least 5,000 years ago) was studied using SNP and microsatellite Y variants. This provided evidence of two geographically distinct populations that may have migrated independently from Asia. It also implied that dingo populations have been compromised by introgression from male, but not female, domestic dogs, because they bear the signature European Y haplotype, but not domestic dog mitochondrial lineages. This has consequences for managing particularly the populations near growing urban areas that are also impacted by culling and habitat loss. In the same way, Y chromosome markers have shown that randy male camels from domestic populations compromise the genetic integrity of endangered wild camel populations.

Y chromosome markers have also been useful in studying and managing endangered populations of marsupials. The small kangaroo tammar wallaby became extinct in mainland Australia but survived as isolated island populations. The species has a promiscuous mating system, with most matings by a dominant male. In an island population, most males shared a single Y chromosome haplotype, attesting to inbreeding problems in this small island population and pointing to strategies for captive breeding to maximize genetic variability.

Thus, Y chromosome markers, with their special male-limited transmission, have many uses in tracking and managing mammal populations, including our own.

3.17 Conclusions and Remaining Puzzles

We now have detailed molecular pictures of the human X and Y chromosomes. They are both even more peculiar than we had suspected from classic genetic analysis.

The X chromosome, though it looks relatively normal and contains a fair density of genes, contains a great excess of genes that have some function in intelligence, or reproduction, or (strangely) both. Particularly striking are regions in which cancer-testis genes are amplified. Its extreme conservatism of gene content, and even order across mammals, demands a special explanation, and I will examine this in Part II.

The Y chromosome is downright weird (Figure 3.10). Its content of repetitive DNA, particularly the apparently useless heterochromatic regions, is far higher than any other region of the genome. It is organized into distinct regions with different evolutionary origins that bear genes with different histories and functions. The pseudoautosomal regions it shares with the X have extraordinary properties of pairing and recombination against all odds. The male-specific region of the Y contains only 27 genes, although many of these genes have been duplicated or further amplified into extraordinary palindromic loops.

Like many things in nature, it is difficult to comprehend these weird structures in terms of their function. To make sense of sex chromosomes, we must look to evolutionary forces that shaped them (Chapter 4).

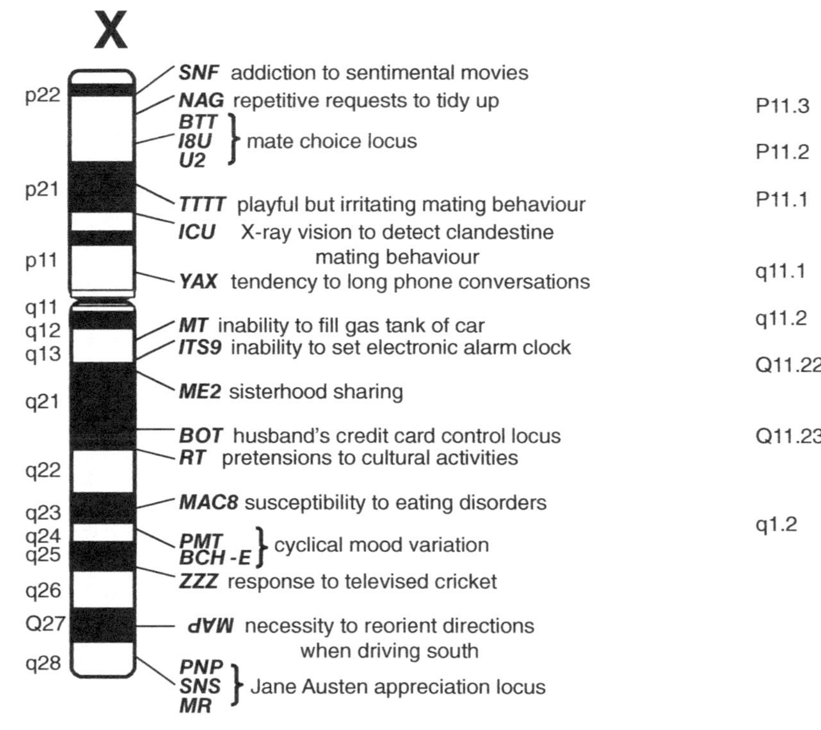

Figure 3.10 Popular view of genes on human sex chromosomes. Banded X and Y chromosomes labelled with imaginary genes that account for stereotypical female and male traits. X genes are all my own, but Y genes include some suggested by colleagues over decades.

FURTHER READING

Classic Papers

Ross M, Grafham, DV, Coffey, AJ, et al., 2005. The DNA sequence of the human X chromosome. *Nature* 434: 325–337

Skaletsky H, Kuroda-Kawaguchi T, Minx PJ, et al., 2003. The male-specific region of the human Y chromosome is a mosaic of discrete sequence classes. *Nature* 423: 825–837

Weissenbach J, 1988. Mapping the human Y chromosome. *Philosophical Transactions of the Royal Society of London. Series B, Biological Sciences* 322: 125–131

Reviews and Research Articles

Colaco S, Modi D, 2018. Genetics of the human Y chromosome and its association with male infertility. *Reproductive Biology and Endocrinology* 16: 14. doi: 10.1186/s12958-018-0330-5

Godfrey AK, Naqvi S, Chmátal L, et al., 2020. Quantitative analysis of Y-Chromosome gene expression across 36 human tissues. *Genome Research* 30(6): 860–873. doi: 10.1101/gr.261248.120

Hallast P, Ebert P, Loftus M, 2023. Assembly of 43 diverse human Y chromosomes reveals extensive complexity and variation. *Nature* 621: 355–364 https://doi.org/10.1038/s41586-023-06425-6

Hughes JF, Skaletsky H, Pyntickova T, et al., 2010. Chimpanzee and human Y chromosomes are remarkably divergent in structure and gene content *Nature* 463: 536–539

Jobling MA, Pandya A, Tyler-Smith C, 1997. The Y chromosome in forensic analysis and paternity testing. *International Journal of Legal Medicine* 110: 118–124

Lahn BT, Pearson NM, Jegalian K, 2001. The human Y chromosome, in the light of evolution. *Nature Reviews Genetics* 2: 207–216

Mangs AH, Morris BJ, 2007. The human pseudoautosomal region (PAR): Origin, function and future. *Current Genomics* 8: 129–136

Maan A, Eales J, Akbarov A, et al., 2017. The Y chromosome: A blueprint for men's health? *European Journal of Human Genetics* 25: 1181–1188

Makova KD, Pickett BD, Harris RS, et al., 2024. The complete sequence and comparative analysis of ape sex chromosomes. *Nature* 630: 401–411 (2024) https://doi.org/10.1038/s41586-024-07473-2

Miga KH, Koren S, Rhie A, et al., 2020. Telomere-to-telomere assembly of a complete human X chromosome. *Nature* 585: 79–84

Rhie A, Nurk S, Cechova M, et al., 2023. The complete sequence of a human Y chromosome. *Nature* 621: 344–354. https://doi.org/10.1038/s41586-023-06457-y

Underhill PA, Kivisild T, 2007. Use of Y chromosome and mitochondrial DNA population structure in tracing human migrations. *Annual Review of Genetics* 41: 539–564

4

Evolution of Mammalian Sex Chromosomes

Sex chromosomes make no functional sense. To understand their peculiarities and how they work, we must understand how they evolved. Evolutionary geneticists just love sex chromosomes. They are atypical in the genome in being represented differently in males and females, and this has consequences in the sex-biased function of genes they bear, their levels of variation and the speed of their evolution.

Sex chromosomes have been the subject of many hard-fought evolutionary battles over many decades, and, in the absence of much data from vertebrates, there have been fierce squabbles about the applicability of, for instance, the 'fast X' hypothesis and 'sexual antagonism'. Now we are compiling enormous banks of data on sex chromosomes of humans, mice and many other mammals, as well as other vertebrates, it is a testing time for evolutionary thought.

We now have the detailed information on sex chromosomes of a range of animals; closely related mammals to spot recent events, more distantly related mammals to explore the early events in our sex chromosome histories, even non-mammal vertebrates and beyond to discover the deepest secrets that unite sex chromosomes of all animals.

As I recounted in Chapter 3, the characterization of mammalian X and Y chromosomes has shown that neither chromosome is in any sense 'normal'. Here I will explore some of the peculiarities of the X and Y properties that offer us insights into the processes that have shaped sex chromosomes in mammals, and in all higher organisms.

4.1 Marsupials Break Ohno's Law to Reveal the Ancient Building Blocks of Mammal Sex Chromosomes

Even the earliest cytological data showed that the mammal X chromosome was exceptional. The eutherian X chromosome is conspicuously constant at about 5% of the haploid length, regardless of whether the genome is divided up into many small chromosomes, or a few large ones. Early genetic studies suggested that gene content, as well as size of the X is conserved between mammal species. Susumu Ohno, an evolutionary cytogeneticist famous for his wild ideas, declared that the gene content of the mammalian X chromosome was absolutely conserved in evolution ('Ohno's Law').

Detailed genetic and physical mapping studies over a wide front confirmed that gene content of the X is, indeed, very conserved among eutherian species, as predicted by Ohno. Most (95%) single copy genes are conserved between species.

Sequencing confirmed that the eutherian X constitutes about 5% of genome size. The fully assembled X chromosome sequences of chimpanzee, cat, cow, pig and sheep all align. Even the mouse, with its rearranged autosomes has an X with almost the identical gene and sequence content, though much rearranged. Surprisingly, gene order as well as gene content on the X is remarkably conserved among eutherians, even between humans and elephants, in which only the centromere has shifted.

This conservation demands a special explanation. Ohno ascribed it to the protection of the X by its involvement in chromosome-wide inactivation (Part II), which would be disrupted by X-autosome translocation, causing sterility in heterozygotes. Minor breaches of Ohno's law are found for genes in or near the PAR, which makes sense since they are not inactivated.

A surprise came with comparative mapping of marsupial X chromosomes. The basic marsupial X is smaller than the eutherian X (about 3.3% of haploid length), but all the classic enzyme markers on human Xq mapped to the X in kangaroos and dunnarts, in accordance with Ohno's Law. However, with the advent of in situ hybridization, it was shown that several genes that are on human Xp were autosomal in marsupials, lying in a block on kangaroo chromosome 5, and dunnart chromosome 3. Sequencing backs this up; the marsupial X aligns to the lower 2/3 of the human X, and an autosomal block aligns with the top.

There are two possible explanations. Either the eutherian X is ancestral and marsupials lost a large chunk early in marsupial evolution, or alternatively, the marsupial X represents the original X, to which an autosomal region was added early in the eutherian lineage. Mapping of human Xq and Xp markers to blocks on separate chromosomes in chicken and fish favoured the second possibility. Thus, the human X is made up of two ancestral segments; the ancient X conserved region XCR that is on the X in all therians, and the X added region (XAR) which is autosomal in marsupials and was added to the X before the eutherian radiation 105 Mya (Figure 4.1).

This dual origin explains several odd features of the human X, since part of it is ancient and part was recently added. For instance, most genes in the ancient XCR have no partners on the Y and are inactivated (Chapter 7), whereas many genes in XAR have Y partners and are not inactivated.

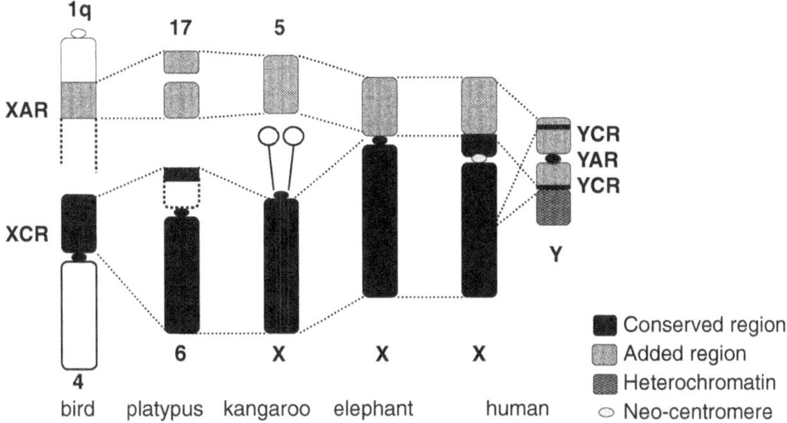

Figure 4.1   Origin of human sex chromosomes. Mapping human X genes in marsupials shows that the X of placental mammals is made up of two ancestral gene blocks. One block (black) is homologous to the marsupial X and represents an ancient conserved X (XCR), the other (stippled) is autosomal and represents a region (XAR) that was added >105 Mya. The separation of these regions by a centromere in the elephant X suggests XAR was added by centromeric fusion. A new centromere took over on the X but not the Y. XCR and XAR are on separate autosomes in monotremes and birds, demonstrating the autosomal origin of our sex chromosomes <166 Mya. The conserved and added regions have homologues on the human Y. The original Y conserved region YCR is represented by only five genes; most of the Y (YAR) originates from the added region.

More refined mapping placed the junction of XCR and XAR above the centromere of the human X. However, the junction occurs right at the centromere of the elephant X. This suggests that X-autosome fusion occurred by centromere to centromere fusion (a very common type of translocation called 'Robertsonian'). The elephant X therefore represents the original configuration. In the ancestor of the eutherian lineage that gave rise to other eutherian mammals, the centromere shifted to its present position, which is conserved between humans, cats and pigs. This was not an inversion, because the order of genes around the elephant centromere is the same as on the human X. The original centromere must therefore have been replaced by a neo-centromere at a new site on the X.

Thus the human X originated from two autosomal blocks of genes that fused about 105 Mya.

4.2 Genes on the X – Mutation, Selection and Sexual Antagonism

We have seen that functions of genes on the human X chromosome are far from the typical mix on the autosomes (Chapter 2). Genes expressed in the brain and in reproduction are disproportionately represented. The X is also particularly rich

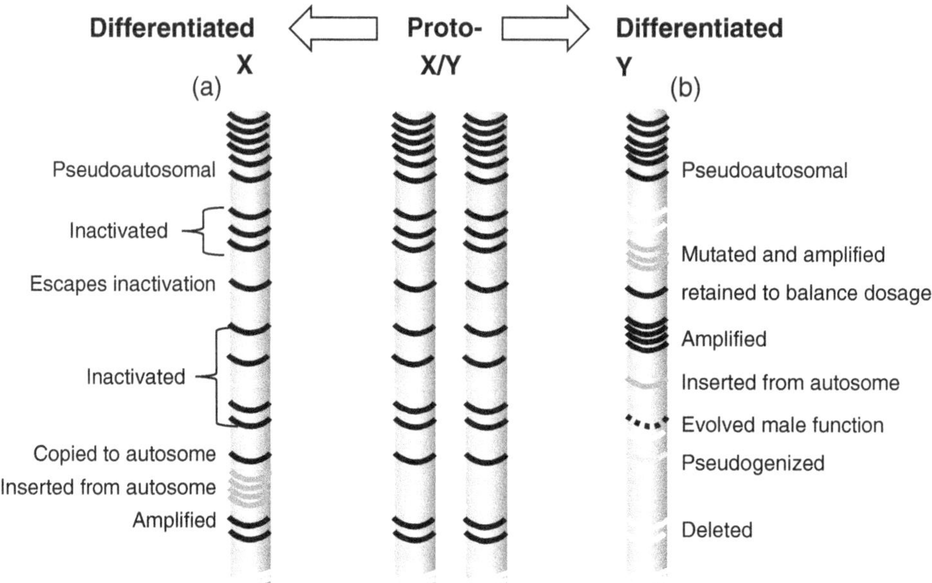

Figure 4.2   Gene evolution on the sex chromosomes. Genes are represented by rings. On the proto sex chromosome, all genes are active (black). Genes within the pseudoautosomal region remain active on both the X and Y. In the X-specific region (a) genes whose Y partners are deleted or specialized are silenced on the inactive X. Genes whose Y partners are retained escape inactivation. Some inactivated genes spawn autosomal copies that are active during meiosis. A few genes are inserted from autosomes and a few genes are amplified. In the male-specific region of the Y chromosome (b) most genes are deleted (white), or mutated/pseudogenized (grey). A few are retained (black) to balance dosage and some are repurposed to a male function (dotted) or amplified.

in 'sex biased' genes that are expressed more strongly (or exclusively) in one sex than the other, and 'sexually antagonistic' genes that confer advantage in one sex but a disadvantage in the other.

Is the unusual make-up of the X a product of bias in the mechanism of gene change – mutation, amplification, movement and retrotransposition? Or have special selective forces shaped the X. The answer is both (Figure 4.2).

An atypical rate and sequence bias of mutation on the X follows from the unequal representation of the X in males and females. Mutation is less frequent in female gonads than male because making eggs takes far fewer divisions than making sperm (4.8). The X chromosome spends more time in the ovary than the testis, so you might expect the mutation rate of genes on the X to be lower than for autosomal genes, which spend equal time in males and females. Indeed, the X does come out lower in comparisons of the rate of substitution at 'silent sites' in DNA (which don't change an amino acid in a coding region) between humans and chimpanzees, mice and rats. This lower mutation might also explain the high GC ratio of the X chromosome because overall, mutation tends to be biased in favour of changes from GC to AT.

Although the gene content of the X is very stable, there are regions in which movement and amplification are rife. Many genes (107 in humans and 149 in mice) lie in 'ampliconic' regions of the X that are prone to amplification. These 'cancer-testis' (CT) genes are expressed in the testis and re-expressed in some cancers. They are conserved in chimpanzees, and seem to have amplified independently in rodents and pigs.

The X chromosome, in both humans and mice, has also acquired more than its fair share of genes copied from autosomes, and contributed a disproportionate number of gene copies to autosomes. This gene traffic is at least three times more frequent than between different autosomes, and involves preferentially genes that are expressed in the testis. A mechanistic peculiarity of the X is ruled out because pseudogenes are not retrotransposed at anything like the same frequency. It is more likely to reflect selection for gene function, and it happened long ago because many of these genes are common to humans and mice.

Such selection is probably the response to another challenge faced by genes on the X chromosome; the X and Y are both inactivated at male meiosis (Chapter 5). It would be fatal if genes like *PGK* with a basic function were inactivated. However, loss of basic functions during spermatogenesis is averted by the evolution of autosomal copies of X genes that are involved in basic metabolism. For instance, X-borne *PGK-A* has an autosomal copy *PGK-B* that is expressed in the testis, and this happened independently in marsupials and eutherians. At least twenty other genes on the conserved XCR have autosomal copies, but only three genes in the added XAR, consistent with its recent recruitment to meiotic sex chromosome inactivation.

A worse problem is that genes on the X are present only in single copy in males. Ohno called this 'the perils of hemizygosity' because any X gene mutation, recessive as well as dominant, is immediately expressed in males. The downside is that deleterious recessive mutations such as haemophilia (and muscular dystrophy, Lesch-Nyhan syndrome, and X-linked mental retardation) as well as colour blindness are always expressed in boys, although their heterozygous mothers are not affected.

The upside is that a new mutation that confers an advantage to males (e.g., making more or better sperm), will be immediately subject to positive selection. A recessive allele will not affect heterozygous females (and there will be no homozygotes as long as it is rare), so even if it confers no advantage, or even a disadvantage to females, it will survive and prosper in the population.

This brings in the concept of 'sexual antagonism'. This refers, not to domestic squabbles, but to the observation that the function of many genes is more advantageous to one sex than the other. In *Drosophila*, there are some male advantage gene variants that are actually lethal in females. Such male-advantage alleles will spread in the population as long as they are not outweighed by their deleterious effects on the other sex. Deleterious effects can be mitigated by modifying gene expression; for instance, many of the male-advantage genes (e.g., CT genes) are expressed only in the testis, so they are silent in females.

These biases might explain the disproportionate numbers of reproduction-related genes on the X. It is easy to see why improved functions in male reproduction might be at an advantage if they lie on the X.

But why are 'intelligence genes' so disproportionately represented on the human X? Perhaps brain-expressed genes with a function in sex-specific behaviour accumulate on the X. An interesting alternative is that females chose partners of high intelligence; who can not only bring home the bacon (or woolly mammoth), but also have time to tell stories and paint on the cave walls. Thus it has been suggested that intelligence is a sexually selected trait with runaway consequences because the alleles that confer intelligence are handed down to females as well as males. This could explain why the human brain expanded so rapidly in the last 100,000 years or so.

Thus opposing forces affect the rate of mutation and the evolutionary clout of the mammal X chromosome. A lower overall mutation rate decreases variation. However, rapid and efficient selection for new alleles that arise (or are imported) on the X, drives faster change in the X chromosome than in other chromosomes.

There is a lot of support for rapid evolution of X genes expressed in male reproductive tissues in several mammals. Comparisons between the human and chimpanzee genomes show X-linked genes have a higher ratio of non-synonymous substitutions than autosomal genes. For instance, sperm proteins encoded by X-borne genes changed faster than sperm proteins encoded by autosomal genes. Many X-linked proteins with functions at all stages of male meiosis showed greater than average variation between mouse subspecies. This 'Fast X' has been called 'the engine of evolution' because it has a disproportionately large effect on evolution and speciation.

So the X is no ordinary chromosome. We must look further into how it became an X and lost most of its partner in males.

4.3 X-Y Homology and the Origin of the Mammalian Y

The idea that differentiated sex chromosomes evolved from an autosome pair was put forward a century ago by HJ Muller to explain *Drosophila* X and Y chromosomes, but applies to all sex chromosome systems.

Muller proposed that the *Drosophila* X and Y chromosomes were originally an autosome pair, but once a male determining gene arose on one partner, that chromosome was confined to males, progressively lost its homology with the X and degenerated rapidly. Evolutionary geneticists Brian and Deborah Charlesworth considered how this would work for how mammal sex chromosomes, and Susumu Ohno adapted this hypothesis to describe the evolution of snake sex chromosomes.

Indeed, there is overwhelming evidence that mammal sex chromosomes evolved from a pair of autosomes. PAR1 regions at the tips of the X and Y are completely homologous, and the overlapping set of single copy widely expressed genes in different mammals (including the sex-determining gene *SRY*) all have X copies (Table 3.1). These Y genes with a copy on the X are the relics of ancient homology.

Some Y genes have X homologues in the XCR that are conserved over all therian mammals. Others (including PAR genes) have X homologues in the XAR. This defines two ancestral blocks on the Y, the conserved region of the Y (YCR) and added region of the Y (YAR). The simplest explanation is that an autosomal region fused to an ancient PAR of the original XY chromosome pair (Figure 4.1). Most genes on the human Y lie within YAR, and only five genes lie in the YCR that was present in the ancestral therian mammal.

Direct evidence that the mammal XY was once an autosome comes from the discovery that human X and Y genes are autosomal in other vertebrates. They cluster in two blocks in birds, reptiles and fish. In chicken, the ancient X and Y conserved region XCR/YCR is homologous to the short arm of chromosome 4 (which is a microchromosome in other birds), and the added region XAR/YAR is homologous to a region on the long arm of chromosome 1. The simplest explanation is that the mammal XY pair was constructed from two ancient autosomal blocks. One became an XY pair before marsupials and eutherians diverged 148 Mya, the other was added after that but before eutherians radiated 105 Mya.

So how did the human X and Y become different? We can quickly see that, though the X has specialized to some extent, it retains its ancestral structure and gene complement. It is the Y chromosome that became different (Figure 4.3). Out of an original 1,000-odd genes, it lost all but 51. Only the five genes within YCR survive from the original XY pair, and 22 from the added YAR. The undifferentiated PAR1 containing 24 genes is simply what is left from the added region. Thus the weird gene constitution of the mammal Y is easily explained by the hypothesis that mammal sex chromosomes evolved from an autosome by the progressive degradation of the Y chromosome.

Despite the inconsistency in gene content, number and activity between species (Table 3.1), it is obvious from the overlapping set of Y-borne genes that the mammalian Y is monophyletic; that is, derived from the same Y chromosome in a common mammalian ancestor. Variation between species amounts to loss or inactivation – or amplification, or acquisition – of different genes in different lineages.

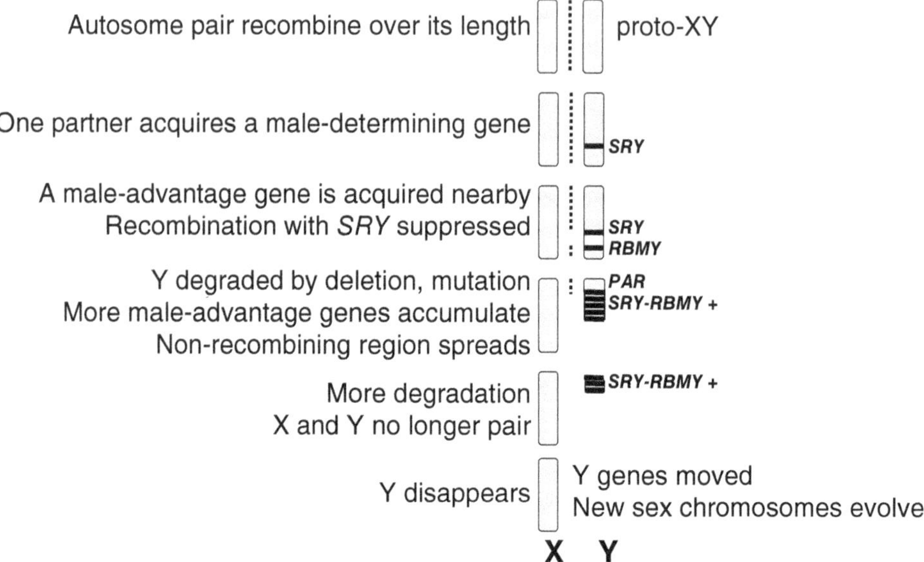

Figure 4.3 Differentiation of mammal sex chromosomes from a pair of autosomes. Differentiation of X and Y chromosomes from the original autosome pair (shown as grey bars) which undergoes recombination along its length (represented as dotted line between them). Differentiation begins when a male-determining gene (e.g., *SRY*) is acquired by one homologue. If a male-advantage gene (e.g., *RBMY*) evolves nearby, there is selection for loss of recombination in the region between them to preserve a male-specific package. Within the non-recombining region (white) mutation and deletion rapidly remove most active genes. The non-recombining region expands as other male-advantage genes accumulate, and Y degradation proceeds until just a small pseudoautosomal region (PAR) remains. This can be lost if alternative means of X-Y segregation evolve. The entire Y may be lost if a rival sex gene evolves on another chromosome.

The secret to understanding this apparently quixotic variation was the realisation that the Y chromosome is essentially a degenerating X, and that the extent of degeneration can vary in different lineages. Now we just need to understand when this happened and how.

4.4 Monotremes Provide Time Zero for Y Degradation

When did mammal sex chromosomes begin to differentiate? Because birds and reptiles have completely different sex chromosomes, the dawn of mammalian sex chromosomes was always assumed to be after the divergence of mammals and reptiles 310 Mya. This estimate had to be drastically revised when it was discovered that monotreme mammals have sex chromosomes that share no homology with those of therian mammals.

The egg-laying monotremes are an ancient branch of mammals that diverged from therians about 180 Mya (1.14). They are an important outgroup to therian mammals, and many attempts were made over the years to determine how their sex chromosomes worked. This should be easy – just line up karyotypes of male and female platypuses, and spot the heteromorphic XY pair in males. However, when this was done with platypus, there were not two chromosomes without a partner in males, but *ten*. Its cousin the echidna (spiny anteater) had nine. At male meiosis these multiple sex chromosomes seemed to line up in a chain, head to tail.

The bizarre situation was resolved by flow sorting and identifying individual chromosomes by 'chromosome painting' (Section 1.9). The ten unpaired chromosomes in male platypus turned out to comprise five X chromosomes (present in two copies in females and one in males) and five Y chromosomes (present only in males). The five Xs were all different; likewise the five Ys.

How on earth could meiosis deliver sperm that carried either a male or a female sex chromosome complement, and not some crazy mixture of Xs and Ys? Chromosome painting showed that at meiosis, these ten sex chromosomes lined up always in the same order X_1–Y_1–X_2–Y_2–X_3–Y_3–X_4–Y_4–X_5–Y_5. They were joined at meiosis by crossovers that occurred in homologous regions (nine PARs) between the tips of the short arms of X_1/Y_1, the tips of the long arms of Y_1/X_2, the tips of the short arms of X_2/Y_2 …. Painting with chromosome-specific probes showed that out of thousands of sperm, none had a mixture of Xs and Ys. All the X chromosomes must therefore have segregated to one pole and all the Y chromosomes to another, producing two kinds of sperm, having either all five Xs (female determining) or all five Ys (male determining). (Figure 4.4).

The real shock came from comparative mapping, which showed that genes that are on the human XCR all mapped to the sixth largest autosome in platypus. Chromosome 6, present in two copies in both sexes, looked much like the marsupial X, with the same genes in much the same order. Yet it had two copies in both sexes, and was not part of the sex chromosome chain.

While mapping and sequencing genes on the ten platypus sex chromosomes revealed no homology to the sex chromosomes of therian mammals, unexpectedly, they had strong homology to *bird* sex chromosomes, which are quite different (Section 4.14). There is no *SRY* gene, and its X homologue *SOX3* maps to platypus chromosome 6, along with other human X genes. The best candidate for sex determination in the platypus is a gene *AMH* (homologous to the human Anti-Müllerian Hormone), which lies on the tiny platypus Y_5.

This result implied that the sex chromosomes of therian mammals, including human, got their start a mere 180 Mya, after therian mammals diverged from monotremes. This is the new time zero, when an autosome became the new mammalian sex chromosome pair.

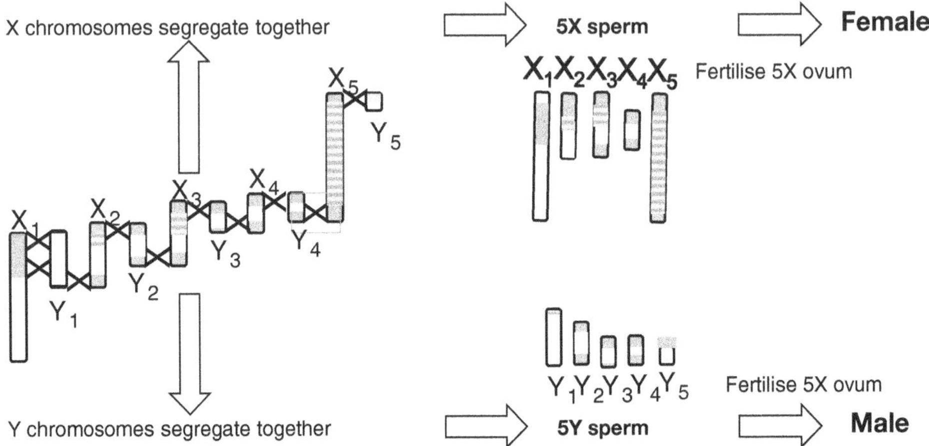

Figure 4.4   The bizarre sex chromosomes of the platypus. Male platypuses have 5X and 5Y chromosomes. The ten sex chromosomes line up in a chain at male meiosis and are held together by crossovers (cross shapes) between the PARs of X and Y chromosomes (shaded). At anaphase the 5Xs all segregate to one pole and the 5Ys to the other. They form two kinds of sperm carrying 5X, or 5Y chromosomes. Females have two copies of the 5X chromosomes and make eggs carrying 5X. On fertilisation of 5X ova, 5X sperm determine female and 5Y sperm determine male offspring. Platypus sex chromosomes share considerable homology with chicken sex chromosomes (striped regions).

4.5 Y Chromosome Degradation

Y chromosome degradation took place in several steps (Figure 4.5). The first step of sex chromosome differentiation was the acquisition of a new sex-determining gene by one member of the ancient autosome pair. The ubiquity of *SRY* in therian mammals qualifies it for this role. The initial event was the evolution of *SRY* from an ancient gene *SOX3* on the original autosome (Chapter 10).

But what happened next to set the process of Y degradation in motion? Ideas centre on suppressing recombination between two genes that are both required for making males or females. Nei suggested in 1969 that sex determination requires changes in two genes on the same chromosome, and Brian and Deborah Charlesworth later proposed that you need mutations in two linked 'antagonistic' genes, one conferring male sterility and the second female sterility. However, this applies largely to plants that evolve separate sexes from scratch; most animals adopted sexual reproduction hundreds of millions of years ago and are merely turning over the genes and chromosomes that accomplish this.

Most ideas invoke loss of recombination between the new sex-determining gene and a second gene on the same chromosome. There are many claims that this second locus must be a sexually antagonistic gene, good in the sex specified by the sex-determining gene but bad in the other sex, However, tight linkage with any

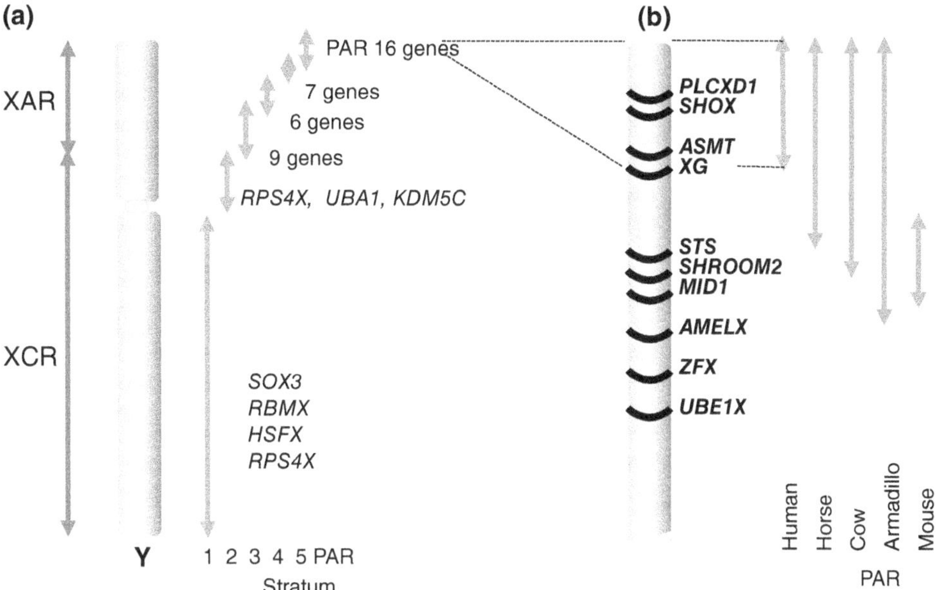

Figure 4.5 Evolutionary strata on the Y chromosome. (a) The human X chromosome can be divided into 6 regions ('strata') on the basis of the divergence between X and Y gene pairs. Stratum 1 (with the greatest divergence) is the oldest, and the Y retains only a few genes in humans and mice. Stratum 5 (least divergence) is the youngest, and contains proportionately more genes on the Y, many X partners of which escape X inactivation. PAR genes show no divergence between X and Y. Strata 1, 2 comprise the X conserved region XCR which is shared by marsupials and placental mammals. Strata 3, 4, 5 and the PAR comprise the added region XAR that is autosomal in marsupials. (b) The PAR boundary has been pushed to different extents in different mammals. The human PAR boundary splits the XG gene. In horse and cow differentiation has progressed only to *STS* and *SHROOM2*, and in armadillo only as far as *AMEL*. In mouse the terminal region has been translocated to autosomes, leaving a tiny PAR that extends to *MID1*.

gene that provides an advantage in this sex should be selectable. For instance, suppression of recombination between *SRY* and a gene important in spermatogenesis (maybe *RBMY*?) would ensure that this gene is active only in males and not wasted on females.

The loss of recombination between the X and Y in the region between the sex-determining and a male advantage gene would have severe consequences for this region. An overwhelming body of data in flies and other insects as well as mammals, shows that in the absence of recombination, deleterious mutations and repetitive sequences accumulate, and active genes are incapacitated and finally lost.

There are many ways in which recombination can be suppressed, including loss of homology by deletion, insertion of repetitive sequence or inversion so the sequences no longer match up. Did this process happen by gradual erosion of sequences around the differentiation front? Or did it occur in fits and starts as large regions were exempted from recombination with the X?

We can examine this process by looking at the front of Y degradation, where the pseudoautosomal region meets the male-specific region. Degradation has proceeded at slightly different rates, so that several mammals have PARs that are longer than the human PAR1 and contain more genes conserved from the original pair. The long elephant PAR contains 40 genes at the tip of the X, the 6.8 Mb bovine PAR has 33, whereas the 2.7 Mb human PAR contains only the most distal 24. Genes such as *STS* that are active in the elephant and horse PARs are represented by pseudogenes in the male-specific region of the human Y (Figure 4.5).

The process by which the PAR was rolled back can be investigated by looking at the sequence surrounding the boundaries in different mammals. There is no sign of gradual attrition at the boundary of the human PAR1, which is in the same place in great apes and Old World monkeys, although an *Alu* element was recently inserted at this site on the great ape Y.

However, there is evidence of recent variation of the PAR boundary that is consistent with gradual erosion. Duplications and deletions within the human PAR have been detected in some mammal families. Sequencing around the PAR boundary in wild mice and *Mus* subspecies reveals structural diversity and copy number variation, including insertion of 400 bp of unique X sequence in one species. Comparisons of the sequences around the boundary in cattle and their relatives also provide some evidence for gradual movement of the PAB, leaving 'tidemarks' of homology. The process is not always unidirectional: for instance, the PAR boundary in cattle was rolled backwards by illegitimate X-Y recombination in newly differentiated regions.

A particularly nice study of how degradation proceeds uses the African pygmy mouse, whose X and Y chromosomes both fused with an autosome only a million years ago. This greatly extended the PAR, and G-band identity reveals no rearrangement. Surprisingly, however, homologous pairing within this 'neo-PAR' is not normal; synapsis and DNA repair is delayed, and there is little recombination over a large swatch nearest the fusion point. This means that even in the absence of inversion, recombination can be suppressed very rapidly over a large region.

But there is strong evidence from many mammal groups that degradation on a larger scale occurred in fits and starts as large regions of the Y rearranged and lost homology with the X. A large inversion occurred early in human evolution to move *SRY* from its original position near *RBMY*, creating a new pseudoautosomal boundary. More recently several inversions of the human Y have occurred near the pseudoautosomal boundary, one that bisected the *XGA* gene that straddles the human pseudoautosomal boundary.

Once homologous genes on the X and Y stop recombining, and once dosage is compensated, Y alleles are free to evolve either specialities of their original function (subfunctionalization, such as somatic and gonad expression patterns), or entirely new functions (neofunctionalization).

Several Y genes (like *SRY* and *RBMY*) acquired vital functions in male reproduction, which ensured selection for their retention on the Y. Comparisons with their partners on the X shows that the highly conserved ancestral genes (*SOX3* and *RBMX*) had general functions in transcription and RNA processing; both are expressed in the brain and testis (fitting their apposition as 'brains-and-balls' genes). Similarly, in marsupials *ATRY* has forsaken *ATRX's* conserved role in chromatin modification and taken on a testis-specific role. Sequence comparisons imply that they are bastardized versions of their erstwhile homologue. *RBMY*, and marsupial *ATRY*, have exon structures similar to their X-borne homologues *RBMX* and *ATRX*, but their introns have been puffed up by the insertion of repetitive sequences.

Yet other genes seem to be on the Y for no particular reason – perhaps they have simply not yet met the forces of degradation. The tooth enamel (amelogenin) gene *AMELY* might be one of these; marooned on a degenerating Y, and conferring no particular male advantage (although big gnashing teeth have been suggested), it has been inactivated or eliminated from the Y in several lineages.

Several novel genes have been recruited to the Y chromosome in different mammals. They can come from anywhere, either transposed by the insertion of DNA sequence or retrotransposed by the insertion of DNA copies of processed RNA. *DAZ* appears to have arrived as a copy of an autosomal gene with very conserved functions in fertility (even in *Drosophila*!). Remarkably, a novel gene on the horse Y (*ETSTY7*) originated from a parasite, became testis-specific and presumably useful (to males anyway), then hugely amplified. Not all Y genes were selected for their effect on spermatogenesis; the olfactory genes recruited to the pig Y may have been selected because smell is important to boars (as well as to truffle hunters).

An outstanding feature of Y genes in all mammals is their propensity to be amplified. Amplification may be selected simply to boost the rate of transcription of useful male-advantage genes. However, my explanation is exactly the opposite. If vital Y genes are damaged by mutation so that they are transcribed less, or make a less active product, there will be strong selection to make more of them to increase their activity (Figure 4.6). Further amplification can then be seen as a desperate race to stay afloat. There are strange situations that support this model; for instance, a species of spiny rat has 100 copies of an *SRY* gene that has lost much of its capacity to bind DNA.

The extraordinary variety in gene survival, specialization, amplification and subfunctionalization has a very considerable stochastic element. There is a lot

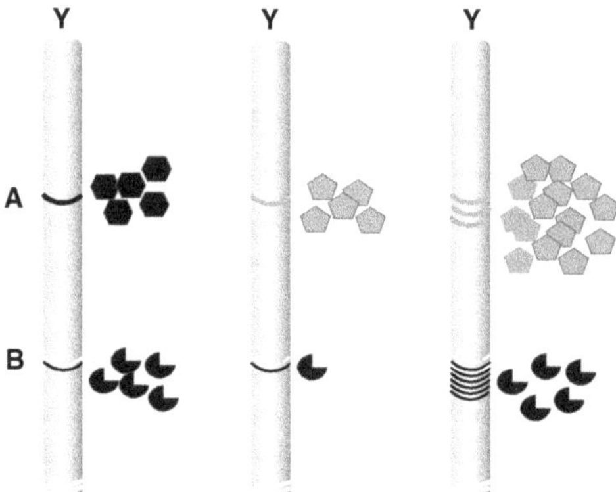

Figure 4.6   Selection for gene amplification on the Y. Gene A makes a vital protein product (black hexagons). A mutation renders it less active (grey pentagons). Amplification of gene A produces more copies of the compromised protein to make up for its lower activity. Gene B suffers a mutation in its upstream regulatory region (white) that lowers its rate of transcription but leaves the protein intact. Amplification restores the amount of protein product.

going on for the Y – inexorable forces that degrade it countered by selection for male-specific function or balancing dosage of essential genes.

4.7 The Decline and Fall of the Y Chromosome

The degradation of the ancient mammal YCR is nearly complete, and the degradation of the recently added region YAR is in its final stages. Will the human Y be degraded to death any time soon (Figure 4.3)?

It would help us to predict the future of the Y if we knew the kinetics of the degradation process. Mammal sex chromosomes are old, and the trail has gone a bit cold, so most of our information comes from other systems. Studies of Y degradation in a *Drosophila* species in which a neo-Y was formed about a million years ago by a Y-autosome fusion show that half the new genes are already inactivated, and repetitive sequence and retroelements have accumulated throughout. This occurred also in the pygmy mouse, suggesting that the first phase of degradation of new Y chromosomes is extremely rapid.

The initial rapid phase of degradation of the mammal Y probably occurred twice; once in the ancient YCR, then again when an autosome was fused to the PAR in eutherians, forming a neoPAR. The presence of evolutionary strata on the mammal Y suggests that this process occurred several times as large chunks of the Y lost recombination with the X. Modelling the process in the evolutionary strata of the human Y suggests that degradation of S1, S2 and S3 was initially rapid then levelled off, leaving just a few genes that had acquired vital functions or were extremely dosage sensitive. Recent strata 4 and 5 retain more of the original genes, but their inactivation or deletion in one or other lineage suggests that they are on the way to being dumped.

So is the human Y chromosome doomed to extinction? Knowing the start time of degeneration (166–150 Mya), the numbers of genes in the original ancestral blocks (about 1,000), and the number that survive today (about 51 in the male-specific region and the PAR), we can calculate that the rate of loss over the last 150 Mya was about 6.3 genes per million years. At this rate the last 51 active genes on the Y-specific region and PARs might be expected to be lost in about 8 million years.

My similar back-of-the-envelope calculation of the fate of the 'wimpy Y' (Figure 4.11) in 2002 caused quite a furore. Feminists wrote books about the disappearing Y. Supporters of the Y chromosome mounted a spirited defence, averring that the human Y has been particularly stable over the last five million years, while the chimp Y lost several genes.

The Y supporters proposed that gene conversion within palindromic loops could correct mutations and save the Y. Conversion is a mechanism by which the DNA sequence of one allele is copied onto the other during repair. The only problem with this idea is that gene conversion is usually not fussy about which allele to copy, and could equally well substitute a mutated version for a normal allele (a 'double or quits' strategy that could alter the kinetics of degeneration but not its overall rate).

A recent study of meiosis presents another intriguing possibility – that the Y chromosome must persist because it bears a meiotic 'executioner' gene. This gene (perhaps *ZFY*) is required for meiosis and is expressed very specifically at an early stage. Along with other genes on the Y, it is silenced by MSCI; if it remains active, meiosis fails. Thus, there is powerful selection for this gene to remain on the Y.

An alternative idea from David Page's lab is that the Y chromosome contains a gene (again *ZFY* is fingered) that evens up the expression of many autosomal genes.

It was also argued that the Y contained genes vital for reproduction, and their loss would render men sterile. However, almost all the genes on the human Y are gone in one species or another, sometimes relocated on an autosome, or their function subsumed by autosomal copies or other gene family members. There is almost nothing left of the original mouse Y chromosome. Only nine genes survive in the 2 Mb X-ancestral region on the tiny short arm. Knockout phenotypes reveal that of these, only two (*Sry* and *Eif2sy*) are critical for male determination and function.

And in several rodent species – pygmy mice, gerbils, voles – the PAR has completely differentiated. Given that loss of the PAR completely messes up meiosis in mice and leads to infertility, how do they do it? No synapsis or chiasmata are observed, but X and Y do pair, side by side and form a sex body that persists through pachytene. This isn't enough for mutant mice that lack synapse proteins, but it permits segregation in achiasmate rodents.

Marsupials met this problem even earlier since they did not benefit from fusion of the X and Y with an autosome that bolstered the PAR. There is no homologous X-Y chromosome pairing at male meiosis in marsupials; instead, sex chromosomes are confined by a plate (made of the protein that ensures chromosome pairing in other mammals) that ensures that the X and Y chromosomes are distributed to different daughter cells. And in some marsupial lineages, the Y chromosome is lost from somatic cells in males (and the inactive X from somatic cells in females), implying that the Y lacks genes with a critical function in somatic cells.

Even more significantly, there are two lineages of rodents that have independently lost their Y chromosome and are busy evolving entirely new sex chromosomes. Two of the three species of the mole vole *Ellobius* lack a Y chromosome entirely and have no *SRY*. In *Ellobius lutescens*, males and females both have a single X chromosome, and in *Ellobius talpinus* both sexes have two X chromosomes, which look identical, but do not pair fully at male meiosis. A third species, *Ellobius fuscocapillus*, has an XX female:XY male sex chromosome complement and a Y-borne *SRY* gene.

What about other genes on the *Ellobius* Y? Of the other 16 genes on the mouse Y, only two genes (*Zfy, Eif2s3y*) survive in both Y-less *Ellobius* genomes, as well as *Ssty*, a gene amplified on the mouse Xq. Interestingly, they have been transferred to the *Ellobius* X, so the proposed meiotic executioner *Zfy* is still subject to meiotic sex chromosome inactivation.

A similar situation is enacted by three species of spiny rat (genus *Tokudaia*), all highly endangered and living on small islands in Japan. Two species have XO males and females. In these Y-less species, gene mapping showed that several Y genes have been transposed or retrotransposed to other chromosomes, including the X (which bears *ZFY*). The third spiny rat species is XX female: XY male, but the sex chromosomes are weird. The X and Y chromosomes are both fused to an autosome, bolstering the PAR and perhaps avoiding the elimination of the Y.

In Y-less mole voles and spiny rats, another gene elsewhere in the genome must have taken over the sex-determining function, and could trigger differentiation of new sex chromosomes, as I describe in Chapter 10 (Section 10.2).

Thus, degradation of the Y chromosome seems to be inexorable; the question of its survival appears to be *when*, not *if*. The human Y may be only temporarily stable; the mouse Y is almost gone, and there are rodents that have lost the Y and are experimenting with new sex chromosomes.

4.8 The Evolutionary Forces That Shaped the Y Chromosome

What evolutionary forces drove the degeneration of the Y chromosome? Again, we must distinguish between generation and selection of variation. Is there evidence for an increased mutation rate? Is there any reason why we might expect purifying selection might not work to expunge mutants from the Y? It turns out to be both. Indeed, things are stacked up against the Y (Figure 4.7).

There is certainly higher variation for genes on the Y than for their partners on the X. Comparisons of sequences between mouse species show differences in divergence rate for the X and Y copies of several X-Y shared genes. The X borne copy is highly conserved between species, whereas the Y borne copy shows extreme sequence variation. For instance, *SOX3* is highly conserved in all vertebrates, whereas its Y paralogue *SRY* is highly variable. Divergence rates of *ZFY* are greater than *ZFX* in many species.

There are reasons why genes on the Y are more prone to genetic mishaps, reflecting differences in male and female germ cell biology (Section 5.2). Cells that give rise to spermatogonia go through 301 pre-meiotic mitoses, many more than do oogonia (Section 4.5), so there is more opportunity for genetic errors to occur in the male germline. The Y is unique in going through meiosis only in the male, never the female germline (Figure 4.7a). 'Male-driven mutation' is recognized in humans: sequencing around a *de novo* mutant gene shows that mutation usually occurred on a chromosome inherited through the father's sperm, a strong male mutation bias.

However, a higher mutation rate does not explain why damaging mutations, deletions and insertions are rife on the Y. Why doesn't selection get rid of them? The answer is that selection doesn't work very well on the Y chromosome.

The most obvious problem the Y chromosome faces is that it no longer has anything to recombine with – it is genetically isolated. Other chromosomes (including the X in a female) can recombine mutation-free bits into a pristine mutation-free chromosome. The Y cannot. Since back mutation is rare, once mutation has occurred on a particular Y, it is stuck there forever (Figure 4.7b).

Absence of recombination means that the Y chromosome is at the mercy of 'genetic drift', the random loss of alleles (Figure 4.7c). In a population of Y chromosomes, mutations will accumulate independently; some Y chromosomes will carry several mutants, others fewer. There will initially be some Y chromosomes with no mutation; if this class is lost (e.g., if men with a perfect Y have no sons), it can never be regenerated. Then the next least mutated Y may be lost and so on. This has been likened to the turning of a ratchet ('Muller's ratchet' after its proponent in 1918).

Selection for and against new mutations on the Y can act in a perverse way. For instance, a degraded allele at one locus can 'hitchhike' along with a useful new allele at a neighbouring locus. The absence of recombination within the male-specific region of the Y means that if a favourable mutation occurs on a particular Y chromosome, selection of this new variant will carry along

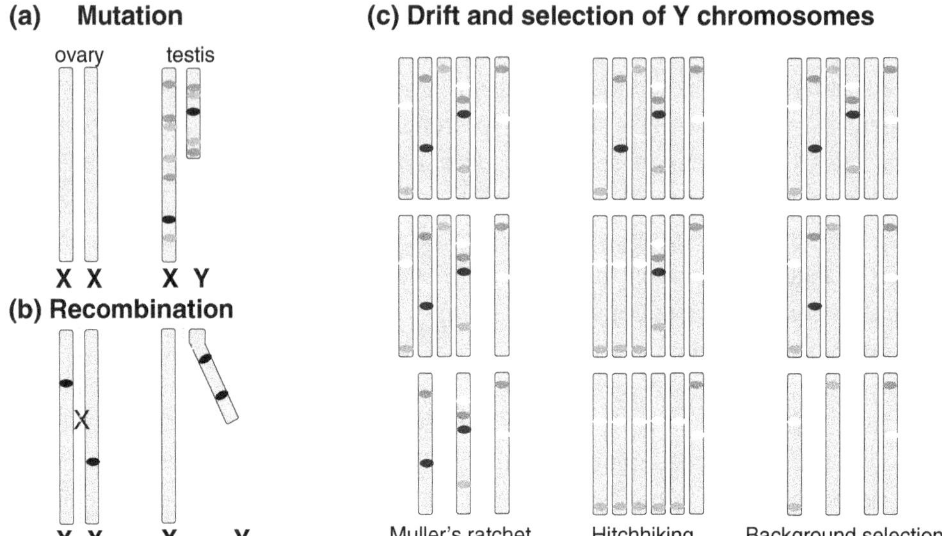

Figure 4.7   Degeneration of the Y chromosome; the perils of increased mutation, lack of recombination, drift and selection. (a) Mutants are represented by oval shapes of different colours, the darker representing more damaging mutations. Mutation is much more frequent in the testis during spermatogenesis than it is in the ovary during egg development. This hits the Y chromosome disproportionately because it is always in the testis. (b) The X chromosome pair undergoes recombination over its full length (dotted line). Thus, if the two X chromosomes in a female carry different mutations, recombination between them can generate an unmutated X in the progeny. However, the Y chromosome at male meiosis has no partner over most of its length and is stuck with the mutations it bears. (c) Three mechanisms of Y degradation resulting from genetic drift. *Muller's ratchet*: A population of males will have Y chromosomes with various degrees of mutation (represented by ovals, darker colour indicating severity). Initially the population will contain a Y chromosome with zero mutations. If this unmutated Y is lost from the population, it can never be regenerated. The next least damaged Y, bearing one mutation may subsequently lost, leaving the population of Y chromosomes with two or more mutants. *Hitchhiking*: occurs when a beneficial mutation (white oval) increases fitness of its bearer and this Y will spread in the population. Any other mutation, deleterious or not (darker grey) on the same chromosome will hitchhike to success, increasing in frequency. *Background selection* is the opposite; a beneficial mutation on the same Y as a strongly deleterious mutation will be lost in the population.

the whole Y that bears it, regardless of mutations or deletions in other, less critical, genes. This combination ('haplotype') sweeps through the population at the expense of Y chromosomes with other haplotypes. The opposite, too can occur; a beneficial new mutation can hitchhike to its death if it lies on the same Y chromosome as a strongly detrimental mutant ('background selection').

'Selective sweeps' fix different combinations of inactivated and degraded alleles in different species, producing the rather haphazard patterns we observe.

Thus, the degeneration of the mammalian Y may be the result of several processes unique to this chromosome. A high rate of variation, lack of recombination, genetic drift, and hitchhiking can account for the loss of active genes and the accumulation of pseudogenes and amplified sequences.

These processes are not really unique to the mammal Y, but general to sex chromosomes of all descriptions in all organisms with sexual reproduction.

4.9 Birds Reveal Common Rules for Sex Chromosome Differentiation

Birds have long been known to possess differentiated sex chromosomes. They are completely different from the mammal XY and work in the opposite way, but comparisons with the mammal XY establish general evolutionary principles.

Bird karyotypes contain nine large chromosomes ('macrochromosomes'), and many tiny 'microchromosomes' (Section 1.15) and are almost invariant between species. Bird sex chromosomes are recognizable under the microscope. But when you line them up in a karyotype, you find that it is the female, not the male, that has a sex-specific chromosome. We call this a ZZ male:ZW female system of female heterogamety.

Ohno originally suggested that the bird ZW pair is homologous to the mammal XY, but gene mapping showed that they share no genes. Rather, the bird Z is homologous to regions of human chromosomes 9 and 5, and the human X is homologous to chicken chromosome 4p (XCR), and a patch of chicken chromosome 1 (XAR). (Figure 4.1)

There is no *SRY* on the bird ZW pair and *SOX3* lies on chromosome 4p along with other human XCR genes. The bird W chromosome was long expected to harbour a female-determining gene, analogous to the male-dominant *SRY* of mammals, but no such gene was ever found. Instead, differential dosage of a gene *DMRT1* on the Z (but not the W) chromosome controls sex determination in birds (Section 10.5).

Birds provide a textbook demonstration of the evolution of sex chromosomes from an autosome. The Z chromosome is extremely conserved in size and gene content even in the flightless ratites (ostrich and emu) that were first to diverge 72 Mya. This dates a single common origin of the bird ZW system. However, the size and gene content of the female-specific W chromosome varies between two extremes. In ratites, the W is nearly as long as the Z chromosome and they recombine within a PAR that covers most of their length (52 Mb of the 87 Mb). In contrast, the putative songbird PAR comprises less than 1 Mb of a tiny W microchromosome.

Homology between the Z and W implies that the W chromosome is a degenerate Z, just as the mammal Y is a degenerate X, and differentiation has proceeded for a shorter or longer time. The large ratite W shares the same genes with the Z over most of its length. Genes on the shorter W chromosomes of chickens and songbirds all have homologues on the Z from which they obviously diverged. Thus, the bird W degenerated from a homologous autosome pair to a ratite-like minimally differentiated pair, then to the extreme differentiation seen in chickens and songbirds. Thus, the process by which the W chromosome degenerated parallels that of the mammal Y (Figure 4.3).

Bird ZW chromosomes, like the mammal XY, are ancient and tell us little about the first stages of ZW differentiation. However, a recent (20–40 Myr) fusion of an autosome to the ancestral sex chromosomes in the common whitethroat provides an opportunity to compare genes on the neo-sex chromosomes. Neo-W genes were more diverged and less expressed than their counterparts on the neo-Z, in line with our ideas of sex chromosome differentiation.

Genes shared by the bird Z and W show different degrees of divergence, and seem to fall into three 'strata', representing three events which supressed recombination. Comparing bird W sequence to reptile outgroups reveals inversions within the W that would have expunged recombination over large regions, but the newest (pseudoautosomal) boundary may have been created by insertions of highly repetitive sequence.

The W chromosomes of emu, chicken and many songbirds have been sequenced, so we now have an inventory of genes that survived on the W. As for the mammal Y, loss is greater in the older W strata. The W chromosome does not seem to have been 'feminised,' in contrast to the evolution of many spermatogenesis genes on the mammal Y. Most W genes have copies on the Z and are expressed widely, so they appear to be housekeeping genes whose dosage must be kept equal between the sexes. Although W degradation occurred independently in different lineages, the set of W genes is very conserved, implying independent selection of genes that are particularly dosage sensitive. Gene amplification is not a feature of the W chromosome, for only one gene is present in multiple copies.

W chromosome degradation in birds seems to be slower than that of the mammal Y. In ratite birds, it has barely begun (we don't understand why), and even in chickens and songbirds a greater proportion of genes have survived on the bird W than on the mammal Y. This may reflect the lower mutation rate on the female-specific W because it is confined to the ovary, whereas the male Y is at higher risk in the mammal testis.

Many studies have searched the W in vain for female-advantage and sexually antagonistic genes. However, genes on the Z are significantly male-biased, reflecting the higher frequency of the Z in males. The ratio of nonsynonymous (coding) base substitutions to synonymous (silent) substitutions in orthologous genes in different species provides some evidence of a 'fast Z'.

Birds are famous for their sex differences, including ostentatious male plumage, songs and mating rituals. Sexual selection is rife in birds (think of the peacock's tail, beloved of peahens but of little practical use). There has been longstanding interest in sex-biased genes on sex chromosomes, particularly those with stronger male expression in gonads and the brain. However, a whole genome association study detected sexually biased expression of genes all over the genome. The role of sex chromosomes in sexual selection and evolution was investigated in many bird lineages by estimating how fast genes evolve on the ZW compared to the autosomes in birds with different mating systems. Faster change was observed in species such as bird of paradise in which a few gaudy males monopolise many females.

Thus, the bird ZW system of female heterogamety with dose-dependent sex determination is the exact opposite of the mammal XY system. However, nearly all the features of mammal XY evolution are mirrored in the bird system, including degeneration of the sex-specific W chromosome in fits and starts as large regions are removed from recombination, the retention of dosage-sensitive housekeeping genes on the W, and biased gene content due to the different representation of the Z and W in the two sexes.

4.10 Snake Sex Chromosomes – Another Independent Origin

Whereas mammals and birds each have an extremely conserved systems of genetic sex determination (GSD), reptiles have an enormous variety of sex chromosomes and sex-determining genes. Some reptiles do not have sex chromosomes at all, but rely on environmental cues to determine sex of hatchlings (Environmental Sex Determination, ESD, Chapter 10).

Different reptile groups display a variety of genetic sex-determining mechanisms, including male heterogamety (XY systems) and female heterogamety (ZW systems), and species with sex chromosomes differentiated to greater or lesser extents. There is striking variation even within groups, implying that sex chromosome systems can evolve very rapidly in reptiles. They present excellent systems for investigating sex chromosome evolution.

Sex chromosome variation occurs despite an extremely conserved basic karyotype. Reptiles share with birds several large normal-looking chromosomes, and many tiny microchromosomes, which have fused in some lineages. Sequence alignments reveal astonishing conservation.

One of the oldest groups are the snakes, comprising about 3,800 species in four major families that started diverging 91 Mya. Most snakes share a ZZ male:

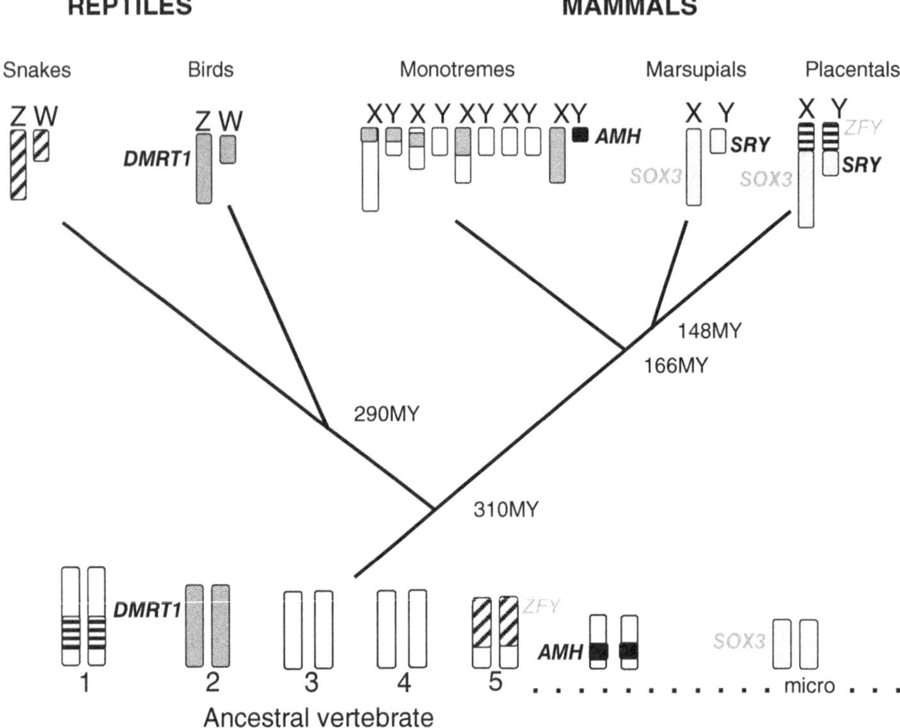

Figure 4.8   Different regions of a conserved vertebrate genome became sex chromosomes in different lineages. The phylogeny of higher vertebrates is shown, with divergence dates at nodes. The karyotype of an ancestral mammal can be inferred from extremely conserved genomes of mammals, birds and reptiles. Ancestral chromosome 5 (diagonal stripes) became the ZW pair in snakes, and chromosome 2 (dark grey) containing *DMRT1* became the ZW in birds. A chromosome region (black) containing *AMH* was fused to bird-like sex chromosomes in monotremes. A microchromosome containing *SOX3* (pale grey) became sex chromosomes in therian mammals, and a region of ancestral chromosome 1 (horizontal stripes) was fused to it in placental mammals.

ZW female system, which led Ohno to propose how vertebrate sex chromosomes evolved from an autosome pair.

Oddly, Ohno was right in his hypothesis, but dead wrong about snakes. Early cytology suggested that different snake families have the same ZW sex chromosome pair (the fifth largest), but the W was degraded to different extents, from minimal in pythons and boids, partial in colubrids to extreme in elapids. Ohno considered these represented steps in the progressive degradation of the W.

Gene mapping, and now sequencing, confirms that the Z chromosome is conserved between many snake species. The conserved snake Z is much the same size as the bird Z, and was long thought to have evolved from the same ancestral sex chromosome. However, comparative mapping showed that the snake Z shares genes with an arm of chicken chromosome 2, and the bird Z with genes on snake chromosome 2. Thus, snake sex chromosomes evolved independently from a different ancestral autosome than either the bird ZW or the human XY (Figure 4.8).

Sequencing genomes of species from each snake family clarified the relationships between sex chromosomes of the three families, but defied cytological predictions. Elapids, indeed, had highly differentiated W chromosomes that were often larger than the Z, but full of highly repetitive DNA. The W showed homology with the Z over a small PAR at one end. But the colubrid Z and W, despite their cytological similarity, showed homology only within a small (7 Mb) terminal PAR.

Pythons and boas were a puzzle, for no sequence difference at all between males and females could be detected on chromosome 5. Subsequently, it was discovered by old fashioned linkage mapping that boids have an X and Y system with homomorphic sex chromosomes, and that chromosome 5 is a plain old autosome. This suggests that the boid sex-determining system evolved recently.

Thus the three mammal groups, birds and snakes, all have sex chromosomes that originated from different autosomes in the karyotype of the common ancestor 310 million years ago (Figure 4.8).

4.11 Reptile Sex Chromosomes – Dizzying Variety and Independent Evolution

Lizards present amazing variability in GSD, as well as temperature-sensitive sex, and sometimes both, as well as occasional parthenogenesis. Early cytology showed that different groups of lizards had almost every possible type of sex chromosome system, including XY and ZW systems and many variants. Some groups, such as the iguanas and skinks, have XY male:XX female systems, whereas other families have either XY or ZW systems, or both. The degree of XY and ZW differentiation is also very variable.

DNA sequencing of the green anole revealed many genes present in half the dosage in males as females, suggesting an XY system, and comparison with birds revealed orthologues on chicken chromosome 15. Few genes lie on the anole Y chromosome, denoting an old and highly differentiated sex chromosome pair. As in mammals, several Y genes have acquired male-specific functions, while their X homologues maintained their original function. This system is ancient, for at least some X chromosome markers are shared by the iguanids, which diverged 108 Mya.

In varanid lizards, there have been several flip-flops over just 17 Mya. The model lizard species *Pogona vitticeps* (the Australian central bearded dragon) has ZW microchromosomes whose genes show homology to chicken chromosomes

17 and 23. In gekkos there are at least two non-homologous ZW chromosome systems, within which the Z maintains its size and morphology, but the W assumes several different forms in different groups. Gekkos are very variable; one species has an XY system with homology to the bird Z, others have homology to several different chicken autosomes. However, skinks (comprising 1,700 species with a huge diversity) have extremely conserved XY chromosomes with homology to chicken chromosome 1; their ancient origin (85 Mya) means they are as stable as the bird ZW, even though they are poorly differentiated.

Turtles have many different sex chromosome systems, including XY and XO and several variants. Gene mapping in different species shows that these systems are non-homologous, originating from different autosomes and presumably containing different (unknown) sex-determining genes. Marine turtles have temperature-sensitive sex determination.

Accumulation of heterochromatin seems to be an important factor in the differentiation of reptile sex chromosomes. The snake W chromosome is loaded with several highly repeated sequences, including Bkm, a female-specific repeated sequence that was once thought to determine sex (Section 8.9). The sex chromosomes of even closely related varanid lizards are distinguished by their different contents of repetitive sequences, ranging from small blocks to the whole W chromosome. For instance, the W microchromosome of the central bearded dragon is recognizable only by its female-specific repetitive sequences. Accumulation of repetitive sequence may denote an early stage in reptile sex chromosome differentiation.

Thus, different XY or ZW systems evolved independently many times in reptiles, and from many different autosomes. Turnover was facilitated by transitions between genetic and environmental sex, as I shall explain in Chapter 10.

4.12 Frog Sex Chromosomes – Why Do They *Not* Degrade?

Amphibians all show GSD and there is no evidence for an environmental influence. However, most amphibian species have no obviously differentiated sex chromosomes. Why not?

Amphibians diverged from amniotes 350 Mya. There are nearly 7,000 species in four orders, but frogs and toads are by far the most numerous. They have large chromosomes and genomes, which has made cytology more attractive than genomics. However, chromosome-scale sequences are now available for a few frogs, and even the axolotl with its enormous (45 Gb) genome.

Frogs have a few, large chromosomes and very conserved karyotypes, and much cytological work has been done on the 13 splendid chromosome pairs in many species. Morphologically distinct sex chromosomes are rare: XY or ZW pairs can be morphologically identified by subtle cytological differences in only a few species. For instance, the poisonous cane toad ravaging Australia has a female-specific W chromosome identified by its extra dollop of heterochromatin. Some neotropical frogs have wildly heteromorphic sex chromosomes, and even complexes that rival the 5X5Y system of the platypus.

Though cytology fails to distinguish them, it is easy to determine which pair are sex chromosomes, and which is the heterogametic sex, by detecting sex linkage of protein or DNA markers in frog crosses. The extraordinary finding is that the chromosome that shows sex linkage may be different; among frog species representing 220 Myr of evolution, 19 turnovers were detected, even between closely related species. For instance, among 28 frog species that diverged only 3–55 Mya, there have been at least 13 changes in the sex chromosomes, mainly between XY systems. Five different chromosomes figure in XY systems and two in ZW systems; changes were not random, but often refashioned the same ancestral chromosome.

This raises the question of whether some chromosomes make better sex chromosomes than others. Gene mapping provided a simple answer: these five chromosomes all carried one or more genes that are sex-determining in many vertebrate species (Figure 10.4).

Comparative mapping of four frog species, and short-read DNA sequencing of 20 frogs confirms that the sex chromosomes of even closely related frogs evolved from different ancestral chromosomes, and that they are little differentiated. The group therefore presents opportunities for studying how sex chromosomes can turn over in evolution.

There are different sex chromosomes even within the same species in two famous frogs. Detailed comparisons of the Japanese wrinkled frog revealed populations with ZW and XY systems living on different small islands, and hybrid zones on a third. Remarkably, the XY and ZW sex chromosome pairs derive from the same ancestral autosome. The same situation has been recreated in laboratory populations of *Xenopus tropicalis*. Although the wild species has an XY system with a male-dominant gene on the Y, lab populations may have Z, W and Y chromosomes in which YZ, YW, and ZZ are males and ZW and WW are females; again, the same region of Z, W and Y chromosomes is sex-determining.

Thus frogs present the opposite puzzle to mammals and birds; we must ask not why the sex-specific chromosome (Y or W) degrades, but why it does *not* degrade. The answer seems to lie in their rapid turnover; the 'Fountain of Youth' hypotheses has that a sex chromosome pair is replaced before it has time to differentiate.

Thus frogs contribute greatly to our understanding of how sex chromosomes arise and what contributes to their longevity. I will consider possible explanations in the following section.

4.13 Sex Chromosomes of Fish; Identifying Ecological Drivers

Fish, comprising one third of all vertebrate species, constitute a very ancient class. Evolutionary differences within fish are as deep as the division between mammals and fish.

Fish show a great variety of sex-determining systems, including XY, ZW and many variants. A few rely on environmental triggers. Some species are hermaphroditic and many species regularly change sex during their life cycle, or when presented with social stimuli (Chapter 10). Only about 10% of the fish karyotyped have morphologically distinguishable sex chromosomes, and there is rapid turnover. Thus, fish are useful for studying the events in sex chromosome initiation and the early stages of differentiation.

Traditional methods – cytological observation and genetic crosses – established several important principles in model fish species. For instance, the medaka and its ricefish relatives share the same karyotype with no heteromorphic sex chromosomes. But sex linkage was demonstrated to seven different chromosomes in different species.

DNA sequencing is now revealing unprecedented insights into the processes of initiation and differentiation of sex chromosomes in many fish groups. Sequencing can detect even tiny initial changes that are out of reach of cytology. For instance, the sturgeon's ZW system (important to the caviar industry), depends on a 16 kb female-specific region. Surprisingly, this has been stable for 180 Myr, contradicting the inevitability of W chromosome degradation.

Even such tiny differentiated regions can contain the next step of sex chromosome differentiation. For instance, XY chromosomes in herring are differentiated by a 230 kb male-specific region, which contains only three genes, one which could induce testis and two that affect spermatogenesis. This illustrates the beginning of recombination suppression in the region around a new male determining gene that contains male advantage genes.

DNA sequencing confirms that the extent of differentiation between sex chromosomes is extremely variable, even within fish groups, and there is no consistent relationship with time of divergence. For instance, comparing read depth in males and females in a group of closely related live-bearing fish showed that they share an XY system with a conserved X chromosome. However, the degree of Y degradation and the proportion of non-recombining region vary from almost nil to almost complete. The rate of degeneration may be very different in the same Y of closely related species; for instance, the enormously variable cichlid fish that live in African lakes show no correlation between divergence time and degree of sex chromosome differentiation.

The process of XY differentiation has been examined closely in guppies, which show an astonishing variety of Y chromosomes in closely related species, and even populations of the same species. Surprisingly, sequence comparisons of Y chromosomes with different degrees of divergence show no evidence of large-scale inversions; suppression of recombination seems to have been a linear rather than a stepwise process.

Comparisons of sequence changes in genes on fish X and Y chromosomes, and on homologous autosomes in closely related species, identified evolutionary strata and divergence dates. In the stickleback group, there are at least three different XY pairs and a ZW pair that correspond to different linkage groups. Three species with the same XY system derived from an ancestral chromosome have different degrees of Y differentiation. Sequencing the Y reveals at least three evolutionary strata that differ in their retention of active genes and accumulation of transposable elements. One species sports a fusion with an autosome, on which genes are mutated but not yet lost; this neo-Y has six identifiable strata.

One rather common route to establishing a region in which recombination is suppressed seems to be the hybridisation with a different species with structurally different chromosomes that do not pair during meiosis. Another strategy is to load it up with repetitive sequence that disrupts pairing. This appears to have occurred in a widespread tropical fish species complex in South America which boasts three different sex chromosome pairs differentiated by rearrangements and variable accumulation of repetitive sequences among populations in pools isolated by waterfalls.

Some fish groups provide opportunities to explore, not only the process of sex chromosome evolution, but the ecological drivers. The classic case is guppies, in which males sport an orange spot to advertise their maleness; such a spot would put brooding females in danger. The problem was evidently averted by a rearrangement that put the orange spot gene on the Y chromosome near to the male determining locus and suppressed recombination between them.

There are many other examples of reproductive traits linked to sex chromosomes. An extreme is a South American freshwater (poeciliid) fish in which males come in five different colour morphs with very different body size and mating behaviours. These morphs are inherited directly from father to son, and sequencing revealed five different Y haplotypes that control five different reproductive strategies.

Cichlids have undergone rampant evolution of many characters in thousands of species and populations adapted to microhabitats over several lakes and rivers in east Africa. Many species show extreme sexual dimorphism in size, colour and behaviour, suggesting sexually antagonistic genes at work. There is an extraordinary diversity of cichlid sex chromosomes, almost all young and little differentiated, which have undergone rapid change. It seems likely that sex chromosomes have played an important part in adaptation to these different environments by fixing different male advantage variants, which originate by virtue of their proximity to novel sex-determining genes.

Thus, fish sex determination encompasses almost every kind of system, and they can turn over rapidly. Genomic studies of many systems are greatly expanding our understanding of how sex chromosomes evolve, and several dearly held concepts of an orderly, time-dependent sex chromosome differentiation have to be abandoned in favour of an 'anything goes' hypothesis.

We can now see that the stodgy, stable sex chromosomes of mammals and birds are something of an exception among vertebrates. Sex determination in reptiles, amphibians and fish is much more variable and flexible. Comparisons between the vertebrate classes enable us to approach some of the deepest secrets of sex chromosomes.

4.14 Sex Chromosome Turnover

The variety of sex chromosomes in reptiles, amphibia and fish imply frequent turnover of sex chromosomes. How are new sex chromosomes chosen, and how do newly minted sex chromosomes take over from the old?

Are some chromosomes particularly suitable for the task of determining sex? There has been much speculation about this question, and it has been proposed that chromosomes with a high density of sexually antagonistic genes and a low density of dosage sensitive genes are most suitable.

The finding of several groups of frogs and fish with virtually identical karyotypes but different sex chromosomes led at first to the anarchic view that anyone can be a sex chromosome. However, as data accumulated it became obvious that within species groups, particular chromosomes were favoured. For instance, among 33 frog species, seven of their ten shared chromosomes were linked to sex in two or more species. However, this is far from a random assignation, as chromosome 1 determines sex in nearly half of the species, probably because it bears two genes that are sex-determining in many species (Figure 10.6). Similarly, among 79 cichlid species, 12 of 23 shared linkage groups showed sex linkage in one or more species but four linkage groups predominated, even among less closely related species.

Is this identity by descent? Or are there limited options for fashioning new sex chromosomes? These questions became even more interesting when homologous genome regions were found to be sex-determining in distantly related animals, such as fish and platypus, birds and toads.

There is some evidence that it helps if the new sex chromosome contains few genes that are dosage sensitive and could cause problems on a degenerating Y or W. But largely, particularly dosage-sensitive genes just have to take their chance, either remaining active on a degraded Y, or blocking its degradation.

There is little evidence that novel sex chromosomes must contain a high density of sexually antagonistic genes. A study of cichlid fishes found no evidence that sexual antagonism is a driving force for sex chromosome turnover. Similarly, studies in frogs favours neutral processes (drift) and deleterious mutants. It seems that all that is needed to seed a novel sex chromosome is acquisition of a sex-determining gene. We will see in Chapter 10 that making new sex genes is surprisingly easy. Genes that can evolve a male or female-advantage variant are likely to pepper the entire genome so should not confine the choice.

So how does a new sex chromosome pair take over? This is difficult if the sex-specific chromosome is highly degraded because if the Y/W has lost many active genes, YY or WW genotypes are likely to be inviable. Also, the specialized gene content and dosage compensation of the X or Z may make it impossible to resume their life as an autosome. In fruit flies, an ex-X chromosome, the tiny chromosome 4, is stuck with a low gene content and no recombination.

Yet turnover of differentiated sex chromosomes has happened, for instance, in Y-less mammals and boid snakes. There must be a period in which old and new sex chromosome systems fight it out, and hybrids cope with duplications and deficiencies that result in subfertility. For a new system to have a selective advantage, it must be more efficient than the old; perhaps it replaces one which has degenerated to the extent that it no longer works very well. Spiny rats, in which an XY species has a severely degraded Y, might represent this situation. Humans might not be far behind.

Turnover of minimally differentiated sex chromosomes must always be easier, and this is why we see so many turnovers in lower vertebrates and so few among mammals and birds. Turnover follows the acquisition of a rival sex-determining gene on an autosome or the shifting of an old sex-determining locus (or a copy) to a new location (Chapter 10). However, it can also follow rearrangement, like fusion of sex chromosomes to an autosome that provides a new sex-determining gene.

There are several examples of takeover by a new sex chromosome in natural populations. For instance, in France an XY system is taking over from a ZW system in toads, and a new chromosome system seems to be taking over in the sea bass. There are some particularly interesting situations in which different populations of the same species have two, or even three different sex chromosome systems, as in the Japanese wrinkled frog.

In reptiles and some fish, GSD can be subverted by environmental factors. For instance, the offspring of ZZ sex reversed female dragon lizards incubated at high temperatures all lack a W chromosome; sex is specified by temperature until a new sex-determining gene arises to define new sex chromosomes. Rapid flip-flops between GSD and TSD speed up turnover; a flexibility not available to homoeothermic mammals and birds.

An accidental experiment shows how rapidly an old ZW system can be replaced. Zebrafish have been captive bred as a genetic model in several laboratories for more than 30 years. Lab populations seemed much the same, but some labs reported sex linkage to chromosome 16, others to chromosome 5. The problem was solved by the discovery that wild zebrafish had a perfectly good ZZ male: ZW female system. Captive breeding had unwittingly selected WW females and discarded the Z chromosome. What we are seeing now is the rapid emergence of different, novel sex chromosomes. Similarly, in the wild the frog *Xenopus tropicalis* is XY male dominant, but in lab colonies there are YZ, YW and ZZ males, and ZW and WW females.

Turnover may also be the only way out of a sticky situation. For instance, doubling the chromosome number in *X. laevis* made a bigger frog but messed up the original sex-determining system seen in its diploid relative *X. tropicalis*. This favoured the evolution of a retroposed gene fragment that suppressed maleness, defining a new W chromosome (Section 10.8).

With our gathering understanding of the birth of new sex chromosomes, is it possible now to make our own? Classic experiments in fruit flies established that selection for a bristle variant only in males led to rapid accumulation of female-deleterious variants in the region. In the platy swordtail, a popular aquarium fish, thirty years (100 generations) of repeated backcrossing to another species with a different sex chromosome system, and selection for particular pigment variants, produced a large female-specific region. The sex chromosome of one species was fused to an autosome in hybrid fish, leading to a new sex chromosome.

Thus, rapid sex chromosome turnover in fish, frogs and many reptiles is proof of the ease of changing over as long as sex chromosomes are little differentiated. Rapid turnover probably accounts for the ever-young frog and fish sex chromosomes, but we still cannot explain why some taxa opt for differentiation and others do not.

4.15 Sex Chromosomes and Speciation

Does sex chromosome turnover have a special role in dividing one species into two? Species are defined biologically as reproductively isolated populations unable to mate to produce fertile offspring. Speciation may occur in populations that have become isolated in some way, most often geographically (allopatry). Or more interestingly, speciation may occur within a population (sympatry).

The role of chromosome change in speciation has been hotly debated for decades. Early cytological studies of species complexes, especially insects, revealed that closely related species often had different karyotypes, and hybrids between them were unable to undergo meiosis and make gametes. In the mouse, heterozygotes for X-autosome translocations and inversions are generally sterile.

The major mammal lineages (monotremes, marsupials and eutherians) have markedly different sex chromosomes. The changes in sex chromosomes (fusions and a complete turnover) are likely to have posed reproductive barriers that could lead to speciation (Figure 4.9). Monotremes have bizarre multiple X5Y5 system that resulted from serial translocation or fusion of ancestral sex chromosomes with three autosomes, with a fourth (different in platypus and echidnas) being added independently in the two families. Each of these fusion events would have caused meiotic mayhem in a hybrid between the original and the translocated forms.

Therian mammals forsook the ancestral system entirely for a new XY system; again, incompatibility between the ancestral and new system would have led to aberrant sex ratios and infertility. The addition in eutheria of an autosomal region to the ancestral therian XY would also have led to hybrid infertility that may have triggered divergence of eutheria. It is tempting to propose that these sex chromosome changes drove mammal divergence.

But does chromosome change cause species divergence, or does it follow divergence? Detailed genetic analysis of *Drosophila* favoured the alternative view that the reproductive barrier that first drove a wedge between incipient species was the

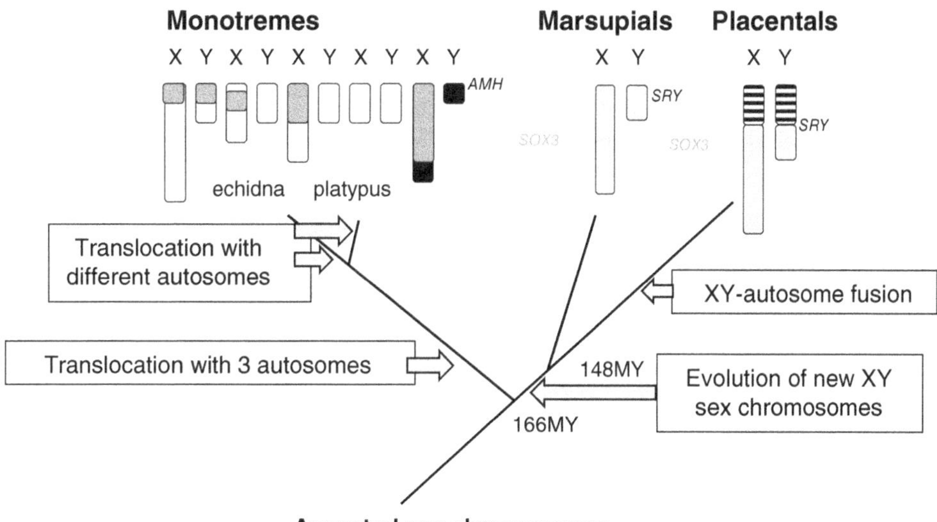

Figure 4.9   Sex chromosome change and mammal divergence. The three mammal lineages differ in their sex chromosomes. Each change (boxed) would have posed a reproductive barrier that could have initiated species divergence.

result of gradual accumulation of mutants in geographically isolated populations, and that chromosome change was secondary. Many claims were made to have identified 'speciation genes', mutations in which drove this isolation. These hybrid sterility genes are concentrated on the X chromosome, providing an alternative explanation for a strong sex chromosome effect on speciation.

Sex chromosomes may present a particular problem to hybrids because necessarily in the heterogametic sex one sex chromosome (X or Z) derives from one species and the other (Y or W) from the other species. Could infertility be simply a mechanical effect of mismatched sex chromosomes? If changes in PAR of one or both species render X and Y (or Z and W) unable to pair, meiosis will fail. Deletions and insertions in this region can disrupt pairing and lead to the complete failure of meiosis in humans and mice.

Alternatively, mutations in Y alleles may occur so that they no longer complement the X copy of a dosage sensitive genes in a related species, and dosage imbalance follows. Dosage differences would be exacerbated by X chromosome inactivation and upregulation of loci on the X when the Y partner is lost (Chapter 5). For instance, if one parental species retains alleles of a gene on both sex chromosomes, but the other parent has lost the Y allele, upregulated the X allele and joined the X inactivation system, dosage differences up to 3:1 between male and female hybrids would result.

Sex chromosomes could also be particularly important for interposing a reproductive barrier between populations because they bear so many genes involved in reproduction. For instance, the mammals X and Y contain more than their fair share of fertility genes and genes with sex-biased expression. Sex chromosomes also have higher rates of mutation (the rapid degradation of the Y/W as well as the 'fast X/Z'), and so diverge faster than autosomal genes.

Hybrids between species show different degrees of infertility and inviability. Interestingly, this is often different in male and female hybrids. J. B. S. Haldane noted decades ago that in systems of male heterogamety (XY systems like mammals) the male was always the more affected, but in systems of female heterogamety (ZW systems like birds) it was always the female. 'Haldane's Rule' that infertility and inviability are more severe in the heterogametic sex screams for a role for sex chromosomes in speciation.

At least in mammals, the Y chromosome may well hold the secret to male infertility in species crosses. The Y has degenerated faster than other chromosomes. It is likely that deletion or inactivation of a dosage sensitive gene in the male-specific region of the Y in one parental species, in the absence of compensatory changes to expression of genes on the X in the other would cause developmental problems which are manifest by inviability or particularly sterility, since some Y genes are critical for sperm production (Figure 4.10).

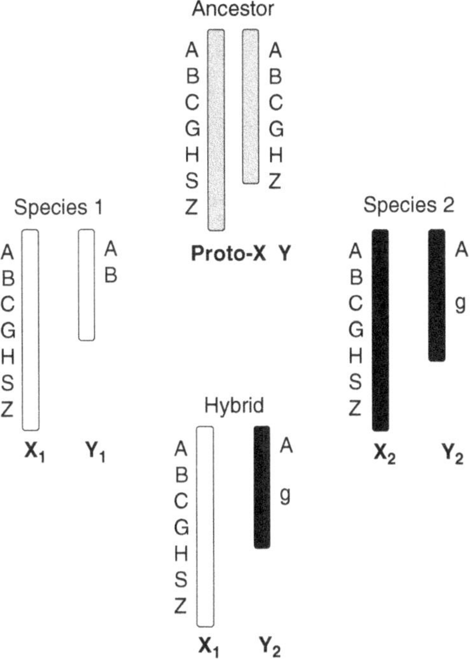

Figure 4.10   Sex chromosome evolution and Haldane's Rule. An ancestral species with a young XY system has minimal Y degradation (genes A-Z). After divergence into species 1 and 2, the X chromosome (X_1 and X_2) is conserved between species but Y chromosome degradation occurs independently so that the genetic constitution of the Y_1 and Y_2 is different. For instance, the dosage sensitive gene B is retained in Species 1, but lost in species 2, with dosage compensation restored by X inactivation of the X allele. In species 1, gene G has been lost from Y_1, whereas in species 2, it has acquired a male-specific function (g). In a hybrid between Species 1 × 2, genes in X_1X_2 females will be balanced, but in males from reciprocal crosses, they will both be unbalanced. For instance, in X_1Y_2 hybrids, gene B is present in only a single dose in males and is not compensated by inactivation of X_1. The function of the male-specific gene *g* may interfere with the upregulated *G* on X_1. The reciprocal hybrid X_2Y_1 will also be impaired by an overdose of B and the absence of the male-specific g. Since many genes on the Y are required for male fertility, male, but not female species hybrids, may be infertile.

A causative role of sex chromosome change in speciation is supported by detailed studies of fish, frog and reptile groups in which closely related species differ by a sex chromosome rearrangement. For instance, in the nine-spined stickleback, a novel XY system arose along with a large inversion that imposes hybrid sterility. In the threespine stickleback, geographically separate populations are distinguished by an X-autosome fusion, and GWAS shows that the neoX contains genes involved in seasonal and behavioural traits important for reproductive isolation.

There are many examples of the association of rapid sex chromosome turnover and species richness. Neotropical fish in Brazil provide spectacular examples of closely related fish species that differ in their sex chromosomes, and there is a strong association between the rapidity of sex chromosome turnover and species richness of cichlid taxa.

Such associations can never prove that sex chromosome turnover *drives* speciation. So we are still not in a position to choose between the problems of species hybrids with Y (or W) chromosomes that don't match, or the slow divergence of sex chromosomes in incipient species.

4.16 Conclusions

Sex chromosomes provide a wonderful example of the power of comparative genomics. We see that vertebrates, although they share much the same genome, have designated different regions of it to determine sex (Figure 4.8).

We started with an exploration of human sex chromosomes, charting how sex chromosomes of eutherian mammals evolved from two autosome blocks about 150 million years ago and differentiated by degradation of the Y chromosome to its present parlous state. Our view of the human Y chromosome has changed as we discover that, far from being a dominant male-determining entity, or even a selfish element that kidnaps male-advantage genes from the autosomes, it represents the last relics of the ancestral autosome. Truly a wimp (Figure 4.11).

Bird sex chromosome evolution – from a different autosome and entirely opposite – has taken place by parallel processes, producing a gene-rich Z and a degenerate W chromosome.

But moving outward to reptiles, amphibians and fish reveals how unusual, and how confined, are these examples of stable sex chromosomes, locked in by their extreme differentiation. And how extremely flexible other vertebrate sex chromosomes can be, with a high turnover as novel sex-determining genes evolve on other chromosomes and take over, and with innumerable examples of young sex chromosomes that differ by a single gene or a few hundred kb.

We still don't understand why degradation of the sex-specific chromosome is rapid in some lineages, and low – or even absent – in others. But we can see that degradation is not necessarily inexorable, an inevitable consequence of genetic isolation, nor is recombination suppression a necessary consequence of acquisition of a new sex-determining gene. We understand better the role of sex chromosomes in sex chromosome incompatibility in species hybrids, but still don't know which causes what.

We see, too, the huge variety in genes that can accomplish sex determination (Chapter 10), and the huge variety of other genes that confer an advantage (not necessarily sexually antagonistic) that might drive recombination suppression, as

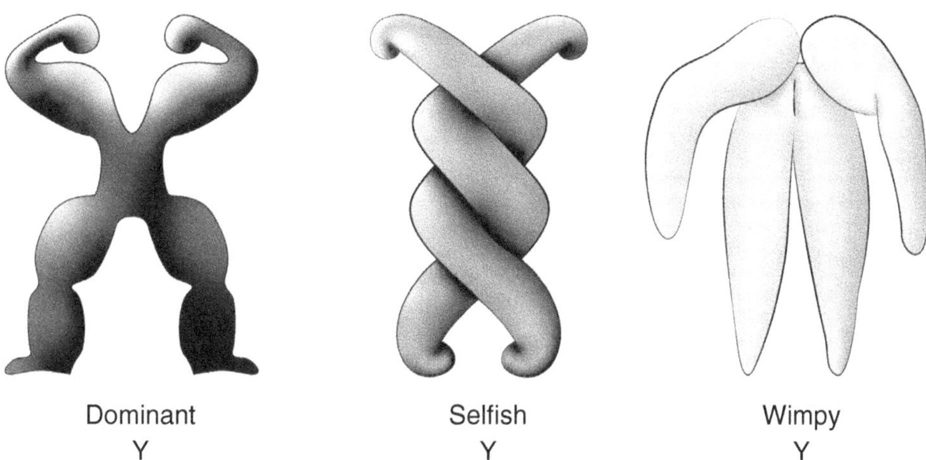

Figure 4.11   My models of the human Y chromosome. The Y is unique in its structure and function, and many evolutionary models have been proposed to explain it. Here are four of mine. The effect of the Y chromosome is to cause male development no matter how many X chromosomes there are, so the Y has traditionally been thought to be a dominant chromosome. Several Y genes appear to have been appropriated from autosomes, leading to a model of a 'selfish Y'. However, investigation of the suite of Y genes shows that they are mostly deleted, or are mutated copies of X genes, supporting a model of a 'Wimpy Y'. Original drawings by Jenny Graves, art by Bronwyn Knight, Knight's Design Cavern, knightscavern.myportfolio.com.

well as the huge variety of responses to problems of differential dosage of other genes on the sex chromosomes (Chapter 6).

Reptiles and fish have an out that we lack – the ability to determine sex by temperature. Birds and mammals, being homoeothermic, are stuck with their XY and ZW systems. Environmental sensitivity confers the ability to switch – suddenly – to temperature sex determination. This wipes the chromosome slate clean in an evolutionary eyeblink – a single generation – and leaves room for novel genetic systems to rapidly evolve.

Until recently, we were limited by the tools of genetic analysis and by tractable animal models; mice and chickens, flies and worms. Now we have powerful new genomic technologies that we can apply to any interesting species you can sequence.

Many questions remain. Does loss of recombination occur between a sex locus and a sexually antagonistic – or merely a male advantage gene? Does degradation of the sex-specific chromosome (Y or W) occur by inversion of large chunks or creeping tidal movements, or both? Does sex chromosome turnover cause speciation or occur after it? My hunch is that the answer to these questions will be 'both – and lots of other things we never thought of'.

FURTHER READING

Classic Papers

Charlesworth B., 1991. The evolution of sex chromosomes. *Science* 251: 1030–1033. doi: 10.1126/science.1998119.

Haldane JBS, 1922. *Sex ratio and unisexual sterility in hybrid animals. Journal of Genetics* 12: 101–109. doi: 10.1007/BF02983075

Lahn BT, Page DC, 1999. Four evolutionary strata on the human X chromosome. *Science* 286: 964–967

Reviews and Research Articles

Bellott D, Page DC, 2021. Dosage-sensitive functions in embryonic development drove the survival of genes on sex-specific chromosomes in snakes, birds, and mammals. *Genome Research* 31: 198–210. doi: 10.1101/gr.268516.120

Cortez D, Marin R, Toledo-Flores D, et al., 2014. Origins and functional evolution of Y chromosomes across mammals. *Nature* 508: 488–493. doi:10.1038/nature13151

Furman BLS, Metzger DCH, Darolti J, et al., 2020. Sex chromosome evolution: so many exceptions to the rules. *Genome Biology and Evolution* 12: 750–763, https://doi.org/10.1093/gbe/evaa081

Graves JAM, 2006. Sex chromosome specialization and degeneration in mammals. *Cell* 124: 901–914

Hayashi S, Abe T, Igawa T, 2024. Sex chromosome cycle as a mechanism of stable sex determination. *The Journal of Biochemistry* 176: 81–95. https://doi.org/10.1093/jb/mvae045

Hughes JF, Page DC, 2015. The biology and evolution of mammalian Y chromosomes. *Annual Review of Genetics* 49: 507–527. doi: 10.1146/annurev-genet-112414-055311

Kitano J, Peichel CL, 2012. Turnover of sex chromosomes and speciation in fishes. *Environmental Biology of Fishes* 94: 549–558. doi: 10.1007/s10641-011-9853-8

Pennell MW, Mank JE, Peichel CL, 2018. Transitions in sex determination and sex chromosomes across vertebrate species. *Molecular Ecology* 27: 3950–3963, https://doi.org/10.1111/mec.14540

Perrin N, 2012. Sex reversal: A fountain of youth for sex chromosome? *Evolution* 63: 3043–3049. https://doi.org/10.1038/srep32874

Setti PG., Deon GA, dos Santos Z, et al., 2024. Evolution of bird sex chromosomes: a cytogenomic approach in Palaeognathae species. *BMC Ecology and Evolution* 24: 51. https://doi.org/10.1186/s12862-024-02230-5

Shayer MIA, Sarre SD, Gleeson D, Ezaz T, 2018. Did lizards follow unique pathways in chromosome evolution? *Genes* 9: 239

Uno Y, 2021. Inference of evolution of vertebrate genomes and chromosomes from genomic and cytogenetic analyses using amphibians. *Chromosome Science* 24: 3–12, 2021

Vicoso B, 2019. Molecular and evolutionary dynamics of animal sex-chromosome turnover. *Nature Ecology & Evolution* 3: 1632–1641, doi.org/10.1038/s41559-019-1050-8

Part II

Activity of Sex Chromosomes

5

Discovery of X Chromosome Inactivation

As I emphasized in Part I, sex chromosomes are a wonderful example of evolutionary Dumb Design. The differentiation of the X and Y chromosomes caused by Y degeneration and gene loss in humans and other mammals presents a number of problems, including difficulty of XY pairing at meiosis, the perils of haploidy in males – and dosage differences for genes on the X chromosome.

In this section, I will discuss the dosage compensation system that ensures fair play between the sexes.

X chromosome inactivation was discovered by Mary Lyon in 1961. It is an extraordinary phenomenon, involving the transcriptional silencing of a thousand unrelated but physically linked genes. It is a whole-X event, involving major cytological and molecular changes. It constitutes the classic example of 'facultative heterochromatinization' in which large blocks of chromatin are genetically active or inactive in different situations. We now call this phenomenon 'epigenetic silencing', because the gene remains unchanged and the silencing is 'above the gene'.

As the most spectacular example of epigenetic control on a grand scale, much effort has been put into studying X inactivation at a molecular level. Fifty years later, we still do not fully understand the intricacies of X chromosome inactivation, but, with the sequencing and characterization of the X, and the beginnings of understanding of a control locus, we can now begin to unravel what is evidently a multilayered control mechanism.

As for the organization of sex chromosomes, our understanding of the control of X chromosome activity depends on comparisons of different animals; mutant mice, human patients with abnormalities of the X chromosome – and of other species of mammals, other vertebrates and even invertebrates, that show major differences in regulation of sex chromosome activity.

In Part II I will discuss the discovery of X chromosome inactivation, how it works and how it evolved. Chapter 5 will describe the original observations and interpretations of what would become the most celebrated example of epigenetic silencing. Chapter 6 discusses the intricate web of molecular change that underpins inactivation. Chapter 7 focuses on what we learn from comparing X inactivation between mammals and other vertebrates, and how X inactivation fits into a general picture of epigenetic dosage compensation.

5.1 The Need for Sex Chromosome Dosage Compensation

Dosage compensation for sex-linked genes makes excellent sense. As I have discussed, the mammalian X chromosome contains between 900-1400) protein-coding genes; Gencode lists 1,144 on the human X. These are involved in a range of housekeeping and specialized functions. The Y chromosome seems to have lost practically all of these genes, retaining only a handful, and often twisting their function toward male-specific purposes.

One obvious consequence of the genetic inequality of the mammalian X and Y chromosomes is that about a thousand genes on the X chromosome are present in two doses in XX females, but only a single dose in XY males (Figure 5.1). This is obviously not fair. Monosomy for even a small bit of an autosome is usually

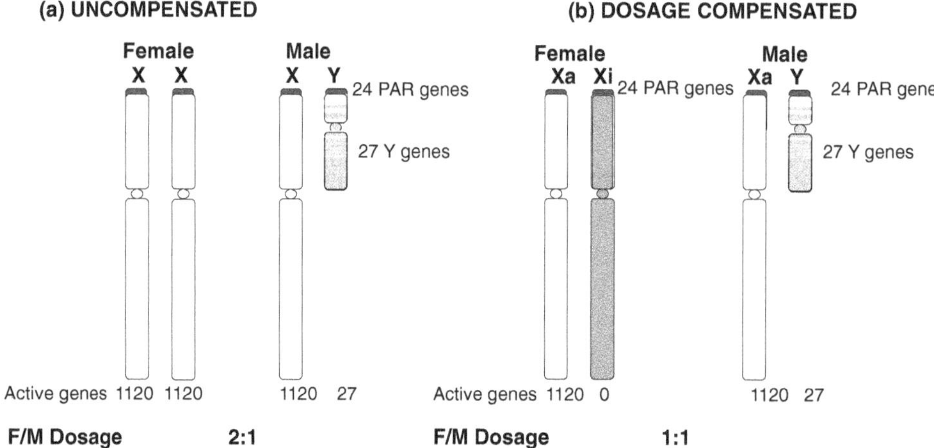

Figure 5.1   Dosage differences and dosage compensation of genes on the human sex chromosomes. (a) Sex chromosomes of female (XX) and male (XY), with shared PAR (black) at the tip. The X chromosome has 1,144 protein-coding genes at the last count (1,120 excluding the PAR); the heterochromatic Y has only 51 (27 excluding the PAR), only a few of which complement X partners. Thus, there are about 1,120 genes that are present in two copies in females and a single copy in males. (b) Dosage imbalance between the sexes is prevented by inactivation of one whole X chromosome (dark grey, Xi) in females so both males and females have a single active X (Xa). However, some genes on the short arm of the Xi escape compensation at least partially (see Figure 5.4).

detrimental, often fatal. So it is not surprising to find that there is a mechanism that compensates for the different gene dosage in males and females by inactivating one X chromosome in females.

But there is another dosage problem to be solved; the relation of X-borne genes and autosomal genes will also be upset if only one X is active. This led Ohno to propose, long ago, that genes on the X are twice as active as autosomal genes.

Dosage compensation systems are known in other vertebrates and even insects with differentiated sex chromosomes (Chapter 7). Dosage compensation was, like practically everything else, first described in the fruit fly *Drosophila melanogaster*. It was an early observation that, despite the 2:1 difference in gene dosage between female and male *Drosophila*, the amounts of X-linked gene products do not differ between the sexes. This gave rise to the idea that the differences in X-linked gene dosage differences between males and females are compensated for in some way.

In mammals, dosage compensation for X-linked genes at the biochemical level was described in the late 1950s with the observation that the activity of the X-borne enzyme locus *G6PD* was the same in human males and females, as well as in patients with abnormal numbers of X chromosomes. But its workings were quite mysterious. In the last 70 years we have gained enormous knowledge of X inactivation, and recognized its value as a model system for epigenetic control of gene activity.

5.2 Discovery of X Chromosome Inactivation

The hypothesis that one X is inactive in the somatic cells of female mammals was first put forward by Mary Lyon in 1961 to account for the blotchy coats of female mice that were heterozygous for alleles of an X-linked coat colour gene (Figure 5.2).

Males were either one colour or the other, so mosaicism was a peculiarity of females with both alleles. Lyon suggested that one or other X chromosome, chosen at random, became genetically inactive in every somatic cell early in embryogenesis. Once inactivation occurred, this state was stable and heritable through somatic cell divisions, giving rise to clones of tissue in which either the maternal or paternal X was active – 'mosaic' phenotypes. For instance, a tabby mouse (or the more familiar tortoiseshell cat) has blotches of fur that express a dark allelic variant from one parent, and blotches that express a lighter variant from the other parent.

Although we humans lack fur, mosaicism is evident in human females heterozygous for X-linked skin conditions; for instance, women who have patches of normal skin, and patches of skin devoid of sweat glands.

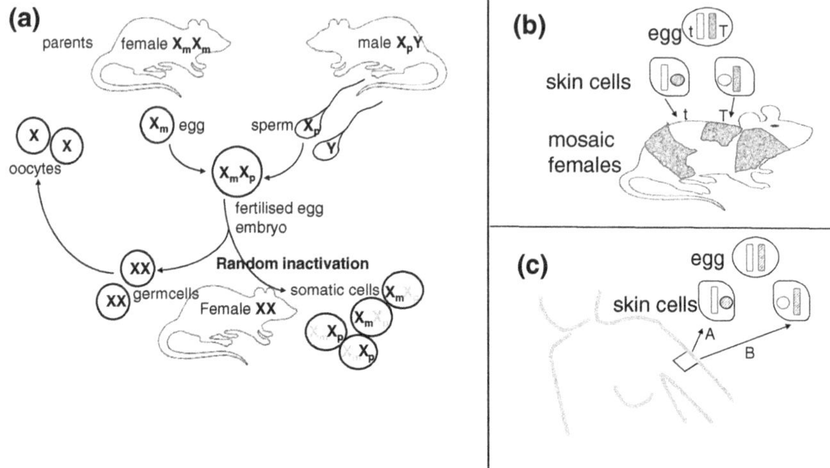

Figure 5.2 Lyon's hypothesis and the evidence for mosaicism. (a) Lyon's hypothesis states that in the somatic cells of female mammals, either the maternal X (X_m) or the paternal X (X_p) becomes randomly inactivated (light grey) in the embryo, and stays stably inactive in cell descendants throughout life. Inactivation does not occur in germ cells, so both X chromosomes are inherited in an active state (bold). (b) In somatic cells of females, either X becomes inactive at random (scrunched up inactive X represented by round shape). If the two X chromosomes bear different alleles of a gene (T/t) that can be detected on the skin, there will be two populations of cells, one expressing the maternal X (with the paternal X inactivated) and one expressing the paternal X (with the maternal X inactivated). Thus Mary Lyon saw that female mice heterozygous for a coat colour gene were mosaics, having patches of fur that expressed the darker allele (T) on the paternal X, or the lighter allele (t) on the maternal X. (c) Human females don't express fur genes, but it can be shown that cells from a woman heterozygous for variants of the enzyme G6PD (A white, B striped) have a mixture of cells expressing A or B but not both.

In the germline, Lyon proposed that inactivation did not occur, or was reversed. In patients with abnormal numbers of X chromosomes, Lyon predicted that there was always only a single active X.

The Lyon Hypothesis neatly accounted for the equivalence of X-linked gene products in males and females, since somatic cells had a single active X in both sexes.

The major tenets of Lyon's hypothesis have been demonstrated time and time again for different X-linked genes in humans as well as mice, but there are elaborations and exceptions which also provide insight into the mechanism, function and evolution of X inactivation.

X chromosome inactivation has been observed in many other species of eutherian mammals, such as rabbits, cats, sheep, cows, and even elephants. Marsupial X inactivation was described a few years later.

The Lyon hypothesis proposed an extraordinary situation – the silencing of a thousand genes on a single chromosome, in only one sex.

5.3 Cytological Changes to the Inactive X Chromosome

The change of activity of one X in female mammals is accompanied by extraordinary changes in its appearance (Figure 5.3).

In fact, long before Lyon put forward her hypothesis, it was known that cells from males and females looked different under the microscope. This was first observed in 1949 by a graduate student, Ewart Bertram, in neurones of cats. Nuclei of human cells regularly had a heterochromatic blob near the membrane in females but not males (Figure 5.3a). This 'sex chromatin' (often called the Barr body after Bertram's supervisor) is apparent in many cell types from human females, but not males. It was used as a 'sex test' to distinguish female from male athletes (a very dubious definition, as we will see).

Even before Lyon had broached her hypothesis, Susumu Ohno proposed that sex chromatin represented a scrunched- up X chromosome in cells from women. This was verified by his observation that the cells of patients with extra X chromosomes had extra sex chromatin bodies to match. The identity of the sex chromatin has since been confirmed by in situ hybridization with specific X chromosome probes.

The conformation of the inactive X within this body has been subjected to many types of cytogenetic and molecular studies over the years. Analysis of the positions of signals from unique sequences on the X suggested that the inactive X assumes a highly organized state, perhaps folded into a hairpin loop by telomere association. Early reports referred to a 'bipartite structure'.

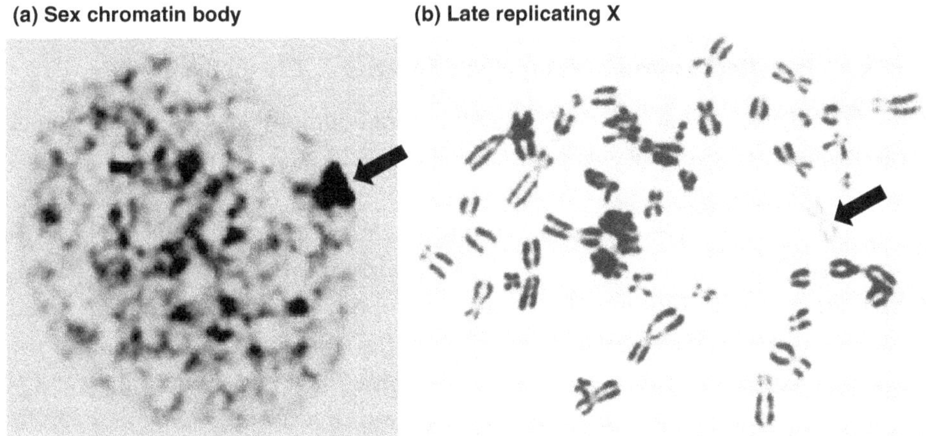

Figure 5.3   Cytological hallmarks of the inactive X chromosome. (a) The inactive X in the nucleus of a human female is visible as the sex chromatin body (Barr body, arrow), a condensed mass often at the periphery of the nucleus. Photograph taken by S. Ohno and supplied by Dr Christine Disteche, University of Washington, Seattle. (b) The inactive X in a human mitotic spread (arrow) replicates later than other chromosomes, and stains lighter when the cell is fed with BUdR at the end of the DNA synthesis phase.

Ultrastructure of the sex chromatin body shows that it assumes a shape different from that of the active X. At the highest optical resolution, it appears as a lumpy structure with channels between condensed domains.

Differences in condensation may also be revealed at metaphase by differential staining methods, which include chemicals with a completely unknown action (such as a hot solution of potassium chloride), as well as enzymes that preferentially degrade 'open' active chromatin.

The earliest observations of the sex chromatin body noted its unusual position in interphase cells. The active X takes its place among other not-very-gene-rich chromosomes in the outer ring of chromosome territories. However, the sex chromatin body in fibroblasts is nearly always plastered up against the nuclear membrane, almost popping out of the nucleus, wrapped in nuclear membrane.

The position of the inactive X in a cell may differ between tissues or stages of the cell cycle. During the phase of the cell cycle when DNA is synthesized, sex chromatin is often almost wrapped around the nucleolus, the body that surrounds ribosomal DNA clusters and is full of rRNA. We know that both these preferred positions are associated with gene repression. These locations are shared by constitutive heterochromatin (non-transcribed repetitive sequence). Repositioning of a DNA sequence to either of these locations may therefore affect its expression.

Thus, the active and inactive X chromosomes show dramatic differences in size, conformation and position in the cell.

5.4 Replication Timing of the Inactive X Chromosome

Early work using radioactive isotopes and autoradiography also identified one X chromosome that replicated conspicuously late in cells from females but not males (Figure 5.3b).

It is quite easy to assess when a chromosome replicates within the DNA synthesis phase (S phase). A radioactive DNA precursor (tritiated thymidine) is supplied to growing cells (cultured fibroblasts or blood lymphocytes). They are fixed a few hours later, and the position of labelled DNA is detected by covering the microscope slide with a thin layer of photographic film ('autoradiography'); silver grains are clustered above chromosome regions that were making new DNA. Fixation is timed so that mitotic cells must have been just completing S phase when label was available.

It was observed in the 1960s that the two X chromosomes in female Chinese hamster cells labelled asynchronously with tritiated thymidine, supplied late in the S phase. A late-replicating X was seen also in cells (white blood cells or cultured fibroblasts) of female, but not male, mice, humans and many other species of eutherian mammals. My undergraduate research thesis reported the presence of a late-replicating X chromosome in blood cells of female kangaroos, the first indication that X inactivation occurred also in marsupials.

The single X in male-derived cells completed replication along with the autosomes. In female cells, one X completed replication with the autosomes, but the other was conspicuously later. In multi-X subjects, all but one X replicated late, so that the number of late-replicating X chromosomes was the same as the number of sex chromatin bodies.

A direct demonstration that the late-replicating X was the inactive one was made using clones of female mule cells. The replication time of the morphologically distinct horse or donkey X was correlated with the differential expression of horse or donkey G6PD.

In the 1980s, fluorescence labelling techniques replaced radioactive isotopes to mark replicated chromosome regions. Replication timing could be visualized by incorporation of a nucleotide analogue, usually bromodeoxyuridine (BrdU) in place of thymidine, followed by detection using a dye sensitive to BrdU or an antibody. Patterns confirmed the early observations, and provided much higher resolution. The correlation of late replication with inactivity was strengthened by the finding of an early replicating region on the terminus of the short arm of the inactive human X, corresponding to the PAR, which is expected to escape inactivation because it is present in two doses in both sexes.

The asynchronous replication of the active and inactive X chromosomes is apparent at the level of the single gene. Early and late-replicating DNA was differentially labelled by BrdU incorporation, and separated by their differential cutting with restriction enzymes. Genes on the inactive human X were included in the late compartment. Thus, late replication correlates with inactivity at the gene, as well as the cytological level.

It was recognized early that the preferential staining, condensation and late DNA replication were all features of a class of 'facultative heterochromatin', as defined from work on insect systems; chromatin that contained genes that were active in some time or place but inactive in others (Chapter 7).

5.5 Silence of Genes on the Inactive X Chromosome

Lyon's hypothesis that one X chromosome is silenced in females relied on the patterns of expression of visible traits in the mouse, particularly X-linked genes affecting coat colour and fur characteristics. Inactivation was inferred from the mosaic phenotype of heterozygous females.

Gene silencing could be more directly demonstrated by early work on a few well-characterized X-linked enzyme loci. The usual set comprises *G6PD, PGK1* and *HPRT*, encoded by genes on the long arm of the human X, within the conserved region (Section 4.1). In females heterozygous for alleles that encoded enzymes with different charges (isozymes) separable by electrophoresis through a gel, patches of cells expressed one allele or the other, but never both.

Cells from women heterozygous for electrophoretic variants A and B of the enzyme G6PD could be cloned. Clones expressed type A or type B. No cells contained both – this can easily be detected because the enzyme G6PD is a dimer consisting of two chains. If types A and B are present in the same cell, dimers will be of three types, AA, BB – and a heterodimer AB. This has intermediate mobility and is detected as an intermediate band.

Marsupials also showed gene silencing. Females heterozygous for X-borne genes also expressed alleles from only one X. However, marsupial X chromosome inactivation was startlingly different from X inactivation in eutherian mammals, because it was always the gene from the male parent that was silenced. Inactivation was also observed to be incomplete; that is, some genes on the X were not inactivated, at least in some tissues.

It was very important to demonstrate that inactivation is coordinated over a single X. Lyon's hypothesis that inactivation is a whole chromosome phenomenon predicted that all the alleles on one X should be expressed, and all the alleles on the other X should be silent concordantly. This prediction has been borne out by many studies of inactivation *in vivo* and in culture.

In mice, Lyon found that two coat markers (affecting fur colour and texture) were co-ordinately expressed in the mosaic patches on the coats of female mice heterozygous for both.

Coordinate expression was later demonstrated more spectacularly using hybrids between two mouse subspecies (*Mus musculus x M. caroli*), in which the parental origin of three genes could be recognized because the enzymes coded by *G6pd*, *Pgk* and *Hprt* alleles differed by charge in the two species, and so were all distinguishable. Clones from hybrid embryos expressed only the set of *M. musculus* alleles, or alternatively, the set of *M. caroli* alleles.

Genetic studies therefore supported Lyon's hypothesis that genes on one or other X chromosome were silenced.

5.6 Cells Can Count X Chromosomes

Remarkably, cells from female mammals inactivate as many X chromosomes as it takes to achieve balance.

An extension of Lyon's hypothesis was that in cells of patients with multiple X chromosomes, all but a single X was inactivated. The Lyon Hypothesis is sometimes referred to as 'the Single Active X Hypothesis'. It seems that cells recognize if there are more than one X per diploid chromosome set, and inactivate one or more X chromosomes to fit.

Early observations established that XO cells had no Barr body, and no late-replicating X. On the other hand, XXY cells had one, and XXX two Barr bodies and late-replicating X chromosomes. Generally, cells from patients with multiple X chromosomes (*n*) had one fewer (*n*-1) Barr bodies and late-replicating X chromosomes, as you would expect if all but one X chromosomes were inactivated.

Inactivation of all but a single X accounts for the near normal phenotypes of patients with abnormal sex chromosome constitutions. Patients with XXY, XXX or even XXXX are remarkably normal, compared to an extra dose of any other chromosome, which is lethal for all but the gene-poor chromosome 21. And XO females have an extremely mild condition in humans, and are near normal in mice, whereas single representation ('monosomy') of even the smallest autosome causes embryonic death.

What happens in embryos with abnormal numbers of chromosome sets? Embryos with three sets of chromosomes (triploids) have one or two inactive X chromosomes. Embryos with four sets of chromosomes (tetraploids) have two inactive X chromosomes. This suggests some sort of equation between the numbers of autosomal sets and numbers of active X chromosomes.

Counting seems to take place early in development, in the preimplantation mouse embryo. Suggestions were made over many years that some autosomal factor in short supply blocked inactivation of a single X. In polyploids, a double dose of blocking factor could protect two X chromosomes from inactivation. In a triploid, the marginal supply would leave one or sometimes two X chromosomes active, and cell selection would alter the ratios of these cells.

But do cells really have to count? Perhaps X inactivation is a stochastic process that produces every combination of active and inactive X chromosomes, of which only 1X-active cells survive. Stochastic processes seem to be behind many apparently continuously variable phenomena, and the observation of stochastic expression of partially inactive genes (Section 6.3) would support this idea.

5.7 X Inactivation Is Super-Stable

Mary Lyon also predicted that X inactivation, once it occurs in a somatic cell, is very stable. This was suggested by the large size, and the uniformity, of the mosaic patches on female mice heterozygous for coat colour variants. This means that the inactive state must be transmitted at each cell division throughout life; thus, it is somatically heritable.

X chromosome inactivation is extraordinarily stable in culture as well as *in vivo*. Human cell clones isolated from heterozygous women express only one allele of *G6PD* indefinitely, and there is no indication of spontaneous reactivation.

A very low X chromosome reactivation rate could also be inferred from growth of heterozygous cells in selective media. Cells from a woman heterozygous for Lesch-Nyhan disease (deficiency of the X-linked enzyme HPRT) were also remarkably stable. Cultures contain a mixture of cells expressing the X bearing the normal *HPRT*⁺ allele, and cells bearing the mutant *HPRT*⁻ allele. Clones of *HPRT*⁺ cells die in medium containing toxic variants of guanine, and HPRT⁻ clones die in selective medium in which cells must use an alternative biochemical pathway. No spontaneous reactivants were detected, implying an overall reactivation rate for the *HPRT*⁻ allele of less than 10^{-9}.

In the established lines of *Mus musculus* × *M. caroli* cells, which differ at three loci, expression of either the *M. musculus* or the *M. caroli* set has been maintained throughout years of growth. An *Hprt*⁻ variant of one of these lines, which had the mutant allele on the active *M. musculus* X, but a normal allele on the inactive *M. caroli* X, never spontaneously re-expressed the *Hprt*⁺ allele from the inactive *X*, even in the face of stringent selection of many millions of cells. X chromosome inactivation therefore seemed even less reversible than gene mutation.

But there are few examples of somatic reactivation *in vivo*. Reactivation occurs in some disease states, particularly in cancer cells, for instance, human breast cancer. There was an early report that an inactive allele of an X-borne enzyme locus was reactivated in aged mice. There are also mutations that increase the age-dependent probability of reactivation in human cells.

Reactivating genes on the inactive X chromosome was long a practical goal because this could potentially cure X linked disease in females heterozygous for a mutation, for instance, of blood clotting Factor VIII, or neural factors involved in Rett's syndrome. Early attempts to induce reactivation of an inactive allele, using all manner of physical and chemical insults, including mutagens, were generally unsuccessful. However, in the 1980s, treatment with agents that interfere with epigenetic processes produced reactivation, and provided the first clues to the molecular biology of X chromosome inactivation (Section 6.3).

5.8 Some Genes Escape X Chromosome Inactivation

Mary Lyon's prediction that X chromosome inactivation was a whole X event was borne out by early studies on the few genes that affected visible traits or electrophoretic mobility of protein product. However, a wider survey shows that some genes on the inactive X 'escape' silencing and are expressed from both alleles ('biallelic') (Figure 5.4).

Not surprisingly, genes within the pseudoautosomal region (PAR1) of the human X, and *Sts* in the mouse PAR escape inactivation. They do not need to be inactivated, since they have a functional allele on the Y chromosome.

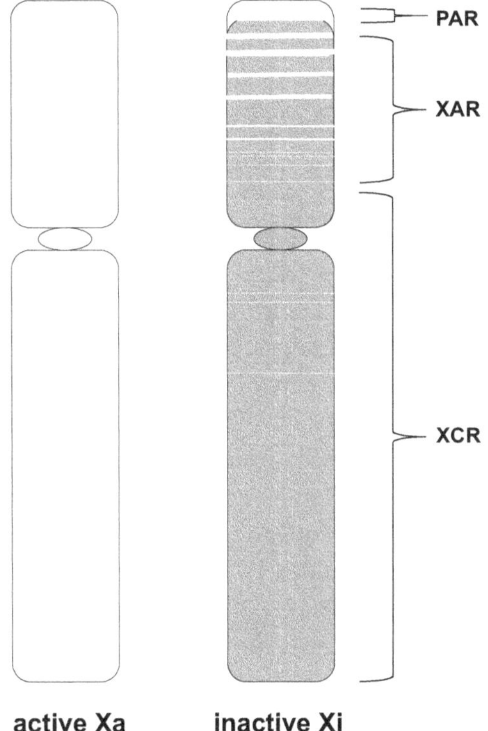

Figure 5.4   Escaper genes on the human inactive X. Active Xa (white) and inactive Xi (dark grey) genes in a human female. Many (270) genes on the human X chromosome escape inactivation fully or partially (white stripes). These genes are concentrated into the added region of the human X, particularly in the regions near the PAR that differentiated from the Y most recently.

Oddly, however, genes within PAR2 at the distal ends of the long arms of the human sex chromosomes are silenced on the inactive X chromosome – but also on the Y chromosome, evidently having evolved yet another kind of dosage compensation.

Several other genes on the human X with active partners on the Y outside the PAR (e.g., *ZFX, RPS4X, UTX, AMELX*), are also exempt from inactivation. Again, this makes functional sense, since these X-Y shared genes do not need dosage compensation, and their two-fold dose may be critical for function. However, it makes less sense that several genes on the X with *inactive* partners on the Y (pseudogenes), such as *STSY* and *KALY*, also escape X chromosome inactivation.

The advent of molecular techniques to examine RNA directly (Section 6.1) revealed that many genes on the inactive human X escape silencing. Expression of human X-borne genes was monitored by assessing their RNA transcripts in mouse-human cell hybrids that retained an inactive human X, or cell lines from patients with skewed X inactivation.

Surprisingly, 150 genes were found to be transcribed from both X chromosomes. Escapers now amount to 20–30% of all genes on the human X. Escaper genes on the inactive X are always less active than their alleles on the active X. Anything above 10% of the active allele is classed as 'escape', and it varies from this up to 70% of the expression from the active X.

Escape seems to be an intrinsic property of escaper genes since they are not inactivated when relocated elsewhere on the X chromosome.

Many escaper genes showed activity in some tissues, but not others. Most remarkably, the extent of escape of some genes was different in different women. A survey of 639 protein-coding genes on the human X found that 80 escape inactivation, and another 93 variably escape. The variability of X inactivation between tissues, and between females adds a new level of variation that could cause differences in morphology, growth or disease states between different women.

Escaper genes are not scattered all over the human X. As shown in Figure 5.1 they are clumped together in escaper regions, which largely lie in XAR, the region of the human X that was added later to eutherian sex chromosomes. Their frequency increases toward the PAR.

In other mammals, too, many genes escape inactivation, and the set of escapers is mostly conserved, at least in primates.

However, escape from inactivation is much less common in mice. RNA expression from genes on the inactive mouse X (in hybrid females with skewed inactivation) detected expression from both alleles in only 17 genes, and variable escape (in different tissues or developmental stages) in another 20. The set of escaper genes in mice and humans overlaps by only 7 genes. The difference between mice and human escape seems to be largely in the size of escaper regions, which contain up to 15 genes in humans but only one or two in mice.

The presence of so many escaper genes on the human X but not the mouse X may explain why XO female mice have a much less severe phenotype than XO women with Turner's Syndrome. In humans, but not mice, two doses of some or all escaper genes may be required for normal development, particularly of the germ cells.

Curiously, the first hint of escape of genes on the inactive X came, not from humans or mice, but kangaroos. Enzyme studies in the 1970s showed that some genes on the inactive paternal X chromosome were partially or even fully expressed (Figure 5.5). This has been confirmed by directly observing transcription of alleles on the active maternal, and inactive paternal X in the opossum, and by RNA FISH in the kangaroo. Partial expression was observed in some tissues, but not others, and was inconsistent between marsupial species. Overall, about 14% of genes on the opossum X escape inactivation, but about 32% escape inactivation in kangaroos.

Thus, genes on the inactive X can escape inactivation to different degrees in different tissues and different species. These exceptions to gene silencing are particularly informative.

5.9 Randomness and Choice, Skewed X Inactivation

One of the more extraordinary features of X chromosome inactivation is that it can affect either X chromosome, and the process is random.

Lyon observed that female mice heterozygous for X-linked coat colour genes usually had similar numbers of dark and light blotches distributed randomly over the body. To explain this large-scale mosaicism, she proposed that the two X chromosomes were randomly inactivated early in embryogenesis when the pool contained few progenitor cells. The predictions that there should be an equal number of patches expressing genes on one or the other X of a heterozygous female have been verified many times in many different systems.

Spectacular confirmation came from constructing mice with a 'blue' gene' (the *lacZ* gene from bacteria) inserted into one X (Figure 5.5). The distribution of blue and white cells could now be directly observed in different tissues and

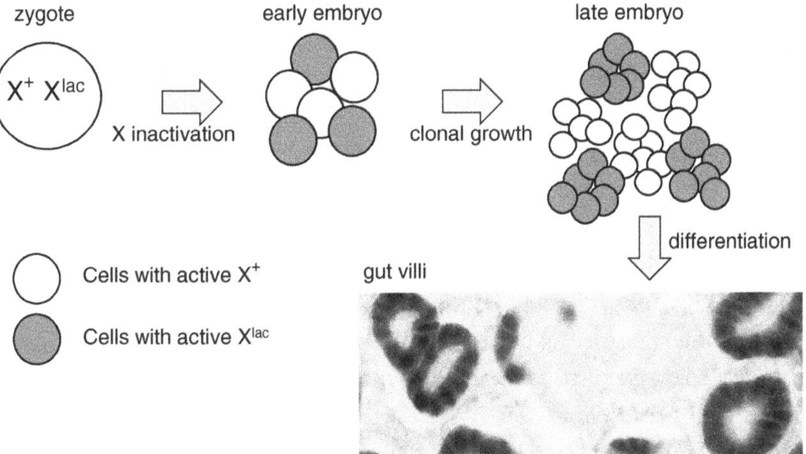

Figure 5.5 Patterns of X inactivation in tissues of female mice heterozygous for a transgene. In female mice heterozygous for a transgene on the paternal X the lac transgene produces a dark blue colour in cells in which the X^{lac} is active, but no colour in cells with an active X^+. Random inactivation occurs in the early embryo, producing a mixture of blue and white cells. As the embryo grows, X inactivation status is maintained through mitosis, so cells multiply to form blue and white clones. These clones undergo differentiation to form tissue with blue and white patches or stripes, according to patterns of cell migration; here they form blue and white rings that will differentiate into gut villi. Photograph supplied by Dr Seong Sen Tan, University of Melbourne.

at different developmental stages. The distribution was consistent with the idea that random X inactivation occurred early in embryogenesis; however, it occurred at different times in different tissues where there was a small pool of progenitor cells. Once inactivation had occurred, it was extremely stable in each cell clone. Coloured stripes and blotches represented patterns of cell movement within the developing embryo.

More recently, and even more spectacularly, mice have been constructed with a gene expressing a red fluorochrome on one X and a green fluorochrome on the other. Again, mosaicism of all tissues was confirmed, and the timing of inactivation in different tissues could be reckoned from the patch sizes and distributions.

The predictions of randomness have also been verified in humans. Cells from women heterozygous for G6PD isozymes were cloned. The proportion of the two types of clones was roughly 1:1, as you would expect if choice of the X to be inactivated were random within a limited cell population.

As for any stochastic event, extreme distributions are sometimes seen. For instance, women heterozygous for mutations of the X-linked gene coding for the red-green visual pigment (conferring colour blindness on half their sons) are occasionally colourblind in one eye. This suggested that the pool of progenitor cells was small, and subsequent analysis showed it may be as low as 8.

Extremely skewed expression ratios (anything outside a 70:30 ratio) are usually due to either a mutation or rearrangement of the X. This can lead to an unbalanced genotype and cell lethality. A deleted or rearranged X that leads to an unbalanced gene complement is invariably inactive because cells in which the abnormal X is active die. Thus, these skewed ratios are the result, not of preferential inactivation, but of cell selection against an unbalanced complement.

In some mice, however, skewed inactivation is genetically determined, and might hint at how the choice is made. More than fifty years ago, mutant mice were described that had heavily skewed distributions. This skewing was shown to be a function of alleles at a locus called the 'X controlling element' (*Xce*). These *Xce* alleles were proposed to form a pecking order of different strengths, so that in a hybrid, the stronger allele renders its X chromosome more likely to remain active. This explains why in females homozygous for one or other of the alleles, X inactivation was random, but in a female heterozygous for two different Xce alleles, the weaker allele is preferentially inactivated. The mouse *Xce* locus was originally mapped by genetic recombination to a site on the X and its location has now been narrowed down to a few hundred kilobases.

Evidence for an *XCE* equivalent in humans is equivocal. Rare families have been described with an inherited skewing of X inactivation. For instance, a family has been reported in which three females heterozygous for blood clotting factor F8 deficiency all suffered from haemophilia.

Both in mouse and human cells, there are epigenetic factors that affect the choice of X, including age-related epimutations. So it is likely that many factors, genetic and epigenetic, conspire, or compete, to alter the chance that one or other X is inactivated.

Thus, the choice of X to be inactivated is itself under genetic control, at least in mice, and this is influenced by many epigenetic factors. One factor that may affect the choice of the X to be inactivated is the parental origin of the X.

5.10 Imprinted X Inactivation

Parent-specific inactivation is known for many autosomal genes (7.14) and is called 'genomic imprinting'. The first evidence for parent-specific gene silencing in mammals came from the discovery that in kangaroos it was always the paternal X chromosome that was inactivated (Figure 5.6).

When marsupial X chromosome inactivation was first detected in the 1960s by the presence of a late-replicating X chromosome in blood cells of female kangaroos, we all assumed that kangaroo X chromosome inactivation would prove to be random like that in mice and humans. However, enzyme studies revealed that it was always the X chromosome from the male parent that was inactive in a heterozygote. For instance, blood cells from a female heterozygous for different *PGK* alleles expressed only the allele from her mother. Paternal inactivation was confirmed by studying replication timing in hybrid kangaroos that had X chromosomes of different sizes and shapes. The late-replicating X was always the one that was inherited from the male parent.

Paternal X inactivation was first thought to be a peculiarity of marsupials. However, in 1975 it was discovered that it was always the paternal X that was late-replicating in rat and mouse extra-embryonic membranes. Isozyme expression confirmed that only alleles from the maternal X were expressed in the trophoblast that gives rise to these tissues.

Enzyme studies confirmed that an early wave of X inactivation occurs in the trophoblast of mouse embryos, and that the inactivated X always came from the male parent. There is evidence, too, that paternal X inactivation occurs in the extra-embryonic membranes of bovine embryos. However, many studies found no evidence of paternal inactivation in humans or other eutherian mammals. Expression of both alleles is seen in the placenta and trophectoderm.

There ensued a long debate as to whether the paternal X was inactivated during male meiosis and simply stayed inactive in the egg. However, alleles in the paternal X were shown to be expressed soon after fertilization and there is a brief 2X-active stage. Thus, the paternal X, though not inactive, retains some kind of 'imprint' from its parental source that marks it for silencing a few cell divisions later.

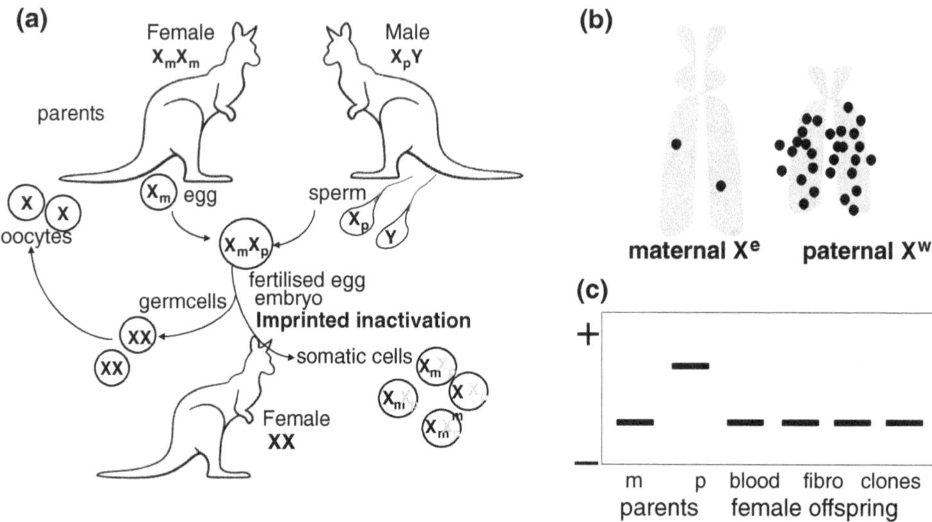

Figure 5.6 Imprinted inactivation in marsupials. (a) In kangaroos and other marsupials, inactivation in the somatic cells of females is not random, but imprinted. It is always the X from the male parent (paternal X, X_p) that is inactivated in the embryo. This was the first evidence of imprinted gene silencing in mammals. (b) Sketch of X chromosome spread from a species hybrid kangaroo in which the two X chromosomes (maternal Xe, paternal Xw) can be distinguished by the presence or absence of a blob of heterochromatin near the centromere. All cells showed late replication of the paternal X, shown by autoradiographic detection of tritiated thymidine supplied during the last part of the DNA synthesis period. (Drawn from Sharman 1971 Nature 230:231). (c) The kangaroo PGK enzyme, encoded by an X-borne gene, has two forms, separable by gel electrophoresis. The mother (m) has the slow form and the father (p) the fast form. Their female offspring are heterozygous at the *PGK* locus. However, only the maternal (slow) form is expressed in blood, attesting to silencing of the paternal X in all blood cells. In cultured fibroblasts the maternal form is expressed strongly, and a trace of the paternal form is seen, in single-cell clones as well as the mass culture. This means that each cell is expressing the maternal form strongly and the paternal form very weakly.

Paternal X inactivation in the extraembryonic membranes could be directly observed in the line of transgenic 'blue' mice (Section 5.9). When the transgene was inherited from the male parent, the trophectoderm in heterozygous female embryos was all blue, confirming the paternal X inactivation in extra-embryonic membranes.

What happens if there is no paternal X in the embryo? It is possible to suck out the sperm nucleus (paternal 'pronucleus') from a fertilized mouse egg and replace it with one from an egg. This produces a 'gynogenetic' embryo with two maternal X chromosomes. There is no early wave of X inactivation in these embryos. Development of extraembryonic membranes is poor and the embryos fail to implant, suggesting that expression of genes from two active X chromosomes may contribute to the eventual death of gynogenetic embryos.

Thus, for some eutherians and all marsupials, the parental source of the X chromosome determines the choice of the X to be inactivated.

5.11 Inactivation Spreads

One remarkable property of the inactive X is that silencing seems to spread throughout the whole X, and even into an autosomal segment fused to the X. Importantly, the signal acts only on regions that are physically joined; we call this '*cis*-acting control' of X inactivation.

An early observation was that when a bit of an autosome became attached to the X ('translocated'), autosomal genes near the fusion point became inactive too. This was first observed as mosaic expression of autosomal coat or eye colour genes in mice. For instance, when the normal allele of the albino gene is shifted from its usual site on mouse chromosome 7 to the X chromosome, it is silenced along with the X into which it is inserted, producing white mutant patches in the fur of heterozygous females.

Genetic inactivation of a translocated autosomal region may be accompanied by cytological changes, such as late replication. The spreading effect into autosomal regions may act over a considerable range, since cytological changes are seen over several G-bands, and genetic inactivation can extend many megabases into the autosomal region in mice and humans.

Silencing of translocated autosomal genes is most effective nearest the fusion point and seems to wane with distance from the X region. However, it is not consistent. X inactivation can even skip over non-inactivated regions to affect genes on the other side. For instance, the insertion of a bit of autosome into the mouse X did not disrupt inactivation of the distal segment, nor did inversion that placed inactivated genes on the far side of the PAR. Thus, the inactivating signal can be transmitted through a non-inactivated region.

The insertion of transgenes into the mouse X provides another way to test how silencing may spread. When large (200 kb) pieces of DNA from an autosome were inserted into the mouse *Hprt* gene; most were subject to X inactivation. Thus, inactivation can spread across 200 kb. However, inserts of some autosomal transgenes, or of an X escaper gene, or a chicken gene were not inactivated, implying that they must lack some element required to respond to the inactivating signal, or contain an element that blocks silencing.

This raises the question – how is the inactivation signal propagated along the enormous length of X chromosome? Let alone into autosomal regions and through insertions? This would seem to require that inactivated loci or regions have some kind of response element. It was suggested that the X chromosome contains 'booster elements' or 'waystations' which renew or amplify the signal along the X, and why it might become weaker with distance from the fusion point.

Perhaps repetitive elements acted to boost the silencing signal along the X? The X chromosome is replete with repetitive elements, and Mary Lyon suggested that these may serve to boost the inactivation signal in cis. Of these, by far the most frequent are Long Interspersed Nuclear Elements (LINEs), in particular the subclass L1.

This attractive hypothesis is supported by the observation that L1 density on the X is double that of autosomes. LINEs are concentrated on the X in humans and mice and even elephants (but not marsupials). LINE density is more concentrated in the older conserved region of the X and less in the region that was recently added to the X in eutherians. LINEs are depleted in regions containing escaper genes that are not subject to inactivation. This hypothesis predicts that spreading of inactivation into attached autosomal regions doesn't get very far because of the low LINE density.

However, the jury is still out on the question of the reality, and the identity of way stations. It is possible that L1 enrichment on the X is simply a reflection of the low recombination rate of the X, rather than its inactivation.

Thus, inactivation spreads in cis to include the whole X except the PAR. It even spreads into autosome regions that are fused or inserted into the X. But the basis of spreading and existence of waystations or booster elements remains controversial.

5.12 X Chromosome Rearrangements Define an Inactivation Centre

The capacity for X chromosome inactivation to spread provided a means of identifying an inactivation centre from which the signal emanated.

A critical observation was that if a mouse X chromosome was broken, only one of the two halves could undergo inactivation and its attendant cytological changes. This gave Mary Lyon the idea that there was a single centre on the X chromosome from which inactivation spreads. Separation of bits of the X from the centre interrupted the inactivation signal, so that genes separated from the centre by fission or translocation are no longer subject to inactivation.

This hypothetical X Inactivation Centre (XIC) would explain the coordinated nature of inactivation, the loss of inactivation on chromosome fission, and its spreading into autosomal regions attached to the X. The observation that inactivation dissipates with distance from the breakpoint suggested that the signal is progressively exhausted.

Just what constitutes the XIC, and how does it propagate its signal along the enormous length of X chromosome? A first step in the identification of the inactivation centre and the investigation of the inactivation signal was the detailed mapping of XIC. The concept of an X inactivation centre predicted that all structurally rearranged X chromosomes capable of being inactivated should share a definable region of the X.

Mapping XIC (Figure 5.7) is made possible by the many cases in which an X chromosome has been internally rearranged by inversion, broken in two (fission), or broken and fused to parts of an autosome as the result of translocation (Section 1.9).

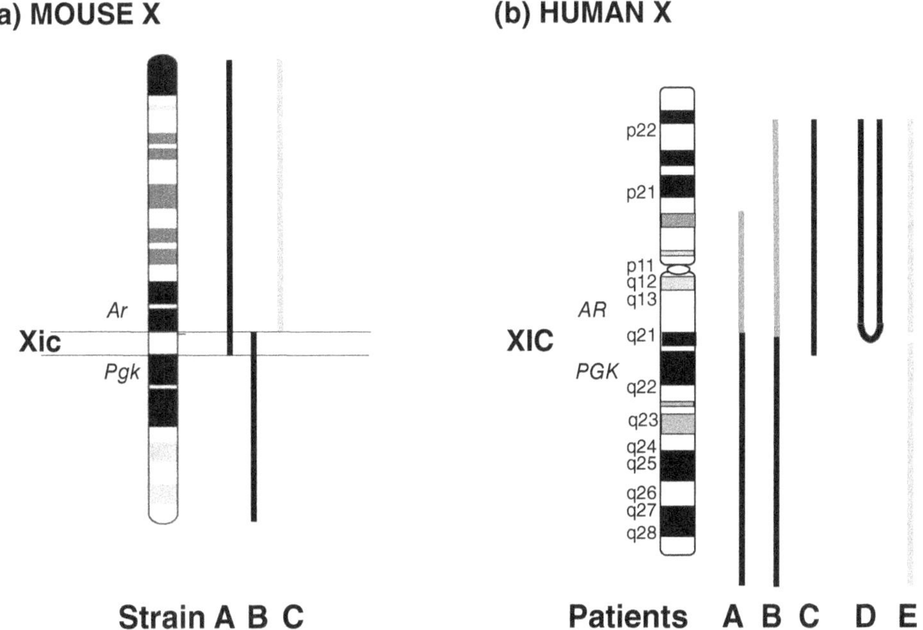

Figure 5.7   Mapping the X inactivation centre XIC on the mouse and human X. (a) The mouse X inactivation centre was mapped by noting inactivation of the X (by differential staining) in strains of mice with partial X chromosomes; strain A has a deletion that does not affect inactivation (black), and B/C are the two halves of an X chromosome split by a translocation; only the bit carrying Xic can inactivate (black), but the bit without the Xic cannot (grey). (b) Partial X chromosomes from patients (A–D) with X chromosome translocations or deletions were isolated in cell hybrids and tested for their activity by assessing replication timing and/or the activity of genes on the X portion. Patients A and B had parts of the long arm of the X translocated to an autosome (dark grey); these parts underwent inactivation (black), so must contain XIC. Patient C had a deletion of most of the long arm, yet the X was inactivated (black), and patient D had an isochromosome composed of two copies of the short arm and the top bit of the long arm (encompassing XIC) joined together, so able to inactivate (black). These abnormal X chromosomes were inactivated (black), so must contain the XIC. The only bit of the X in common between these patients is a small region of band Xq1.3, between genes *AR* and *PGK*, which must contain XIC. This was confirmed by the failure to inactivate (grey) an X chromosome in patient E, who had a tiny deletion within Xq1.3.

A putative mouse *Xic* could be mapped to particular chromosome bands on the X by the position of breakpoints beyond which loci translocated from the X escaped inactivation. The mouse *Xic* region was mapped to a position between the breakpoints for two rearrangements, which lie between *Ar* on the centromere side and *Pgk1* on the telomere side of the mouse X chromosome. This pinpointed *Xic* to a region of about 8 Mb (Figure 5.7).

Independently, the X controlling element (*Xce*) that contained genes that affected the strength of X inactivation, mapped to nearly the same spot on the mouse X.

The human XIC could also be quite precisely mapped with respect to breakpoints of rearranged human X chromosomes. The presence of a single X inactivation centre on the long arm of the human X was originally proposed on the grounds that isochromosomes (duplicated chromosome arms) of the X short arm were never observed. It was suggested that these could not be inactivated because they lacked an inactivation centre, so their presence was lethal. Indeed, tiny non-inactivated marker chromosomes derived from the human X are associated with severe mental retardation and other anomalies. The absence of deletions of the X long arm near the centromere suggested that *XIC* must lie in this region of the human X.

To pinpoint the human X inactive centre, a detailed analysis was made of four structurally abnormal human X chromosomes that were inactivated and must therefore all contain the putative XIC (Figure 5.7). These abnormal X chromosomes were isolated from their normal partners in rodent-human cell hybrids so that their extent could be precisely mapped by noting the expression (or non-expression) of human DNA markers. These abnormal X chromosomes shared only a small (1 Mb) region of overlap on human Xq13 between *AR* and *PGK*, so this region must logically contain XIC in humans as well as mice.

The hypothesis of a *cis*-acting control of X inactivation by a single X inactivation centre (XIC) led to the discovery of a controlling gene within this site (described in Section 6.7). This enormously advanced our understanding of X inactivation.

5.13 X Chromosome Reactivation in Female Germ Cells

Lyon predicted that the X must reactivate during gamete formation, and that inactivation occurs anew in the fertilized egg or embryo. This is basically correct, but the timetable is quite complex, and there are differences between the sexes, and between mammal species (Figure 5.8).

Classic genetic analysis showed that both X chromosomes are passed from the mother to her offspring in an active state. For instance, tabby female mice with patches of normal (dark) and mutant (yellowish) fur produced XY male offspring, half of which had dark fur, and half yellow fur. Even in mice with completely skewed inactivation, and female marsupials with paternally imprinted

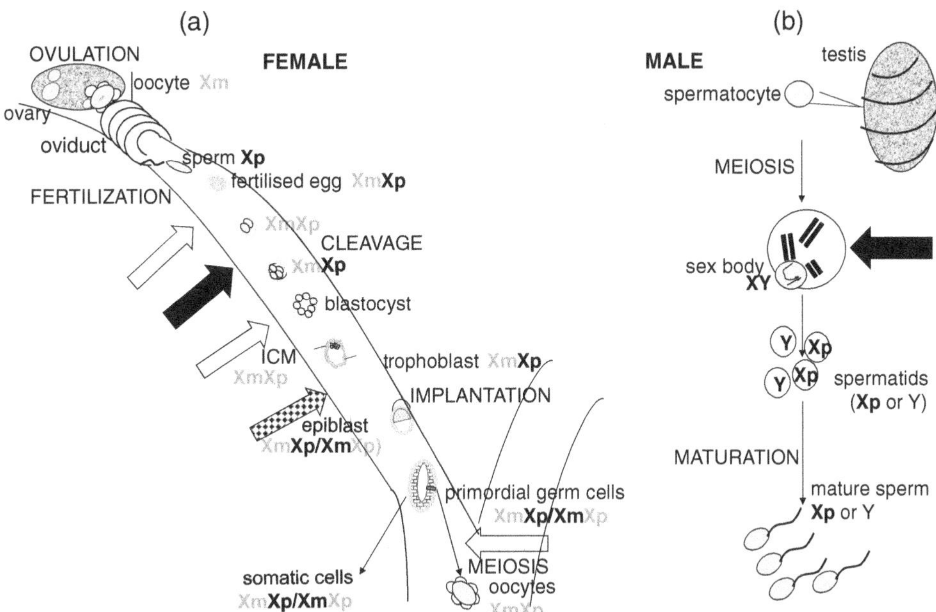

Figure 5.8   Cycle of X reactivation and X inactivation during mouse development. Reactivation (white block arrows) and inactivation (filled block arrow for paternal inactivation and chequerboard for random inactivation) in female mouse development produce active (grey) or inactive (black) maternal X (X_m) or paternal X (X_p). (a) This is a wild ride through the female reproductive tract. Start at ovulation. Oocytes (haploid, with one active maternal X_m) are released from the ovary and pass through the coiled oviduct (Fallopian tube) into the uterine horn (in most mammals, but not mice, the two uterine horns are fused). Fertilization occurs with a sperm carrying an inactive paternal X (X_p). The paternal X is reactivated over the next day or two, so both X_m and X_p are active in early cleavage. At the 4–8 cell stage, the paternal X is once again inactivated. It stays inactive in the cells of the blastocyst (trophoblast) that are destined to become extra-embryonic membranes. In the inner cell mass (ICM) of the blastocyst (which is destined to form the embryo proper), the paternal X is reactivated briefly, then random X inactivation sets in so that all the somatic cells of the developing embryo have either an active maternal or paternal X, and the adult female is a mosaic. The primordial germ cells are initially randomly inactivated, but the X is reactivated before meiosis occurs in the embryo, so that oocytes all have a single active X. (b) In males, meiosis occurs in the testis throughout adult life. The X and Y chromosomes are inactivated (MCSI) during meiosis, forming a sex body, and the paternal X_p stays inactive until after fertilization.

inactivation, half the progeny received and expressed genes on the X that was inactive in the mother.

This could mean that there is a dedicated germline, a population of cells set aside early in development within which the X remains exempt from inactivation. This was originally assumed by Lyon (Figure 5.2) because it was known that in all mammals cells are set aside early in development which will develop into sperm and eggs.

However sensible this idea sounds, it was soon proved to be wrong. Germ cells were shown to arise from a lineage of cells which had already undergone inactivation, necessitating a reactivation step.

It is always difficult to observe what is happening in eutherian embryos because they are tucked inside the mother and hard to get at. Nearly all developmental studies have been done in the mouse, so to understand the cycle of X inactivation, we need some very basic information about mouse embryology.

Development of the mouse embryo is very rapid. All the first stages occur in the Fallopian tube. From the time the mature oocyte is released from the ovary (ovulation), it takes about a day for fertilization to occur to form a zygote. The zygote divides twice in the next day to reach a 4-cell stage at day 2, then over the next day it divides again to make ball of 8 cells.

Then things really start to happen; the embryo hatches out of its covering (the zona pellucida), and by day 4 the cells arrange themselves into a hollow ball (called the blastocyst) with an inner cell mass at one end. The outside cell layer (called the trophoblast) develops into extra-embryonic membranes. The inner cell mass quickly starts to differentiate into an inner core ('epiblast') that will become the embryo proper, and an outside layer that contributes to more extra-embryonic membrane. By day 4.5 the late blastocyst has made it into the uterus, and implants in the uterine wall (where it becomes even harder to get at). The embryo develops rapidly, and pups are born at 19 days.

In male and female embryos, progenitors of eggs or sperm ('primordial germ-cells') arise at about 7 days (40 cells). They originate at one end of the epiblast, a long way from the ridge of cells that will develop into the gonad. They proliferate and migrate forward, right through the hindgut, and into the developing gonad by day 10.5. The genital ridge is initially bipotential and begins its sex-specific development into ovary or testis at day 12.5.

Making haploid eggs and sperm from diploid primordial germ cells requires meiosis, the specialized reduction division (Section 1.6). In female embryos XX primordial germ cells begin meiotic divisions at day 13.5 and are all done by the time the embryo implants.

Early observations of a condensed blob in primordial germ cells first suggested that XX female germ cell progenitors have an inactive X. This was backed up by careful observations on single mouse embryos that the activity of X-linked enzymes was the same in XX as in XY and XO progenitor cells.

The inactive X must reactivate before germ cells reach the gonad because similar enzyme assays showed that activity became twice as high in XX than XO embryos. Even more convincing was the observation that in a female embryo heterozygous for two G6PD isozymes, the telltale heterodimer band that means both alleles are expressed in the same cell appeared at about day 11 in germ cell progenitors, indicating reactivation. More recent studies using DNA SNPs also detect expression of alleles on both X chromosomes before and during female meiosis.

This implies that the inactive X is reactivated during migration of primordial germ cells or their arrival in the gonads of female mouse embryos. Reactivation is a slow process, occurring over days, and is not necessarily uniform between genes, or between cells.

Thus, by the beginning of meiosis, both X chromosomes are active. Meiosis segregates these two active X chromosomes into different oocytes, so each contains one active X and is ready for fertilization.

Studies of human embryos also showed that one X is randomly inactivated in primordial germ cells but not in oocytes. Single-cell RNA sequencing of the transcripts of X-borne genes in human primordial germ cells and throughout meiosis recently showed that some cells still had an inactive X even during meiosis, and were still reactivating it. Thus, reactivation is prolonged and rather heterogeneous in human primordial germ cells.

We conclude, therefore, that female primordial germ cells arise from cells in which inactivation has already taken place. The X is reactivated during germ cell migration into the developing ovary, and both X chromosomes are passed into egg cells in an active state.

5.14 X Chromosome Inactivation in Male Germ Cells

The single X chromosome in XY males is active in all somatic tissues. But it becomes inactive in sperm and must be reactivated in the zygote after fertilization.

It is much easier to study development of sperm than eggs because testes are more accessible than ovaries, and meiosis occurs in adult mice. Many cytological and enzyme studies show that the single X chromosome in XY males is active in the primordial germ cells which develop into spermatocytes (cells that give rise to sperm).

But at male meiosis, the X, along with the Y, becomes heterochromatic and inactive in male germ cells. Meiotic Sex Chromosome Inactivation of the X (MSCI) was first recognized cytologically and later confirmed by analysis of the expression of X-borne genes throughout spermatogenesis.

At male meiosis the X and Y chromosomes are condensed into a special compartment called the sex chromatin body, or sex vesicle. This is seen as a dark body bulging out of cells during the meiotic stage when homologous chromosomes have paired and are condensing.

MSCI is absolutely essential for spermatogenesis. Mice with mutations in the genes that are involved are sterile, because meiosis does not get beyond the first stage.

This X (and Y) inactivation during male meiosis seems to be a special case of a general mechanism to silence chromosome regions that are not safely paired at meiosis. It is called Meiotic Silencing of Unpaired Chromosomes (MSUC). Autosomal regions deprived of their homologue by some genetic accident are also dragged into heterochromatic bodies and undergo the same molecular changes, in female as well as male meiosis. MSUC is probably a general mechanism that protects against non-homologous recombination and rearrangement, damaged DNA or germ cells in which pairing was defective. MSUC may explain why XO females are sterile, because the unpaired X is inactivated and oocytes cannot develop.

MSCI involves the proteins of the synaptonemal complex, which accumulate along the unpaired parts of the X and Y. This seems to be a signal for other proteins to bind, including two that are involved in DNA repair.

Most X-linked genes are silenced during male meiosis, but some genes escape, perhaps by self-pairing sequences that read the same forwards and backwards (palindromes) on the sex chromosomes. In mice, spermatid-expressed genes that are massively amplified on the X may use this method to dodge prolonged inactivation during meiosis.

Indeed, the inactivity of X genes at meiosis poses a problem if they are required for spermatogenesis. One evolutionary solution has been to make autosomal copies of vital genes that can remain active through meiosis, as I discussed earlier (Section 4.2).

In summary, an active X is inactivated in the male germline, and reaches the unfertilized egg in an inactive state. This is completely the reverse of female germ cells, in which an inactivated X chromosome is reactivated. The outcome of all this switching is that XX fertilized eggs have an active maternal X and an inactive paternal X.

5.15 X Chromosome Inactivation during Early Embryonic Development

What happens to these X chromosomes at fertilization and the earliest stages of divisions of an XX zygote (Figure 5.8)?

Lyon postulated that both X chromosomes are active in the fertilized egg, and that inactivation occurs randomly in the blastocyst around the time of implantation, when there are about 100 cells in the embryo. The low number of progenitor cells would explain why mosaicism for X-linked characters is seen as large blotches.

Indeed, fertilized eggs and cells from early cleavage stages from humans, macaques, cats and rabbits lacked cytological signs of an inactive X (no late replication, no sex chromatin). These were first apparent at the blastula stage, just before implantation of the embryo. Studies of the activity of X-linked enzymes bore this expectation out. In human embryos heterozygous for the dimeric enzyme G6PD, the presence of a heteropolymer band revealed that both alleles are active in the same cells in chorionic villi. This seems to be the most common pattern in eutherian mammals.

In mice, too, there is evidence for a 2X-active stage before implantation. Cells of pre-implantation mouse embryos lacked a heterochromatic and late-replicating X. XX female embryos had twice as much enzyme activity as XY males and XO embryos.

But X inactivation in the mouse embryo is more complex than we expected, undergoing an extra, early wave of inactivation that has puzzlingly different properties. As for other mammals, the inactive X delivered by the sperm is reactivated at fertilization. A brief 2X-active stage can be detected by enzyme dosage, and RNA FISH; nearly all blastomeres had two transcription signals, implying that they are 2X-active at this stage.

In rats and mice a first wave of inactivation was detected as early as the 4-cell stage. It is decidedly non-random, inactivating only the paternally derived X, inherited via sperm. All cells of the embryo underwent a wave of paternal inactivation, as shown by the expression of only the maternal allele. Single-cell analysis of alleles of many X-borne genes verifies that silencing of the paternal allele begins at the 4-cell stage. Thus, although the inactive paternal X reactivates at fertilization, it must retain an imprinted mark that makes it prone to inactivation in the early embryo (Section 7.7).

Microassays of enzyme activity showed that extra-embryonic tissues retain paternal X inactivation. But in pluripotent cells of the blastocyst, the paternal X is reactivated once again, and there is another brief period in which cells are 2X-active.

When the embryo implants at day 4.5 and pluripotent cells begin to differentiate, they undergo another wave of inactivation. But this time choice of the X to be inactivated is random.

In marsupials (Section 5.8) all the cells of the XX embryo undergo paternal inactivation, which is retained throughout life. The structure of the marsupial embryo is different from that of eutherian mammals; the blastocyst is a single-cell sheet (unilaminar), and there is no inner cell mass. Instead, large cells accumulate at one end of the blastocyst, then move to make a second layer. Cells again accumulate at this pole and form a third layer, from which the embryo and the extra-embryonic membranes develop. Embryos, as well as extra-embryonic membranes, show paternal X inactivation. There is some evidence that X inactivation occurs early in marsupials, in the unilaminar blastocyst.

Thus, despite expectations that X chromosome inactivation is a uniform and efficient system, we find that different mammals, and different tissues, adhere to quite different timetables. Waves of reactivation and inactivation, random and paternal, occur that are different in rodents and marsupials from that in most eutherians.

5.16 Modelling X Inactivation in Stem Cells

All the interesting changes of state of the X chromosome occur in germ cells and early embryos, and they are intimately linked to differentiation. The inaccessibility of mammalian embryos makes it difficult to study the mechanism of X chromosome inactivation. A big advance has been the development of stem cell lines in which changes of activity occur and can be studied in a Petri dish.

Stem cells are simply cells that have not yet differentiated. They can replicate, and they can differentiate into many types of tissue (called pluripotency). They are present in the embryo and hang around in most tissues throughout life, being vital for growing new tissue and for running repairs.

We can isolate stem cells from embryos, adult tissue (for instance, bone marrow), or induce pluripotency in adult cells by reversing the differentiation process. They can be grown indefinitely in an undifferentiated state under the right culture conditions, then allowed to differentiate into a mixture of cell types.

Embryonic Stem Cells (ESC) are derived from mouse and human embryos by growing blastocyst cells in the presence of feeder cells or protein factors. Originally they were isolated from embryonal carcinomas derived from an ovarian tumour (teratoma) or by implanting an early preimplantation embryo into another body site. ESCs can grow into any type of differentiated cell in the body, but not placenta or extra-embryonic membranes. Adult stem cells can differentiate into a limited range of cell types; for instance, bone marrow stem cells can differentiate into a range of blood cell types, but not neurones or liver cells. Stem cells can also be isolated from the inner cell mass (epiblast), but these have already undergone X inactivation.

We can also produce stem cell lines from mouse primordial germ cells. These Embryonic Germ Cells (EGC) can be induced to proliferate and to become oocytes that undergo the first steps of meiosis (but go no further unless transplanted into a gonad).

It is possible also to induce pluripotency in mouse somatic cells. Normal cells, for example skin cells, are made transgenic for 'pluripotency genes' (the big four are *Oct4, Sox2, Klf4* and *c-Myc)* delivered by a virus. Expression of these genes during cell growth can completely reprogram mouse cells into induced Pluripotent Stem Cells (iPSC).

All these types of mouse stem cells have been used to investigate both X chromosome inactivation and reactivation. Mouse teratocarcinoma lines derived from embryonal carcinomas were the first to be used. Some of these retained two active X chromosomes, detected by enzyme dosage and the expression of both alleles in heterozygous clonal lines. They were useful for studying the kinetics of X inactivation, and for early molecular studies.

Stem cells may also be derived from the trophoblast embryonic lineages of a mouse blastocyst. On differentiation or injection into a blastocyst, these cell lines show paternal X inactivation, so they can be used to study imprinted X inactivation.

ESCs, EGSs and iPSCs derived from embryos, primordial germ cells or somatic cells of female mice usually have two active X chromosomes. On differentiation, one of the two X chromosomes undergoes random inactivation with dynamics similar to those of implanting embryos. Spectacular confirmation of complete reactivation was the reprogramming of mouse cells with a green fluorescent protein transgene on the inactive X. IPSC cells were all green.

Very detailed studies of X inactivation have been done on differentiating mouse stem cells. For instance, allele-specific RNA-seq was used to show that genes were inactivated asynchronously when mouse ESC were differentiated. Escaper genes never showed any evidence of silencing, implying that 'escape' is really escape and not reactivation.

Reactivation of an inactive mouse X can likewise be studied in detail during reprogramming of somatic female cells to a pluripotent state. In mouse iPSCs, X reactivation occurs only after reprogramming is complete, and its steps closely mirror the steps of regaining pluripotency. It results in the complete reactivation of the inactive X, which seems to be obligatory for maintaining the pluripotent state.

Human ESC and iPSC have also been developed and studied. However, their X inactivation and differentiation status can be variable and unstable during culture. Most have already undergone X inactivation. There is usually a lot of variation within a culture. A few lines can undergo X inactivation, but genes on the inactive X chromosome may be partially expressed.

Despite this heterogeneity and instability, human ESC and iPSC cells have contributed some useful insights. RNA-seq shows that cloned cells that have undergone inactivation all express genes from the same X, and in accord with Lyon's predictions that inactivation is stable and somatically heritable, clones stably express genes on one X.

X reactivation in human iPSCs is also asynchronous and unstable, occurring at different times, to different extents, in different lines, and is susceptible to culture conditions. However, the system has helped to clarify the steps of reactivation. Stem cells must contain a transactivating factor that enables reactivation, since fusion with 1X-active cells reactivated the inactive X of differentiated mouse cells. The inactive X lost its late replication characteristic and re-expressed the inactive *Pgk* allele.

Stem cell systems thus provide a valuable model for studying the molecular basis of X chromosome inactivation and reactivation.

5.17 Conclusions

From its earliest invocation, X chromosome inactivation had characteristics that seemed quite magical and defied mechanistic explanation.

Uniquely, inactivation affects a whole chromosome, silencing about a thousand physically linked genes. It can spread along the chromosome, and even into fused autosomal regions, but only to physically joined sequences.

Inactivation can affect either X chromosome, and the same X is active or inactive at different stages, so inactivation must be 'epigenetic' rather than involve changes in DNA base sequence. Inactivation in the eutherian embryo is random, but once the choice of X is made, inactivation is stable and somatically heritable. Inactivation in marsupials, however, always silences the X derived from the male parent, and paternal X inactivation is also a feature of the first wave of inactivation in rodent (and bovid) embryos.

The role of the X chromosome in disease must take inactivation into account. The mosaic phenotype, and its variability, explains the variability in the severity of recessive X-linked mutations in heterozygous women. The rare diseases that affect women only, such as Rett syndrome, are actually an intermediate phenotype of heterozygous females: XY male embryos with the mutant allele die before birth. And knowing now that 15% of genes on the human X escape inactivation, but only 3% of genes on the mouse X escape, explains why XO women have much more severe morphological and reproductive anomalies than XO mice.

Inactivation is tightly linked to differentiation, and is programmed into developmental stages. There are waves of inactivation and reactivation during the life cycle; rather than being uniform and efficient, they may be different in different animals and different tissues. The development of stem cells provides a good model of inactivation and reactivation during mammalian development that can be used to study details of the process at the molecular level (Chapter 6).

Thus, X chromosome inactivation has become perhaps the best model system in which to study epigenetic control of gene activity, as I shall describe in Chapter 6. X chromosome inactivation also has a special place in the sociology of science, in that it seems to have attracted outstanding women scientists, starting with Mary Lyon, and including pioneering luminaries such as Marilyn Monk, Barbara Migeon, Gail Martin, Christine Disteche, Edith Heard, Carolyn Brown and Jeannie Lee.

FURTHER READING

Books

Migeon BR, 2007. *Females Are Mosaics: X Inactivation and Sex Differences in Disease*, New York, Oxford University, 2007

Classic Papers

Lyon MF, 1961. Gene action on the X chromosome of the mouse (Mus musculus *L*). *Nature* 190: 392–373

Monk M, Harper ML, 1979. Sequential X chromosome inactivation coupled with cellular differentiation in early mouse embryos. *Nature* 281: 311–313

Sharman GB, 1971. Late replication in the paternally derived X chromosome of female kangaroos. *Nature* 230: 231–232

Reviews and Research Articles

Berletch JB, Yang F, Disteche CM, 2010. Escape from X inactivation in mice and humans. *Genome Biology* 11, 213

Bermejo-Alvarez P, Ramos-Ibeas P, Gutierrez-Adan A, 2012. Solving the X in embryos and stem cells. *Stem Cells and Development* 21: 1215–1224

Carrel L, Willard HF, 2005. X-inactivation profile reveals extensive variability in X-linked gene expression in females. *Nature* 434: 400–404. doi: 10.1038/nature03479.

Gartler SM, Riggs AD, 1983. Mammalian X chromosome inactivation. *Annual Review of Genetics*. 17: 155–190

Heard E, Disteche CM, 2006. Dosage compensation in mammals: fine-tuning the expression of the X chromosome. *Genes & Development* 20: 1848–1867. doi: 10.1101/gad.1422906

Latham K, 2005. X chromosome inactivation in preimplantation mammalian embryos. *Trends in Genetics* 21: 121–127

Lyon MF, 1999. X chromosome inactivation, *Current Biology* 9: PR235–R237

Rodriguez YM, Borensztein M, 2023. X-chromosome inactivation: a historic topic that's still hot. *Development* 150: dev202072. doi:10.1242/dev.202072

Sun Z, Fan J, Wang Y., 2022. X-Chromosome inactivation and related diseases. *Genetics Research (Cambridge)* 1391807. doi: 10.1155/2022/1391807

Takagi N, Sasaki M, 1975. Preferential inactivation of the paternally derived X chromosome in the extraembryonic membranes of the mouse. *Nature* 256: 640–642. doi: 10.1038/256640a0

6

Molecular Biology of X Chromosome Inactivation

The mysteries of X inactivation – its counting mechanism, the choice of the X to be inactivated, the extraordinarily stable inactivation and spreading of inactivation over immense molecular distances – have attracted intense study of the molecular mechanisms of X inactivation.

Several early proposals were made to account for the stable and heritable change in activity over such a vast expanse of the genome. Differences in binding to non-histone proteins could be set up, and maintained by cooperative binding. Differential DNA methylation was proposed to be maintained by the preference of methylases for a hemi-methylated substrate. Alternatively, differential modification (phosphorylation, acetylation) of protein rather than DNA was suggested. Late replication could be imagined to be a cause, rather than an effect of inactivity if it sequestered one X from transcription factors. Likewise, physical containment of the condensed interphase mass, or maintenance of a condensed superstructure, was proposed to be the cause of inactivation.

It now appears that everybody may be right – the hyperstable eutherian X inactivation seems to result from cooperation between molecular changes at many levels.

6.1 Yes, the Inactive X Is Transcriptionally Repressed

Lyon predicted that inactivation of alleles on the X chromosome results from repression at a transcriptional level. This idea was accepted for decades, because it provided the only logical explanation for the coordinate inactivation of physically linked genes. What else could it be? (Well, in those days we didn't know about the many means of post-transcriptional repression).

However, it was difficult to test this prediction directly, and the sole evidence was the dearth of incorporation of radiolabelled RNA precursors into the sex chromatin body.

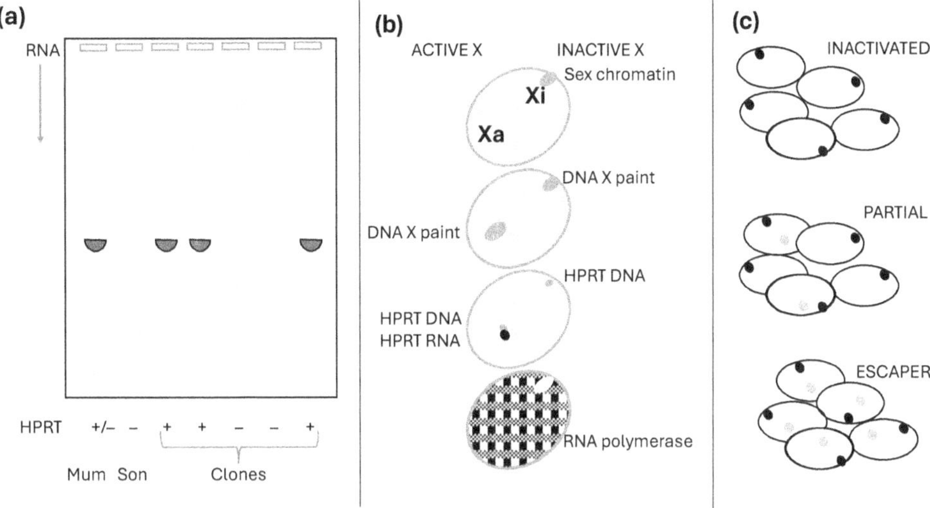

Figure 6.1   The inactive X chromosome is not transcribed. (a) Northern blot analysis of RNA transcribed by white blood cells from a boy with the sex-linked Lesch-Nyhan disease (*HPRT–*) and his heterozygous mother (*HPRT+/-*). Cells from the affected boy show no *HPRT* transcription. Cells from his mother show normal transcription, but clones grown from her single-cells showed either normal or no transcription. In three clones the *HPRT+* allele was on the active X and was transcribed (+) and in two the *HPRT+* allele was on the inactive X and was not transcribed (–). (b) Different ways of detecting the active X (bottom left of nucleus) and inactive X (top right of nucleus). From top: X-specific DNA paint reveals both X chromosomes, cytological stain reveals sex chromatin (inactive X), probing *HPRT* DNA reveals the gene on both chromosomes (grey spot) but probing for *HPRT* RNA lights up the HPRT locus only on the active X (black spot), tagging RNA polymerase stains the whole nucleus except for the inactive X (a hole in the labelling). (c) Probing female cells for RNA transcribed from different genes reveals inactivated, partial or escaper genes. Genes not transcribed on the inactive X produce only one signal, genes partially active on the inactive X show some nuclei with two signals (often one is weaker, shown in grey) and genes that escape inactivation show two signals per nucleus.

It was 25 years before a direct test of transcriptional inactivation could be made at the molecular level (Figure 6.1a). The first evidence came from the (by now rather crude) Northern blot analysis of specific mRNA, done on my sabbatical with Stan Gartler in Seattle. White blood cells from a boy with the Lesch-Nyhan syndrome (caused by deficiency of the sex-linked enzyme HPRT) transcribed no normal *HPRT* mRNA. Cloned populations of cells from his normal mother, who was heterozygous for this condition, were of two types, according to which X was active. One population expressed HPRT (and grew in HAT selective medium); the other expressed no HPRT (and grew in medium containing the poisonous substrate thioguanine). Northern analysis showed that the HPRT+ clones transcribed normal

HPRT message, but the HPRT-deficient clones transcribed no HPRT mRNA. In these cells, the normal HPRT allele on the inactive X was present but was not transcribed into RNA.

Although this verification of Lyon's prediction was critical for our understanding of X inactivation, it hardly caused a stir. Indeed, lofty *Nature* referees rejected the paper (despite Mary Lyon's endorsement) because 'what else could it be?'

Transcriptional repression of genes on the X was confirmed fifteen years later using PCR methods (Section 1.12) to detect transcription of different alleles from minimal material (single embryos or even single cells). Cells from a female heterozygous for allelic variants at one or more X-borne genes were used; often these were SNP markers (single nucleotide polymorphisms detected by sequencing RNA transcripts). DNA copies of 'reverse transcribed' RNA (RT-PCR) could be amplified using primers unique to the gene of interest. What's more, PCR products were differentiated by restriction enzyme cutting that distinguishes the transcripts from the two alleles. Or with judicious choice of primers, different alleles could be amplified as different sized fragments.

To avoid the problem of mosaicism, cells were taken from subjects with one X preferentially inactivated. Another important source was cell hybrids retaining a single active or inactive human X on a rodent background. Primers could be designed to amplify specifically from the human gene. For instance, mRNA from several human X genes was detected in hybrids with an active human X, but not in hybrids retaining an inactive X. In this way, transcriptional inactivation of alleles of many genes on the inactive human X was confirmed.

PCR could be applied to many different tissues, so that inactivation of tissue-specific genes, such as the muscle-specific *DMD* and liver-specific *F9*, could be studied. Silencing of the allele on the inactive X was then documented for many tissue-specific X-borne genes as well as the usual set of housekeeping enzymes.

The conclusion that X inactivation affects a single X was thoroughly vindicated by large-scale transcriptional studies of many allelic differences in human X chromosomes. These showed alleles from the same parent were concordantly transcribed or repressed.

In line with the lack of transcription of one X in human or mouse female cells, fluorescent antibody staining showed that RNA polymerase II (the enzyme that transcribes most genes) is excluded from the region that contains the inactive X, which is visualized using an X-specific DNA paint.

In summary, we now have watertight evidence that genes on the inactive mammal X chromosome are silent because they are not transcribed into RNA.

6.2 Control of X Chromosome Transcription

The new techniques of transcription analysis have been brought to bear on many of the genetic questions that phenotypic and cytogenetic analysis could not answer, such as the balance of X-autosome activity, the timing of inactivation during development and the surprising variation in the level of transcription between genes, between tissues, between individuals, and between species. This has led to a re-evaluation of the molecular basis of transcriptional inactivation.

As I mentioned in Section 5.1, Ohno long ago hypothesized that dosage compensation, as well as ensuring fair play between the sexes, must keep the balance between X-borne and autosomal genes by upregulating genes on the single X in males and the active X in females. Upregulation was first demonstrated by Christine Disteche, who compared the activity of a gene that had been moved from the X to an autosome in a mouse subspecies.

However, when it became possible to measure transcription of all the genes on the mouse X and compare them to autosomal genes, the relationship became muddier, and there were many debates about the suite of genes considered and whether upregulation depends on their ancestral origin. At least for the set of X genes preserved from the ancestral autosome, upregulation has been established; curiously this involves both regulation at the transcriptional and post-transcriptional level, involving several epigenetic mechanisms. Remarkably, the extent of upregulation is finely tuned to the extent of inactivation, so that upregulation is strong in cells with an inactive X, but absent in cells at 2X-active stages of embryonic (and ES cell) differentiation. Indeed, there is evidence that the degree of upregulation depends on the degree of inactivation in different tissues and stages.

Single-cell transcriptomics (isolation and sequencing RNA from single cells) offers a much more detailed account of the inactivation cycle than older studies using mashed up cell populations. This has clarified early events in both mouse and human embryogenesis, and pointed out real differences. As well as the occurrence of paternal inactivation in mouse but not human embryos, there are differences in the timing of lineage specification. X inactivation seems to occur more gradually in human embryos, and they go through a stage at which alleles of lineage-specific genes are coexpressed.

These techniques also document unexpected variation of transcriptional regulation of different genes in adults. Surveys of hundreds of genes in many tissues and many individuals are now available. Sensitive analysis of transcripts in single human cells reveals a core of about 700 genes that are solidly inactive in all tissues, and about 270 that escape fully or partially in one or many of 29 tissues (Section 5.8). And it confirmed that escape could be polymorphic; some escaper genes were completely silenced on the inactive X in some women, but transcribed in others.

In situ RNA hybridization can also determine whether a gene on the X is transcribed from one or both X chromosomes in individual cells. A fluorescently labelled DNA probe to the gene will hybridize efficiently to the site of primary transcription where there are many copies of the unprocessed RNA. Probing an autosomal gene produces two signals in nearly all cells, from both females and males. Probing an X-borne housekeeping gene such as *PGK* or *HPRT* produces only a single signal in XX female cells, as well as in XY male cells (Figure 6.1b).

RNA in situ showed that most genes on the human and mouse (and elephant) X chromosomes were 1X-active in nearly all female (as well as male) fibroblasts according with the observation that most genes are entirely silent on the inactive X. However, partially active escaper genes in all species showed heterogeneity. Rather than all nuclei showing one strong signal and one uniformly weak, the cell population contained different proportions of nuclei with one signal (1X-active) and nuclei with two (2X-active). This was first documented in cells from female kangaroos, which showed partial activity of paternal *PGK* (the first demonstration of partial X activity). The frequency of 1X-active cells was consistent across cultures and experiments, but for different genes it varied from 30 to 90%.

Thus, the different degrees of escape from inactivation in eutherian and marsupial females were not due to ratcheting down of transcription on the inactive X in all cells, but to the proportion of cells that were expressing both alleles. This paints a very different picture of transcription control, depending on the probability, not the rate of transcription. For different genes, the proportion of 2X active cells can be anything between 0 (fully inactivated) and 100% (fully escape) (Figure 6.1c).

The development of single-cell RNA seq (Section 1.12) made it possible to minutely examine the details of transcription, revealing that activity is due to the fraction of time that a gene is 'on' (burst frequency). Partial silencing is achieved by regulating burst frequencies on the X. This is consistent with the observation of stochastic transcription from partially active escape genes in eutherians and marsupials.

Thus, X chromosome transcriptional silencing is far more variable than has been appreciated, and is controlled, not by a rheostat-like ramping down of transcription, but a molecular on-off switch. Upregulation of X activity also involved several posttranscriptional mechanisms.

Observation of partial and variable expression of genes on the inactive X raise many questions. Does dosage difference of X-borne genes contribute to phenotypic differences between men and women? Escaping genes are expressed more highly in women than men, and the extent of this sex bias is different for different genes – and different women.

6.3 DNA Methylation and X Chromosome Inactivation

Great excitement accompanied the discovery that removing methyl groups from DNA could reactivate genes on the inactive X (Figure 6.2).

It was known by the 1970s that DNA could be modified by addition of methyl groups to cytosine-guanosine nucleotide pairs (CpG) in DNA, and clusters of these methylated sites occurred in 'GC-rich islands' near promoters of genes (Figure 6.2a).

Two groups of scientists (Art Riggs in Los Angeles and Robin Holliday and student John Pugh in London) independently suggested that one X was activated by methylation of cytidine residues in DNA. Addition of methyl groups could initially be accomplished by a *de novo* methylase, then differences could be perpetuated by a maintenance methylase with a predilection for sites methylated on one DNA

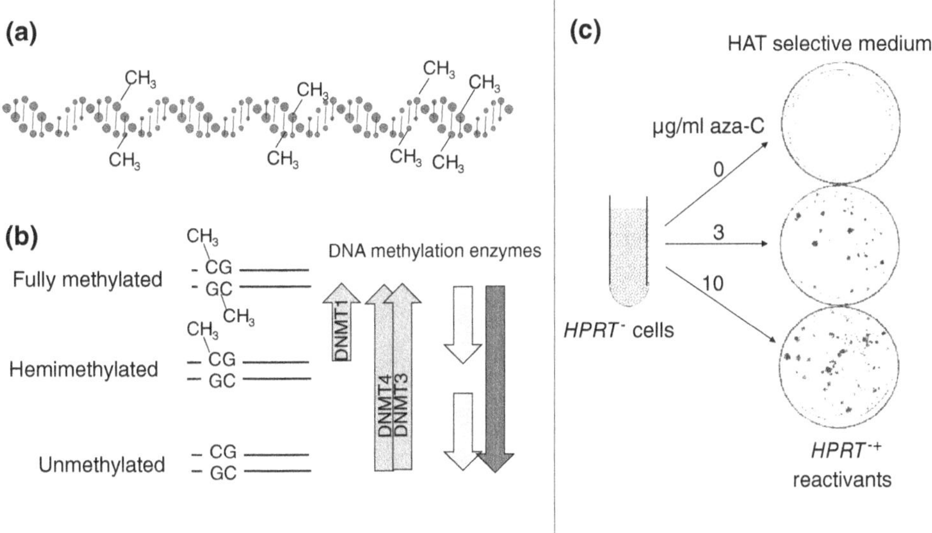

Figure 6.2 DNA methylation and X chromosomes inactivation. (a) DNA double helix with methyl groups (CH₃) enzymatically attached at the 5' position of cytosine residues in the DNA chain. Most methylated sites are CpG dinucleotides. (b) Methylation patterns for fully methylated, hemimethylated and unmethylated DNA. Methylation is established in the early embryo by *de novo* methylases DNMT3 and 4. During DNA replication, DNA methylation is retained by the old strand, and the new strand is methylated by enzyme DNMT1, which recognizes hemimethylated sites. Methylation of a site may be lost passively during replication (light arrows), or by specific demethylating enzymes (black arrow). (c) Experiment showing that removal of methyl groups from DNA reactivates genes on the inactive X. Mouse cells with a mutant *Hprt-* allele on the active X and a normal *Hprt+* allele on the inactive X express no HPRT and will not grow in HAT selective medium. Growth in methylation inhibitor 5-azacytidine causes dose-dependent reactivation, observable as HPRT+ clones growing in HAT medium.

strand but not the other. This was an appealing mechanism for X inactivation, because it explains how differences could be set up and self-maintained.

Methylation was first assessed laboriously by breaking DNA with enzymes that worked only on unmethylated DNA. Bisulfite sequencing made it possible to map methyl groups over larger stretches of genome, and it is now possible to map them over the entire methylome (Section 1.13).

These techniques showed that most CpG sites are methylated in mammalian cells, but, conspicuously, GC rich islands near active promoters are not. During development, waves of DNA demethylation and remethylation occur in the embryo (but not the extraembryonic membranes) as the genome is repatterned to silence different cohorts of genes. Particular genes that are turned on or off in development undergo changes in DNA methylation.

Several enzymes (DNA methyltransferases, called DNMTs) were discovered that transfer methyl groups onto cytosine residues in DNA. Two (DNMT3 and 4) add methyl groups to unmethylated CpG sites ('de novo'), and knockout experiments show that these are critical for life. Another (DNMT1) binds to DNA at sites that are methylated on one strand after DNA replication and adds a methyl group to the new strand. This means that once a CpG dinucleotide is methylated, it continues to be methylated every time the DNA is replicated (Figure 6.2b).

The hypothesis that X inactivation involved differential DNA methylation was triumphantly confirmed ten years later, except that the original hypothesis got it back to front – DNA methylation is associated, not with activity, but with inactivity.

The first demonstration that DNA methylation was involved in silencing the X came from the effects of methylation inhibitors on expression of X-borne genes. A group in Los Angeles used cell hybrids carrying an inactive human X, while I used cells cloned from a hybrid mouse embryo heterozygous for distinguishable alleles of three X-borne genes. In both sets of experiments, treatment with methylation inhibitors (such as the cytidine analogue 5-azacytidine) produced spectacular reactivation of a selectable gene (*HPRT*) and other genes on the inactive X. Reactivation was dose-dependent. It seemed that stripping methyl groups off DNA is all you need to lift the transcriptional repression (Figure 6.2c).

Many observations backed up the conclusion that DNA methylation caused X chromosome silencing; take-up of methylated bases into DNA silenced genes and transfer of the selectable *HPRT* marker worked when it was on the active, but not the inactive X; chemical removal of methyl groups restored its function in transformation. This showed that the repression was a property of the DNA itself.

Moreover, DNA methylation accompanies X chromosome inactivation in the embryo. Mutation of DNA methylases has gross effects on X inactivation; knocking out the *de novo* methylases in mice causes X reactivation in the embryo, but not in the extraembryonic lineage, suggesting it is not required to maintain paternal X inactivation. In humans *DNMT* mutations produce deficient methylation, which causes the rare autosomal ICF syndrome (immune deficiency, centromere instability and facial dysmorphology) in which genes on the inactive X are reactivated.

As the first difference in the inactive X to be established at the molecular level, reactivation of the X by demethylation caused quite a stir. These early experiments were eagerly interpreted as DNA methylation providing a complete explanation of the properties of the inactive X.

Such pleasing simplicity could not last. Firstly, reactivation of genes on the inactive X by methylation inhibitors never occurs in normal diploid cells, but only in somatic cell hybrids and interspecific hybrids. This suggests that some other element of the regulatory network is missing or deranged as a result of a 'genomic shock' event that softened up a step in a multistep mechanism of X inactivation.

Also, detailed examination of methylation in and around cloned housekeeping genes on the X found general changes of CpG methylation at the promoter, but could not nail down a specific CpG site whose methylation was absolutely consistent with inactivation. Differences between X-borne genes, and methylated sites were hard to reconcile with a simple causal relationship. So was the timing of methylation change, which occurred some days after reactivation of the inactive *HPRT* allele in human stem cells.

Thus, methylation cannot be the only, or even the primary molecular difference between the active and inactive X. What other molecular differences distinguish the inactive X?

6.4 Histone Variation and Modification

The basic histone proteins that bind to DNA and give chromatin its nucleosome structure (Section 1.2) had long been suspected of fulfilling more than just a structural role. It was pointed out in the 1950s that anything that disrupted the electrostatic binding of these positively charged small proteins to DNA might regulate its accessibility and activity. Now we know that many modifications are involved in regulating X chromosome activity (Figure 6.3).

Histones are small and highly basic (positively charged) proteins that wrap the negatively charged DNA around 'nucleosomes' (Section 1.2). Each is composed of an octamer, two each of histones H2A, H2B, H3 and H4. Histone H1 acts as a linker, binding to DNA between nucleosomes and regulating higher-order conformation.

Variant histones have been discovered with specialized functions (for instance, in centromere function and heterochromatin structure), and great excitement

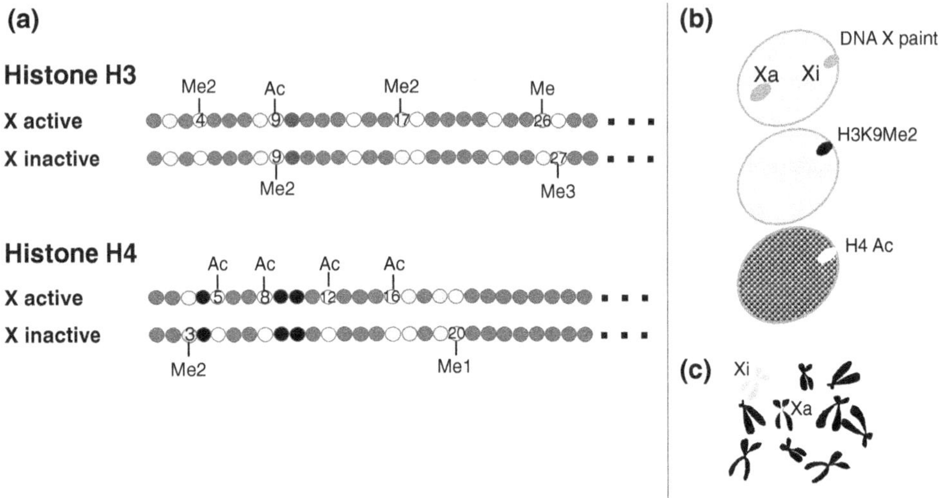

Figure 6.3   Histones and X chromosome inactivation. (a) Tails of histones H3 and H4, showing histone modifications on active and inactive X chromosomes. Amino acids are represented as circles; lysine as an open circle and arginine as a grey circle. Modifications of numbered amino acids are shown above (activating) or below (inactivating) the X; addition of 1–3 methyl groups Me1, Me2, Me3: Acetylation Ac. (b) Nucleus at interphase with active Xa and inactive Xi chromosomes both stained by DNA paint (top). Staining with antibody to H3K9Me2 strongly marks the inactive X (middle) and staining with antibodies to acetylated histone H4 stains all chromosomes except the inactive X (bottom). (c) Sketch of chromosomes stained with antibody to acetylated histone H4. The inactive Xi fails to stain, whereas the active Xa stains like the autosomes.

accompanied the discovery in the late 1990s of an elongated variant of histone H2A. This macroH2A replaces histone H2A specifically on the inactive X chromosome. However, this turned out to be a relatively late event, and knockdown of macroH2A did not prevent X inactivation. Instead, macroH2A seems to have rather general repressive action, being critical for the maintenance of nuclear order and the association of heterochromatin (including sex chromatin) with the nuclear periphery.

Histones can also be modified by enzymatic attachment of chemical groups (methyl, acetyl, phosphoryl) to amino acids in the tails that stick out from the nucleosome, or even in the internal folds that bind DNA (Figure 6.3a). It was known by the 1990s that the density of negatively charged acetyl groups attached to histone was higher on transcribed genes in the yeast and *Drosophila* genomes.

When the distribution of acetylated histone along metaphase chromosomes in mouse and human cells (and even kangaroo) was visualized by staining with fluorescent antibodies, it was seen that one X chromosome was strikingly unlabelled in female, but not male cells. In interphase cells, the sex chromatin and late-replicating (inactive) X was unlabelled by antibody to acetylated histone H4, but labelled with antibody to methylated histone H3 (Figure 6.3b,c). Lysine residues at several sites on the tails of histone H3 and H4 were acetylated on the active X, but not the inactive X.

Tellingly, the timing of histone deacetylation coincided with inactivation and the loss of DNA polymerase II. In undifferentiated human stem cells, the two active X chromosomes were both acetylated, but as X inactivation occurred during differentiation, one X became unstained. This occurred a day before genes were inactivated, and several days before DNA methylation was detected. Thus histone deacetylation seemed to be an early change, suggesting that it is directly involved in transcriptional silencing. Histone modification occurs in all parts of the developing embryo, the extraembryonic membranes as well as the inner cell mass.

Antibody staining, and chromatin immuno-precipitation has revealed many other modifications to the tails of the core histones (particularly H3 and H4) that mark either the active or the inactive X (Figure 6.3a). Some modifications are activating and others repressive. Of particular importance is dimethylation of lysine 9 on Histone H3 (abbreviated H3K9me2). This is an early change, and so spectacular that it can be used to distinguish the inactive X at any life cycle stage. Histone H4 deacetylation is the earliest change detected in differentiating stem cells.

Several other histone di- or tri-methylation marks on H3 and H4 are associated with inactivity of the X – but there are also a few (on arginine or lysine) that distinguish the active X. Addition of phosphoryl and ubiquitin marks have been described on H3, H4 and H2A. Indeed, histone modification has become something of a Pandora's Box; at least 80 different modifications had now been described as the 'histone code', some activating, others repressing. Deposition of active and inactive marks is not uniform along the X chromosome, as shown by the banding patterns produced by specific fluorescent antibodies. Some are mutually exclusive, others overlap.

What is the relationship between histone modifications and DNA methylation? One big advantage of DNA methylation was that patterns could be maintained after DNA replication. It was not obvious how this could be accomplished for histones; however, it appears that at least some modifications are distributed to both daughter strands, and enzymes patch up the holes. The loss of at least one active histone mark is correlated with DNA methylation, suggesting that the super-stable inactivation of the mammalian X chromosome is brought about in a series of steps, involving first chemical changes to histones, then DNA methylation that 'locks' the inactivation in place.

Each of these steps affects the conformation and position of the chromosome, and the accessibility to the transcription machinery.

6.5 Chromatin Conformation of the Inactive X Chromosome

The inactive X is obvious in interphase cells as a dark-staining blob plastered up against the nuclear membrane or the nucleolus (5.2). What does this sex chromatin look like at the molecular level? And are its unique position and conformation causes or effects of its inactivation? At last, we now have the techniques to examine this question (Figure 6.4).

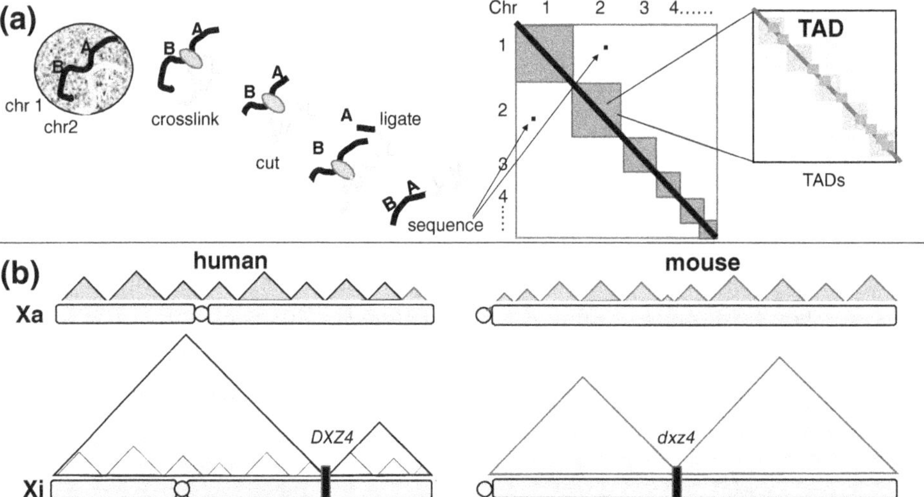

Figure 6.4 Chromosome conformation capture (3C) and architecture of the active and inactive X. (a) 3C works by crosslinking the DNA in a population of nuclei with formaldehyde. DNA sequences that are physically near each other in 3D are covalently bound together. For instance, sequence AB on chromosome 1 is crosslinked to sequence LM on chromosome 2. DNA is purified, and the dangly bits are cut off with restriction enzymes. Then the cut ends are ligated enzymatically, the formaldehyde crosslinks are removed, leaving hybrid molecules that represent the population of crosslinked DNA. These are sequenced, and known sequences arranged in order along chromosomes. A heat map is constructed in two dimensions that represents the frequency of each pair of sequences. Frequent interactions appear as darker areas, rare interactions are white. For instance, the proximity of AB-LM sequences may be represented many times and registers as a strong signal that denotes a frequent association between these sequences in the cell. The most frequent associations are between sequences close together on the same chromosome, explaining the strong diagonal and prominent squares. A blowup of the sequence (inset) shows that each chromosome is represented by small squares, which represent Topologically Associated Domains (TADs) of DNA replication. (b) 3C profiles of the active Xa and inactive Xi chromosomes of humans and mice. Xa and Xi are distinguished by SNP variants. Along the active X, regular small peaks, similar to those of autosomes, denote TADs. However, the inactive X shows two large domains of interaction. separated by sequence DXZ4 that marks a hinge region. This structure completely obliterates TADs on the mouse, though not the human inactive X.

An appealing early idea was that physical containment, or supertwisting within the sex chromatin body might directly restrict access of transcription factors to the entire inactive X.

Indeed, there was early evidence from DNAse sensitivity of chromatin for restricted accessibility of genes on the inactive X. Cutting with DNAses denotes an 'open' structure correlated to gene activity, and failure to cut (leaving a 'footprint' of missing smaller fragments on a gel) with inactive chromatin. 'Footprints' of protected DNA were found near promoters on the inactive human X which were missing from DNA on the active X. Reactivation was accompanied by a switch to nuclease sensitivity as an early event. Thus, the promoters of active and inactive alleles appeared to have a different chromatin configuration.

The position of the inactive X, typical of inactive chromatin, was initially interpreted as a passive process, resulting from being pushed to one side (the molecular equivalent of being 'sent to Coventry' where no one will talk to you). However, it turns out to be the result of binding with many specific factors (Section 6.11).

An enormous step forward in analysing the conformation of the inactive X came with methods to crosslink with formaldehyde bits of DNA that lie close together in 3D. DNA is then cut with restriction enzymes and ligated, forming hybrid molecules that each represent a pair of sequences that are physically close in the nucleus. High throughput DNA sequencing then provides the frequency of each contact (Figure 6.4a).

This chromosome conformation capture (3C) technique provides a cornucopia of data on the associations between different DNA sequences in a chromosome, or even a whole genome. Improvements in selecting DNA-DNA pairs to be analysed have produced refinements (4C, 5C, Hi-C). It is gene-specific and can be made allele-specific by recognising specific SNPs.

The frequency of contacts made by each DNA sequence to all other sequences can be represented as a 2-dimensional heat map. A whole chromosome is represented as a series of interactions of each sequence with every other sequence, in order along the chromosome. Sequences that are close together on the same DNA molecule will be the most prominent, and this shows up as a strong diagonal which provides information on fragment order that is useful for sequence assembly.

More important for analysis of the structures of active and inactive X chromosomes are the interactions between more distant sequences on the same linear DNA molecule which reflect their physical closeness in the interphase nucleus (Figure 6.4b). Two distant sequences on the same molecule that are attached in a loop will provide a strong signal away from the diagonal.

A pattern of rather regular boxes along the diagonal is seen for all the autosomes. This is interpreted as a pattern of large (0.2–1 Mb) Topologically Associated Chromatin Domains (TADs) that represent a series of loops along the chromosome.

These loops are maintained by binding to an 'architectural protein' CTCF (named for a factor binding to sequences of CCCTC in all eukaryotes). Other proteins including cohesin (also required for keeping chromosomes together at division) are involved, as well as other factors that mediate DNA interactions.

TADs don't seem to represent stable associations between particular sequences, but are expunged and re-established at each cell cycle. The TAD structure was shown to be common to all autosomes, and is consistent between different tissues, suggesting TADs are not involved in regulating gene activity.

The active and inactive X chromosomes of females may be separately analysed by using mouse or human cells with a skewed inactivation, and looking specifically at contacts made by DNA sequences on the two X chromosomes that differ by many SNPs.

The active human X gave a 3C profile with TAD structure similar to that of autosomes and the single X of males (in fact, TADs were first identified in studies of X chromosome inactivation). However, the inactive X proved to have an extraordinarily different conformation. TAD structure was weaker, and instead, the inactive X was arranged as two 'superdomains', which represented huge 'super-loops' of DNA (up to 77 Mb) (Figure 6.4b). DNA-DNA association maps of the mouse active and inactive X chromosomes also showed a normal structure for the active X, but two huge superdomains on the inactive X. There were no signs of the TAD structure on the inactive X, meaning that sequences along the X associated at random. A similar pattern was found for the macaque and the cat, so it seems to be a general property of the inactive X.

In all these species, the two domains of the inactive X were separated by a hinge region. The hinge lies within orthologous regions on the conserved old region of the human and mouse X, but because of rearrangements, the human and mouse X divide the chromosome into different sized superdomains. Human and mouse hinges share DNA sequence motifs, including a repetitive locus called *Dxz4*, which binds CTCF. Deletion of *Dxz4* from the mouse X using CRISPR/cas9 editing totally disrupted the bipartite structure and the pattern of histone modification and DNA replication but, surprisingly, did not prevent X inactivation. This counters early speculations that tighter condensation or supertwisting chromatin might itself limit accessibility to the transcription machinery.

So again we have spectacular changes – structural as well as chemical – that seem to reinforce, rather than initiate, X inactivation.

6.6 Late Replication of the Inactive X Chromosome

One of the earliest, and most consistent, observations of the inactive X chromosome was that it replicates its DNA later in the DNA synthesis period than the active X (and the autosomes). Now we can ask, what molecular features do transcription and replication share? Is late DNA replication merely a side effect, a correlate, or could it be a cause of X inactivation?

There is an overall relationship between replication timing and transcriptional activity. At the microscopic level, transcriptionally silent and late-replicating domains are co-located at the periphery of the nucleus. DNA sequence analysis shows that early replicating domains are rich in transcribed sequences, and late-replicating regions are poor.

Early molecular studies revealed multiple origins of replication, but in mammalian cells, most are inactive, and the whole process seems rather leisurely. Replication of different domains initiates at intervals throughout the 6–8 hour DNA synthesis period. The time of initiation seemed to be a property of the domain, explaining why homologous autosomes follow the same replication pattern and patterns are conserved even between species.

New methods for examining DNA replication at the molecular scale are based on DNA sequencing of cells that are synchronized, or sorted according to DNA content. A twofold difference in the read depth of each sequence shows whether a particular sequence has been replicated or not. Early and late-replicating regions can be accurately mapped using HiC, building a picture of replication of the whole chromosome. Since cytological observation of chromosomes is not required, replication can be studied in any tissue at any stage of development.

These techniques show that DNA replication starts from a few origins that are rather randomly primed ('licensed') by recruiting proteins, including helicase that can unwind DNA, which then binds to DNA polymerases. Like RNA transcription, this process is stochastic in the sense that an origin may be active in some cells and not others, with a certain probability.

Active origins then synchronize to replicate larger domains (500 kb–1 Mb), which accord with TADs. These can be mapped to different locations in the nucleus. Early replicating domains tend to be central, and late-replicating domains are often at the periphery, embedded in heterochromatin. Origins are also associated differentially with histone modifications; early origins are marked by active modifications, while late origins are bound with repressive modifications.

Allele-specific sequencing has been used to compare the replication timing of DNA sequences on the active and inactive X. This verifies that the active X replicates along with the autosomes, whereas the inactive X initiates DNA replication much later. Surprisingly, however, sequencing methods show that the inactive X is in a hurry; it replicates more rapidly, from more origins, than the active X and autosomes. This is consistent with the altered conformation of the inactive X, which lacks regular TADS.

How are DNA replication and transcription related? Both transcription and replication are sensitive to chromatin conformation and epigenetic modification, and it seems most likely that, rather than a cause-and-effect relationship, the same set of epigenetic changes that limit accessibility to the DNA of the inactive X alters the availability of transcription start sites, as well as DNA replication origins. Everything is correlated with everything. This does not help us much to distinguish between a causal relationship one way or the other, or a common response to chromatin structure and position.

6.7 Breakthrough – Discovery of *XIST*

The big breakthrough in understanding of X chromosome inactivation came in 1991, with the accidental discovery of the gene that seems to have ultimate control of the process (Figure 6.5).

As I discussed in Section 5.11, cytogenetic studies showed that a single locus on the X chromosome (the X inactivation centre XIC) was required to induce silencing

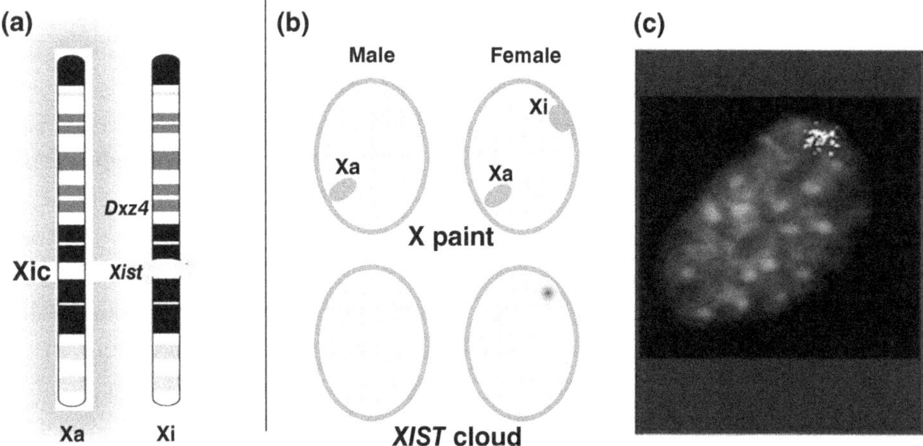

Figure 6.5   Location, transcription and detection of mouse *Xist*. (a) Map of active and inactive X chromosomes (Xa and Xi) showing location of *Xist* within the X inactivation centre (Xic), and the hinge region *Dxz4*. Xa genes are transcribed (pale grey glow), except for *Xist*. Xi genes are repressed except for *Xist*, which is strongly transcribed. (b) At interphase the two X chromosomes in a female nucleus may be identified by in situ hybridization with X chromosome paint. RNA FISH with an *Xist* probe detects a 'cloud' of *Xist* RNA over the position of the inactive, but not the active X. In male nuclei there is no Xist cloud over the single active X. (c) A fluorescent *Xist* probe detects an *Xist* RNA cloud over the inactive X of a female mouse nucleus at interphase. Photograph supplied by Dr Christine Disteche, University of Washington, Seattle.

of the inactive X. Only genes physically attached to this sequence ('*in cis*') could be inactivated. The XIC of mouse and human X chromosomes were flanked by the same genes, so it was expected that cloning and sequencing would soon reveal the secrets of this mysterious region.

But despite herculean efforts, no protein-coding genes could be found within this region. Instead, a strange gene was identified during a search for the human steroid sulfatase gene *STS*. This gene mapped nowhere near the PAR (where *STS* lies), but within the XIC region on the long arm of the human X. The mouse version was identified from the orthologous region on the mouse X.

What marked this gene for frenzied attention was its unique expression pattern. In both species it was found to be expressed in cells from females but never from males, suggesting it is transcribed only from the inactive X (Figure 6.5a). Indeed, transcripts were detected in cell hybrids retaining an inactive human X but not hybrids with an active X. In line with its contrary transcription behaviour, it replicated late on the active X and early on the inactive X.

The gene was called *XIST* for X̲-I̲nactive-S̲pecific T̲ranscript (*Xist* in mouse following rodent nomenclature; I shall use *XIST* except where I refer specifically to mouse).

Features of *XIST* were decidedly odd. On a northern blot (Section 1.11) that separates RNA molecules according to size, human *XIST* was seen to be transcribed into a very large (17 kb) RNA. Mouse *Xist* makes a 15 kb transcript, and also a truncated version that lacks the final 5.6 kb. Mouse and human *XIST* have a similar exon-intron structure except that one mouse exon is included in an intron in non-rodent species. But their exon sequences are very different, sharing only 76% homology, largely in several repetitive regions.

The location of its transcript is the most spine-tingling feature of *XIST*. It stays in the nucleus, unlike messenger RNA, which is shipped to the cytoplasm to be translated into protein. In situ hybridization with a fluorescent *XIST* probe showed that it is localized strictly to one of the X chromosomes, forming an '*XIST* cloud'. The position of the cloud overlaps with the sex chromatin body, representing the inactive X at interphase (Figure 6.5b).

The *XIST* gene was immediately recognized as atypical. Its transcript was the first of many long noncoding (nc) RNA molecules shown to have a function in gene control in mammals.

6.8 Structure of *XIST*

Mouse *Xist* has seven exons and human *XIST* has eight. The very long Exon I of mouse and human *XIST* contains five repetitive arrays (A–D and F), and the very long last exon contains repeated sequence E (Figure 6.6). The A repeat in exon 1,

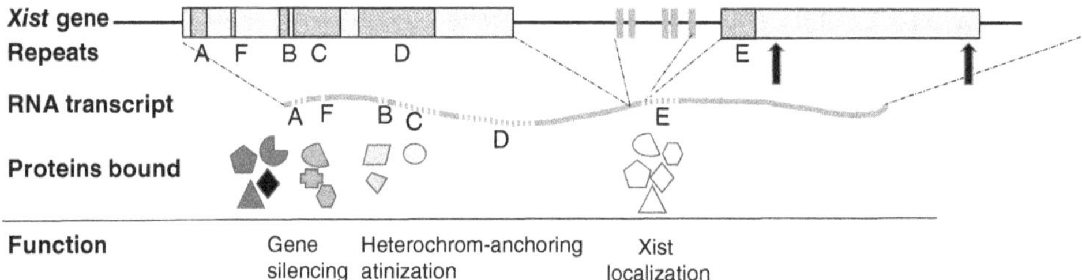

Figure 6.6   Structure of the mouse *Xist* gene and functions of its transcript. Mouse *Xist* has 7 exons, including large exons 1 and 7. It is transcribed into two isoforms (of length 12 kb, and 17 kb, black arrows). *Xist* contains several highly repetitive regions (A-F, stippled) that bind different suites of specific proteins (represented by different shapes and shades, see Section 6.11). Knockouts reveal the functions of these repeat regions.

containing 7–8 copies of a 26-nucleotide sequence, is the most conserved region. Repeat C is specific to mice.

XIST has now been cloned from many eutherian mammals, and mapped to homologous regions of the X. Although its sequence is poorly conserved overall, it is clearly orthologous, with a conserved structure and transcription start site, as well as the repeats.

What does *XIST* do? In every species, *XIST* transcript was polyadenylated and spliced like messenger RNA. But it was evidently not translated into protein, since it lacked a credible open reading frame.

Long nc RNAs usually maintain a complex tertiary structure of a series of stems and loops, and there are many models proposing how *XIST* repeats form modules and functional domains. It is possible that tertiary structure is maintained in different mammals, even though DNA sequence are different, and this might be important for exposing parts of *XIST* transcripts (e.g., A repeats, E repeats) to binding.

The discovery of *XIST* in eutherian mammals heralded a new wave of investigation into the molecular biology of X chromosome inactivation.

6.9 *XIST* Controls X Chromosome Inactivation

Several lines of evidence quickly confirmed that *XIST* was more than just another correlate of X inactivation, but had a vital coordinating role.

Most of this work used mouse embryos, or embryonic stem cells. Many of the human studies were done using stem cells or undifferentiated hematopoietic cells. Especially useful are ES cells that have been engineered to turn on *XIST* expression at will, for instance, by inserting a promoter upstream that could be activated by treating with a drug.

Firstly, *XIST* is transcribed at the right time and place. *Xist* RNA was detected in female, but never male, mouse embryos. *Xist* is expressed as a single signal exclusively on the paternal X at the 2-cell stage, and accumulates on this X about a day before the first (paternal) wave of X inactivation. Later, transcription of both the maternal and paternal *XIST* alleles were detected just before the onset of random X inactivation in the inner cell mass.

It is different in human embryos. In the absence of early paternal inactivation, *XIST* forms a diffuse cloud initially over both human X chromosomes (and even over the single X in males) during cleavage. It zeroes in on the inactive X only later.

In germ cells, too, *XIST* was expressed appropriately. Its expression in the adult mouse testis correlated with the inactivation of the X during spermatogenesis. *Xist* expression in sorted foetal oogonia ceased at the time at which reactivation occurred.

XIST expression was also correlated with X inactivation in mouse and human embryonal stem cells. *XIST* was silent in undifferentiated ES cells that have two active X chromosomes. Transcription began about two days after differentiation, at the onset of X inactivation.

Thus, *XIST* expression is an early event in X chromosome inactivation in mouse embryos, germ cells and human ES cells, and its timing correlates strongly with X inactivation. The complete concordance of *XIST* expression with X inactivation suggests a common mechanism for X inactivation in all types of cells, and is consistent with the hypothesis that *XIST* regulates X inactivation.

The best evidence that *XIST* is the overall controller of X chromosome inactivation is what happens when the gene is inserted in the wrong chromosome, or is knocked out.

When *Xist* was introduced into the genome of XY (male) mouse ES cells, it integrated into many different sites on autosomes. The undifferentiated ES cells didn't seem to notice, but when they were allowed to differentiate, *Xist* was transcribed and made clouds at these autosomal sites. Spectacularly, *Xist* transcription silenced genes up and downstream of the insertion site, and the region became late-replicating. In early mouse embryos, *Xist* transgenes underwent paternal inactivation. In human cells, too, an *XIST* transgene inserted into autosomes silenced genes near the insertion site.

Two early mouse *Xist* knockout experiments confirmed that *Xist* expression was necessary for X inactivation. The *Xist* promoter and a bit of Exon 1 was deleted from one X chromosome in XX mouse ES cells. *Xist* could be transcribed only from the other X chromosome, and this was the only chromosome inactivated in differentiated cells. This implies that RNA transcripts of an intact *Xist* inactivate the X chromosome that bears it ('*in cis*').

Xist could also be deleted from the genome of mice. Deletion didn't affect XY males; they were perfectly healthy, and even made sperm. Female mice with the deleted *Xist* on the maternal X chromosome were viable too, but X inactivation was completely skewed toward inactivating the paternal X that bore a normal *Xist* gene. When the male parent provided the deleted X, all the female pups died, showing that a paternal X that could not be inactivated was lethal. This shows that *Xist* expression from the paternal X is necessary for paternal inactivation in the trophectoderm, and that paternal X inactivation is vital for survival of females.

A surgical deletion of only the *XIST* gene was made in human XX leukaemia cells using CRISPR/Cas9 genome editing. This obliterated *XIST* transcription when the stem cells differentiated, causing global reactivation of genes on the inactive X.

The conclusion is that if *XIST* is knocked out from one X chromosome in embryos or stem cells, transcriptional repression cannot occur on that X chromosome. This is so for paternal as well as random inactivation.

Once cells have differentiated, X inactivation becomes independent of *XIST*. No reactivation was observed in cell hybrids with an inactive X when *XIST* was deleted. Thus, continued *XIST* expression is not required to maintain the inactive state.

In summary, transcription of *XIST* seems to be required to initiate all types of X inactivation in eutherians.

6.10 How Does *XIST* Work?

How can *XIST* possibly control the activity of a thousand genes, and affect the gross cytological properties of such a huge region of the genome?

XIST cannot code for a protein because it contains no open reading frame. It must work through its nc RNA transcript. It exerts its effect in cis; that is, only on the X on which it lies.

XIST is transcribed, and the long RNA transcript is spliced. In mice, the *Xist* RNA stays in the nucleus, and about 2,000 molecules rapidly coat the X chromosome from which it is transcribed. The coating is seen as an 'RNA cloud' at interphase by hybridising with fluorescence labelled *XIST* probe (Figure 6.5c). *XIST* is enriched first at its own locus, then molecules bind at 50 other gene-rich sites and spread out from these with the help of anchoring proteins. *XIST* RNA stays bound to the inactive X throughout the cell cycle, at least in mice.

After *XIST* binding comes the exclusion of RNA polymerase, observable as a black hole after antibody staining (Figure 6.1). This is followed by the transcriptional silencing of X-borne genes, coupled with their late replication. At this early stage the inactive X loses its TAD structure, forms two giant domains and becomes tethered to the periphery of the nucleus and/or the nucleolus. Then occurs a whole symphony of modifications to the core histones of the coated X chromosome; loss of active histone marks and deposition of inactive histone marks. Variant histone macroH2A is recruited; and de novo methylation of CpG pairs occurs in promoters of genes on the inactive X.

Do different regions of *XIST* accomplish different steps in inactivation? Targeted deletions were made in different regions of mouse *Xist*, and their effects on different functions were observed (Figure 6.6). As discussed, mouse knockouts missing the *Xist* promoter transcribed no *Xist* RNA, and the X carrying the deletion could not be inactivated. Deletions of other parts of *Xist* permitted the transcription of at least some aberrant *Xist*, and formation of an *Xist* cloud; however, many genes on the inactive X remained active, making an RNA FISH signal even in the midst of the *Xist* cloud.

Deletions of the A repeats in exon 1 of the mouse Xist had the most spectacular effect. They did not affect *Xist* coating of the X, but they destroyed the 3D conformation of the inactive X, and permitted genes to be transcribed. Thus, the A repeats, though not required for *Xist* binding to the inactive X, are critical for chromatin conformation and gene silencing.

In contrast, deletion of the E repeats in Exon 6 affected *Xist* coating of the X, but this also needed C and F repeats in other regions of *Xist* exon 1, which seem to act additively. When C repeats were deleted, *Xist* RNA lost its tethering to the nucleation center on the inactive X.

Functional domains of human *XIST* were deduced by gene editing using CRISPR/Cas9 in a human leukaemia cell line. As for mouse *Xist*, deletion of the promoter prevented *XIST* transcription, obliterated the *XIST* cloud and permitted transcription of genes on the inactive X. Deletion of non-promoter regions reduced the transcription and produced aberrantly spliced transcripts, or affected their stability. Then eight different regions of *XIST* were deleted, and the expression of genes on the X monitored by RNA FISH and RNA-seq. As for mouse *Xist*, deletion of A and B repeats caused aberrant conformation and allowed some transcription from the inactive X, while repeat E was crucial for localization of *XIST* RNA to the X.

DNA methylation of *Xist* sequences is required for their initial activity; their methylation is one of the earliest and most unambiguous changes that distinguish the active and inactive X. *XIST* RNA is also methylated at 78 adenosine sites, many in the A repeats, and this happens very specifically as the X undergoes silencing when ES cells are differentiated. Knockout of the RNA methylases that do the job results in defective silencing.

In summary, different regions of the *XIST* RNA play different roles in the complex business of inactivating the X. These roles are similar but not identical in humans and mice.

6.11 Protein Partners in X Chromosome Inactivation

Proteins that bind to *XIST* have been identified by 'proteomic screens', using several different methods. In one, protein is crosslinked with *XIST* RNA. The complex is isolated from cells, and probed with 'antisense' sequences complementary to *XIST* that hybridize to *XIST* RNA. Bound proteins are then identified by mass spectrometry. This captures all RNA-binding proteins, including those that have nothing specifically to do with silencing.

Clever genetic screens using *XIST* transgenes were much more specific. For instance, an autosomal *XIST* transgene in an XY male ES cell line was made to be inducible by incorporating a promoter that is activated by a drug. When it was turned on, it silenced the chromosome that it resided on, and the cell died. This allowed selection for anything that disrupted *XIST* function and enabled the cell clone to survive.

To discover proteins that are involved with initiating and establishing X inactivation, embryos or ES cells were used. To capture proteins involved in maintaining X inactivation, differentiated cells were used. A useful way to assess the effects of knocking out these proteins was to hook up *XIST* to reporter genes, whose transcription is easily assessed.

Taken together, all these screens identified 500 proteins that bind to *XIST*. These include many proteins involved in RNA splicing and stability, kinases and proteins in well-known signalling pathways, some of which could play a specific role in X inactivation as well as a general role. Some (e.g., ATRX) have functions in heterochromatin stability. A specific function in X inactivation was validated for more than 90 *XIST*-binding proteins by observing the activity of reporter genes. Here I discuss some of the proteins with major effects on crucial steps of the inactivation pathway in mice.

Firstly, what factors turn on *XIST*? Before the X is inactivated in the undifferentiated cells of the inner cell mass, pluripotency genes (including *Oct4* and *Sox2*) are expressed, and these specifically inhibit *XIST*. *XIST* is activated only when they turn off during stem cell differentiation. This control is mediated by a protein encoded by the nearby *RNF12* gene that binds to motifs on the *XIST* gene. *RNF12* is repressed by pluripotency factors, so is turned on when ES cells differentiate, and in turn, turns on the *XIST* gene, setting off the whole X inactivation pathway (Figure 6.7).

The next step is coating of one X chromosome with *XIST* RNA. This does not, as was initially thought, depend on hybridization of *XIST* RNA sequences to complementary base sequences in the *XIST* gene. Instead, it requires two major protein factors. One is a transcription factor called YY1 ('yin-yang'), which tethers *XIST* to its site of transcription. YY1 can bind to both DNA and RNA. It brings together *XIST* repeats C and F, and acts as a bridge between *XIST* RNA and the X to be

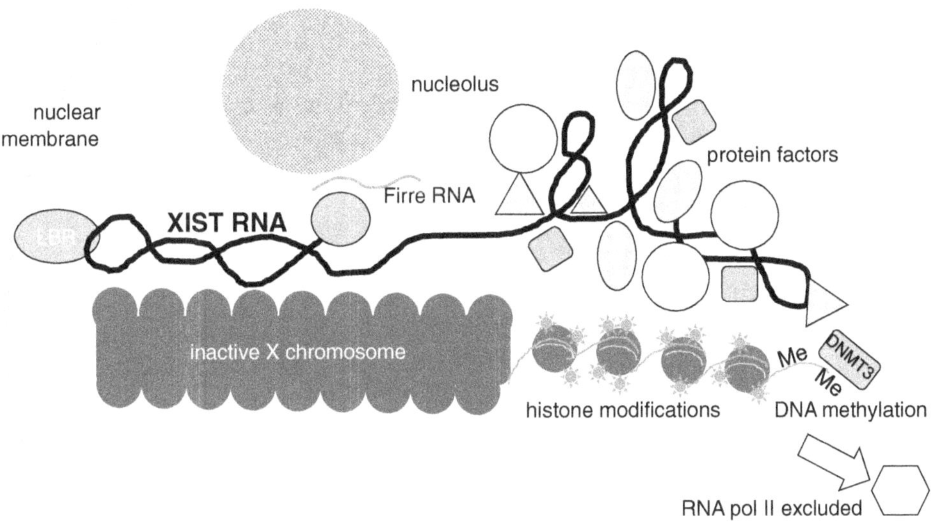

Figure 6.7   Molecular biology of X chromosome inactivation. Simplified diagram of the interaction of inactivating mechanisms. XIST RNA coats the inactive X and interacts with a protein factor (LBR) that stick it to the nuclear membrane, and another ncRNA (Firre) that sticks it to the nucleolus. Other protein factors that bind to different regions of the *XIST* RNA insert variant histones into nucleosomes, catalyse methylation of DNA (DNMT3), and modify histones (remove active marks and insert repressive marks) that alter the conformation of the inactive X and exclude RNA polymerase II, repressing transcription of genes on the inactive X.

inactivated. Coating also required several nuclear scaffold factors ('matrix' proteins) that bind to *XIST* domains. Coating the whole X seems to be accomplished by spreading within the 3D matrix from the limited regions of the X that directly bind a molecule or two of *XIST* RNA.

Tethering the inactive X to the nuclear periphery requires several proteins that bind to *XIST* and to the nuclear membrane. The nuclear membrane consists of a lipid bilayer supported by an inner lamina that contains the structural protein lamin. Binding of lamin to the *XIST*-coated X requires a transmembrane protein called LBR (Lamin B receptor) that binds *XIST* RNA at three sites, including repeat A. Knocking out *LBR* results in defective silencing. Tethering of the inactive X to the nucleolus requires binding of *XIST* RNA with a different factor.

The altered configuration of the inactive X chromosome is a dramatic sign of its inactivation, and there has long been speculation as to how this change is wrought at the molecular level. Ideas have included the fetching picture of an X strung so tightly between sites on the nuclear membrane it cannot loop into TADs. Or physical constraint that is exacted by accumulation of repressor complex proteins that form a local 'phase separated' concentration of protein aggregated around *XIST*. More likely is that *XIST* RNA countermands proteins that maintain the TAD domains. There is evidence that four *XIST* repeats (Figure 6.6) bind to an enzyme that is part of a chromatin remodelling complex that enforces the normal TAD structure, and this suppresses TAD formation.

Transcriptional silencing is the whole point of X chromosome inactivation. Silencing requires a huge protein called SPEN, one of a big family of 'Split end RNA binding proteins' first found in *Drosophila*. SPEN is recruited to the inactive X by binding to *XIST* repeat A. It then binds to other corepressors and recruits (or activates) the enzyme that strips acetyl groups from histone 3, the earliest histone modification. This shuts out RNA polymerase II (Pol II), initiating the creation of a silenced compartment. Knocking out SPEN allows Pol II to reactivate genes on the inactive X. SPEN also repels architectural proteins CTCF and cohesion, which inhibits TADs and creates megadomains.

A vital step in X chromosome inactivation is the recruitment of epigenetic silencing factors. This is mediated by highly conserved enzyme complexes called <u>P</u>olycomb <u>R</u>epressive <u>C</u>omplexes, PRC for short. PRCs are known in many systems (polycomb refers to extra sex combs on the front legs of flies, which is how they were first detected). PRC1 is recruited to the inactive X via a protein that binds to two *XIST* repeats, and PRC2 is recruited by the activity of PRC1. Several components of these complexes are specific enzymes that add or remove small groups from the amino acid tails of histones and endow them with repressive or activating function (Section 6.4).

PRC1 seems to start off the process of folding the inactive X into its unusual 3D structure. It then binds to another factor, SMCHD1 (discovered in mice with deranged imprinting), to fold up, origami-wise, the inactive X into its two superdomains.

A late event in X chromosome inactivation is DNA methylation, which appears to lock in transcriptional silencing. This requires de novo transmethylases DNMT3 and 4 (Section 6.3), as well as other protein factors.

Thus, *XIST* RNA exerts its multiple functions in the localizing, 3D reorganization and transcriptional silencing of the inactive X through its interactions with a plethora of protein partners (Figure 6.7).

6.12 Long Noncoding RNAs in the X Inactivation Centre

As if this were not complex enough, the X Inactivation Centre (XIC) also contains several other nc RNAs which control *XIST* expression (Figure 6.8). This explains

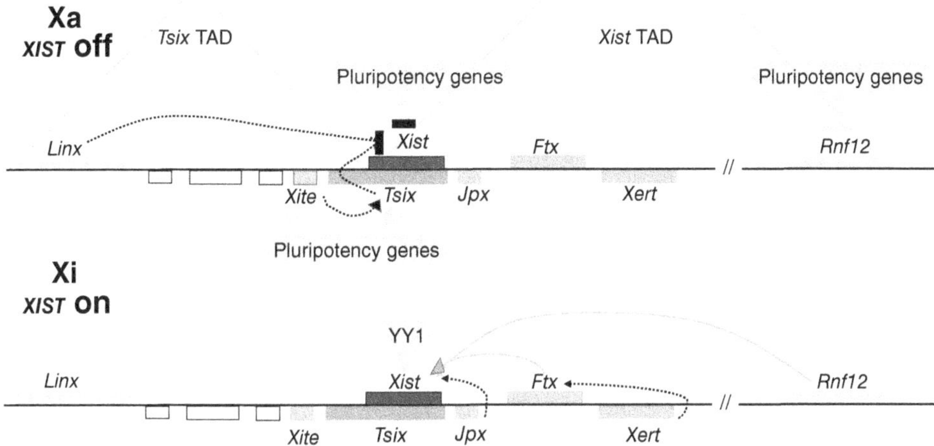

Figure 6.8   Control of *Xist* expression by long ncRNAs. Control of *Xist* expression in mouse X chromosome inactivation depends on many factors located in two TADs (represented by triangles) that contain *Xist* repressors (*Linx, Xite, Tsix*) and *Xist* boosters (*Jpx, Ftx, Xert*). Activation is represented by arrows and repression by blocking bars. On the active Xa, *Xist* is turned off in cis by RNA transcribed from noncoding (nc) genes *Tsix* (antisense to *Xist*) and *Linx*, as well as by pluripotency genes active in undifferentiated cells. *Tsix* is upregulated by nc gene *Xite*, as well as by pluripotency genes. On the inactive Xi, *Xist* is turned on when pluripotency genes are switched off at differentiation. Inhibitors *Tsix* (and *Xite*) are inactivated, and activators *Jpx* (nc) and *Ftx* (probably protein-coding and boosted by nc *Xert*) upregulate *Xist*. Direct binding occurs *in trans* by protein encoded by *Rnf12* and transcription factor YY1.

why sequences 450 kb either side of the XIC are required to accomplish random X inactivation.

Overlapping *Xist* in the mouse XIC, was a 40 kb transcript, cutely named *Tsix* (which I pronounced 'T-6' until reprimanded by its discoverer, Jeannie Lee, who mandates 'Tsi-X'). It is 'antisense' to *Xist*, transcribed from the opposite strand from a promoter outside *Xist*. Spliced and unspliced transcripts run the whole length of *Xist*. Like *Xist*, the *Tsix* RNA has no coding potential – it is another long ncRNA.

Tsix is a specific repressor of *Xist*. *Tsix* and *Xist* transcription are mutually exclusive, so *Tsix* activity blocks *Xist* transcription, and therefore blocks inactivation of the X that bears it. Like *Xist*, *Tsix* transcripts act strictly in cis, so *Tsix* transcription demarcates the active X. *Tsix* knockouts and transgenes verify that it negatively regulates *Xist*; it silences the *Xist* promoter by DNA methylation and histone modification. *Tsix* activation is the earliest event that demarcates the active and inactive X chromosome in mice.

Tsix knockout leads to preferential inactivation of the X lacking normal *Tsix*. Upregulation of *Tsix* does the opposite, leading to preferential activity of the X possessing the upregulated *Tsix*. *Tsix* is activated by pluripotency factors, so it is turned on in undifferentiated cells, blocking *Xist* expression. It is also repressed by the protein product of *Rnf12*, which is repressed by pluripotency factors. Both have the effect of boosting *Xist* expression.

That isn't all. There are at least seven nc genes in the 100 kb critical region of the mouse Xic. They are arranged in two groups that have opposite effects, one which boosts *Xist*, and the other that boosts *Tsix*. These groups lie in separate Topologically Associated Domains (TADs) (Figure 6.8).

In the *Xist* TAD lie two other nc genes, *Jpx* and *Ftx* upstream of *Xist*. Both seem to boost *Xist* production and X inactivation; yet another nc gene *Xert* may boost *Ftx*. In the other (*Tsix*) TAD lies a nc gene called *Xite* (for X inactivation Intergenic Transcription Element), which lies downstream of *Xist* and is transcribed from the same strand as *Tsix*. *Xite* contains enhancer-like elements and acts like an enhancer of *Tsix*, repressing *Xist* and the activity of the X on which it is active. Another smaller enhancer and another nc gene that also boosts *Tsix* expression lie downstream of *Tsix*, and there is yet another nc gene (*Linx*) further away in the same TAD that represses *Tsix*.

Another nc gene *Firre* lies on the X. It is far from the XIC, but is also transcribed only from the inactive X. The long nc *Firre* RNA binds with *Dxz4* RNA on the inactive X at the hinge between the two superdomains, and is important in tethering the inactive X to the nucleolus.

There are also autosomal nc genes that are important in regulating X chromosome inactivation in mice. Curiously many of these are clustered at both ends of chromosome 1, and in the orthologous region of human chromosome 19. These genes may be co-regulated and may even interact physically in 3D with *Xist* and the inactive X.

Human X chromosome inactivation does not involve the same set of ncRNAs. Some of the mouse ncRNAs have no human orthologues, and some orthologues act differently.

There is a *TSIX* orthologue in the human X inactive centre, in the same relative position, but it is shorter and overlaps *XIST* only minimally. *XIST* and *TSIX* can be expressed concomitantly from the same X, and there is little evidence that *TSIX* represses *XIST*. Potential *XIST* activators Jpx and Ftx also have human orthologues, but there is no sign of *LINX, XITE* or other *TSIX* enhancers.

Human cells were also discovered to have a different long nc RNA. *XACT* (for X-active coating transcript) lies on the X far (40 megabases) from the X Inactivation centre. It makes a huge (>250 kb, and largely unspliced) transcript.

XACT RNA coats the active X in early embryos and stem cells. Initially, *XIST* and *XACT* are expressed from both X chromosomes in all cells (even in males) and form diffuse clouds over both X chromosomes. Then *XACT* homes in to a single X in female cells. *XIST* homes into the other X, and the two signals don't overlap. This suggested that *XACT* and *XIST* compete, and the function of *XACT* is to protect the potentially active X from *XIST* binding and inactivation.

Because mouse and human X inactivation involve different sets of ncRNAs, it would be salutary to look for ncRNAs in the X inactivation centres of other mammals. There is great scope to discover other ncRNAs (and great scope for cute naming, e.g., *XCES* for activators and *XPUNJ* for repressors). Particularly interesting would be to investigate proteins and ncRNAs in marsupials, in which not even *XIST* exists.

Long nc RNAs in and outside the XIC are therefore engaged in tugging and tethering the X chromosome to the silencing compartment, and recruiting a variety of protein cofactors that bind it in a stable active or inactive state.

6.13 How Does It All Work?

If this chapter is complicated, it is because mechanisms of X chromosome inactivation are unbelievably intricate. I have tried to cover the wide sweep of it rather than get lost in the details.

It has been difficult to sort out X chromosome inactivation because it is multilayered. Everything is correlated with everything, so it is hard to discern what causes what.

When Mary Lyon put forward her single active X hypothesis in 1961, one of the compelling features was its simplicity; all it would need was a mechanism to protect one X from being repressed by some kind of molecular change. DNA methylation seemed, in 1980, to offer such a mechanism. However, reactivation of the X by inhibition of methylation occurs only on inactive X chromosomes in interspecies crosses and cell hybrids. This lead me to propose a two-layered control. It is more like ten!

Now we find that there are several silencing mechanisms; DNA methylation, modification of several histones, specialist histone variants, ncRNAs, physical constraint and altered conformation, and these are mediated by a bewildering array of genes (I have introduced only a few). Some act *in trans* via their RNA or protein products; others in cis via ncRNAs.

There have been many attempts to build models to show how these factors all interact to silence one X. In the absence of a colour panel, I can offer only a very simplified diagram, but I hope it will leave you with the conviction that X chromosome inactivation is a complex process that provides backups upon backups.

How does such a model explain some of the unique features of X inactivation? From the start Mary Lyon highlighted some peculiarities of X chromosome inactivation that had no obvious explanation in 1961: the extraordinary stability of X inactivation but the ability to reactivate in germ cells, the cytological transformation of the inactive X, the remarkable ability of the cell to count the number of X chromosomes it needs to be active, the inactivation of a whole X (bar a few escapers) and its ability to spread in cis along a huge chromosome, and the choice of paternal or random inactivation, and its propensity to skew.

Now we know about DNA and histone modifications, the several ncRNAs and innumerable protein factors, the TAD structure and 3D megadomains, can we explain these features any better?

The hyperstability of X chromosome inactivation is readily explained in molecular terms by the multiple mechanisms that collude to silence genes. Coating with *XIST*, tethering to a silent corner of the nucleus, accumulating repressive and losing active histone marks and DNA methylation all combine to repress genes.

The position and odd conformation of sex chromatin at last have satisfying molecular explanations in the binding and inhibition of factors that support the normal looped TAD structure. We now know of proteins that remodel the inactive X into two huge superdomains and tether it, through *XIST* RNA, to the nuclear membrane and the nucleolus.

The cell's mysterious ability to count the number of X chromosomes to inactivate to achieve balance suggests a unique factor or binding site that protects one X (per diploid complement) from inactivation. Perhaps the human *XACT* gene, which coats the active X, fulfills this role. This or another gene on the X could interact with autosomal gene(s) in such a way that excess of one or the other is registered. The protein-coding gene *Rnf12* and the noncoding gene *Jpx* in the mouse Xic have been suggested to be able to sense X/autosome ratios.

The remarkable spreading of X inactivation through 150 megabases of DNA containing about 1,000 genes now becomes understandable in terms of *XIST* coating that is nucleated at a limited number of sites and spreads throughout the megadomains in 3D. This combines early ideas of physical restraint with our knowledge of ncRNAs which tend to linger at the point of transcription (unlike messenger RNAs, which are shipped out of the nucleus to be translated in the cytoplasm). The recruitment of physically adjoining regions to the silent domain accounts for spreading of inactivation into autosomal regions in X-autosome translocations.

Another mystery flagged by Lyon is how the X to be inactivated is chosen. Random X inactivation could be explained by the balance of *Xist / Tsix* expression; in the absence of an imprint, whichever X is first to express one will inhibit the other. Skewed inactivation occurs in mouse hybrids with *Xce* alleles that differ in duplications and rearrangements. *Xce* maps close to the inactivation centre, and it seems likely that variants affect *Tsix* transcription, leading to different levels of *Xist* binding, since knockdown of *Tsix* leads to skewed inactivation.

Paternal X inactivation, too, can be understood in terms of altered expression of the same factors. In mice this early wave of imprinted inactivation is initiated by expression of *Xist* by the paternal X. Expression of *Tsix* and its enhancers from the maternal X, enforced by a maternal supply of the protein SMCHD1, keeps maternal *Xist* silent so the maternal X stays active. The imprint on the paternal X may have been left by the inactivation of the single X at spermatogenesis (MCSI), though this does not depend on *XIST* expression.

6.14 Conclusions and Lingering Questions

Exactly how transcription of genes on the inactive X is repressed is still being worked out, as part of a much wider picture of epigenetic modification of gene action in many systems. Escape of some genes on the inactive X seems to be wired into these regions, which continue to be transcribed even when moved into an inactivated region. Barriers to the spreading of inactivation into these regions seem to be posed by clusters of the insulator elements CTCF and YY1, which may protect regions from inactivation by isolating them into a separate chromatin loop.

Developmental timing of X inactivation and reactivation is accounted for by control of *XIST* transcription by pluripotency factors. Downregulation of these during differentiation lifts repression of *XIST* and its accompanying ncRNAs. *XIST* and its boosters and repressors, are expressed before the onset of random silencing in eutherian mammals. X reactivation in germ cells, and in induced stem cells, seems again to be a response of transcription of nc RNAs to pluripotency factors.

In summary, most of the unique features that Lyon highlighted have an explanation in the interactions of long ncRNAs and specific proteins with chromatin, initiated by changes in pluripotency factors.

There are two surprising conclusions from our consideration of X chromosome inactivation.

The first is how complex it is. Surely three silencing methods is a bit over the top? Not if dosage compensation of the X is critical for survival and reproduction, as would appear to be the case.

The other surprising conclusion is that though the silencing of one X in female therian mammals is ubiquitous, the phenotype, and especially the identities and actions of the molecules that make it happen, are astonishingly diverse. The most striking example is the identity of the long ncRNA that is central to inactivation; in eutherians this RNA is transcribed from the *XIST* locus, whereas in marsupials, a long ncRNA with similar properties is transcribed from a completely different locus (Section 7.4). Confusingly, some of the ncRNAs that boost *Xist* or *Tsix* production in mice are absent from the X inactivation centre of other mammals. Another example is paternal X inactivation, which occurs in the extraembryonic membranes of rodents and cows but no other eutherians, and is ubiquitous in marsupials. What is the point?

As for many baffling biological questions, the answers are likely to be found in evolution rather than function. Mammals are not alone in moderating dosage of genes on sex chromosomes. In Chapter 7 I will examine X inactivation and dosage compensation in more distantly related organisms. This will allow us to gauge how important dosage compensation is, see how the X inactivation system evolved, and how it shares elements with other dosage compensation systems for sex chromosomes and autosomes.

Importantly, the answers to these questions inform the many other systems, increasingly recognized, in which genes are transcriptionally activated or silenced by epigenetic mechanisms, as well as systems in which the parental origin of blocks of genes leaves an 'imprint' to be recognized in subsequent activity or silencing.

FURTHER READING

Classic Papers

Brown CJ, Ballabio A, Rupert JL, 1991. A gene from the region of the human X inactivation centre is expressed exclusively from the inactive X chromosome. *Nature* 349, 38–44

Graves JAM, Gartler SM, 1986. Mammalian X chromosome inactivation: testing the hypothesis of transcriptional control. *Somatic Cell and Molecular Genetics* 12, 275–280

Jeppesen P, Turner BM, 1993. The inactive X chromosome in female mammals is distinguished by a lack of histone H4 acetylation, a cytogenetic marker for gene expression. *Cell* 74: 281

Riggs AD, 1975. X inactivation, differentiation, and DNA methylation. *Cytogenetics and Cell Genetics* 14: 9–25

Reviews and Research Articles

Cooper DW, Johnston PG, Graves JAM, Watson JM, 1993. X-inactivation in marsupials and monotremes. *Seminars in Developmental Biology* 4: 117–128

Cotton AM, Chen C-Y, Lam LL, 2014. Spread of X-chromosome inactivation into autosomal sequences: role for DNA elements, chromatin features and chromosomal domains. *Human Molecular Genetics* 5, doi: 10.1093/hmg/ddt513

Deng X, Ma W, Ramani V, et al., 2015. Bipartite structure of the inactive mouse X chromosome. *Genome Biology* 16: 152

Disteche CM., 2012. Dosage compensation of the sex chromosomes. *Annual Review of Genetics* 46: 537–560. doi: 10.1146/annurev-genet-110711-155454

Fang H, Disteche CM, Berletch, 2019. X inactivation and escape: epigenetic and structural features. *Developmental Biology* 7: 219. doi: 10.3389/fcell.2019.00219

Furlan G, Galupa R, 2022. Mechanisms of choice in X-chromosome inactivation. *Cells* 11: 535. doi: 10.3390/cells11030535

Galupa R, Heard E, 2018. X chromosome inactivation: a crossroads between chromosome architecture and gene regulation. *Annual Review of Genetics* 52: 535–566.

Jacobson EC, Pandya-Jones A, Plath K, 2022. A lifelong duty: How Xist maintains the inactive X chromosome, *Current Opinion in Genetics & Development*, 75: 101927, https://doi.org/10.1016/j.gde.2022.101927.

Loda A, Collombet S, Heard E, 2022. Gene regulation in time and space during X chromosome inactivation. *Nature Reviews Molecular Cell Biology* 23: 231–249. https://doi.org/10.1038/s41580-021-00438-7 Report of molecular details.

Lu Z, Carter AC, Chang HY, 2017. Mechanistic insights in chromosome inactivation. *Philosophical Transactions of the Royal Society B: Biological Sciences* 372: 20160356

Malcore, R.M., Kalantry, S, 2024. A comparative analysis of mouse imprinted and random X-chromosome inactivation. *Epigenomes*, 8: 8. https://doi.org/10.3390/epigenomes8010008

Poonperm R, Ichihara S, Miura H, 2023. Replication dynamics identifies the folding principles of the inactive X chromosome. *Nature Structural & Molecular Biology* 30: 1224–1237

7

Evolution of X Chromosome Inactivation and Dosage Compensation

In this chapter, I will discuss how X chromosome inactivation might have evolved in mammals to compensate for the progressive loss of alleles from the Y chromosome, a strategy that makes perfect sense. An evolutionary view also makes sense of some discordant observations, and some puzzling differences between species. However, we must recognize that evolution does not necessarily choose the most obvious or efficient solution.

I will show how we can exploit genetic differences between species in the expression and the molecular mechanism of X inactivation to explore how the several layers of control could have evolved to build the extraordinarily stable silencing of the inactive X in eutherian mammals.

Then I shall describe how dosage compensation works (or doesn't work) in other vertebrates that have differentiated sex chromosomes; monotremes, birds, snakes, and even frogs and fish. And ask the question – how essential is dosage compensation anyway? Are there other ways of coping with dosage inequities brought on by sex chromosome differentiation?

I will then describe briefly some of the best-known dosage compensation systems, in our invertebrate models, revealing many ancient silencing mechanisms whose components we recognize in fish, flies and humans.

Finally I shall go way beyond my brief to point out that these mechanisms are there in all sorts of other systems in which epigenetic silencing has been selected for, constituting an ancient molecular toolbox full of ways to regulate the activities of genes, or of whole regions of the genome.

7.1 Y Chromosome Degradation and Evolution of X Inactivation

If we consider that dosage compensation is a must for genes on the X that have no partner on the Y, we would conclude that Y degradation drives X inactivation (Figure 7.1).

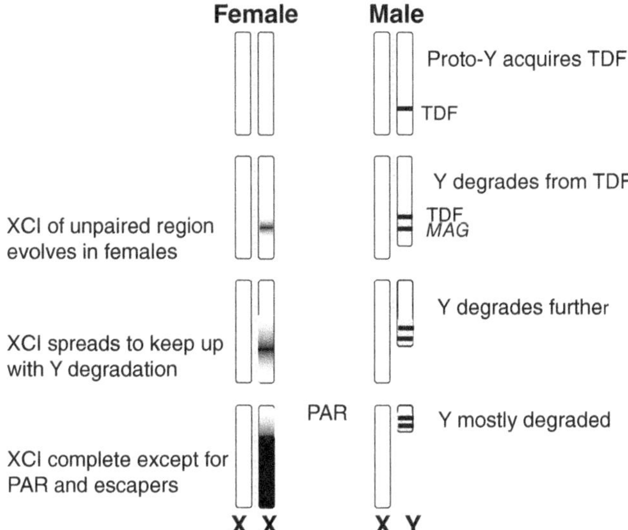

Figure 7.1 X inactivation follows Y degradation. An autosome pair becomes sex chromosomes when a testis determining gene (TDF) is acquired by one partner (here the proto-Y). Initially in males the proto-Y pairs fully with the X (grey dots). The Y begins to degrade (pattern) when other male-advantage genes (MAG) evolve near TDF and this region no longer pairs with the X. In females, X chromosome inactivation (XCI) evolves in the unpaired region (shaded) to restore dosage with the Y. As the Y degrades further, XCI spreads into regions with no Y partner. Ultimately, nearly all of the Y is degraded, leaving a small pairing region (the PAR at the top), and XCI spreads across most of the X (black), except for the PAR and escaper genes near the top.

Sex chromosomes of all animals get their start when a new sex gene evolves on one member of an autosome pair. For mammals, this meant the evolution of a testis-determining gene, which defined the chromosome as a Y and set it on its path of self-destruction (Section 4.6).

As we saw, comparisons between mammal groups show that there is practically nothing left of the original therian Y chromosome; only 4 of about 700 genes survive in eutherian mammals. However, in placental mammals (but not marsupials), the addition of a large autosomal region to an ancient pseudoautosomal region provided another few hundred genes, and the Y degradation process then began working to degrade these too. The human Y retains only 2.6 Mb at its tip that is homologous to the X. This pseudoautosomal region (PAR) contains 28 protein-coding genes that need no compensation.

The degradation process has been running independently in different eutherian mammal lineages, leaving PARs of different sizes but overlapping gene contents. These differences go some way to explaining differences we see in the genes that are inactivated. For instance, several genes are pseudoautosomal in horses but have lost their partners in humans; these genes are inactivated in humans but not horses (Figure 4.5). Makes sense – they don't need to be.

Our observations support a model in which chunks of the Y are progressively rearranged and removed from recombination with the X. Within these non-recombining regions, mutations and deletions inactivate Y genes, leaving their partners on the X unpaired and subject to dosage differences between XY males (one copy) and XX females (two copies). This sets up selection for dosage compensation of unpartnered X genes.

The process of Y degradation is rather jerky, so we might expect that incorporation into the X inactivation system might be jerky too.

Mammalian X chromosome inactivation is a whole-chromosome system, and we know that inactivation spreads into non-inactivated regions, even into bits of autosome fused to the X. So when a region of the X loses its Y homologue, it could simply be recruited into the X inactivation system. This process would be like the tide gradually coming in, washing over gene after gene.

However, there are genes within these inactivated regions that escape inactivation. At least 15% of genes on the human X are arranged in escape regions, protected by insulator elements CTCF. These regions lie mostly within the recently added region of the X, and their density increases as you go toward the pseuodautosomal region. They might simply represent regions that have been recently unpartnered, but have not yet been recruited into the X inactivation system.

In mice, few genes on the X escape, and these are arranged singly and protected by insulators. Why the difference? Perhaps mice simply represent a more advanced stage of Y degradation and recruitment of unpartnered X genes into the X inactivation system. As the tide of X inactivation has come in, many escape regions have succumbed, and the single escaper genes, making a last stand, are all that is left of the larger escape regions we see on the human inactive X.

We conclude that X inactivation has largely followed Y degradation, a jerky process which involved rearrangements of the Y that produced nonhomology over large regions and selected for recruitment of unpartnered X regions into the X inactivation system.

But could it be the other way around? Perhaps inactivation creeping along a partially differentiated X chromosome selected for loss of the partner genes on the Y. However, on this hypothesis you might expect to find genes that are inactivated on the X even though they retain active partners on the Y that complement gene function. There is no firm evidence of such genes.

7.2 Dosage Sensitivity of Genes and Evolution of Inactivation

This simple picture of Y degradation setting up selection for recruitment of unpaired X regions into the X inactivation system is complicated by the different roles of X genes. Some genes are very sensitive to twofold dosage differences, but for others it really doesn't matter. This may account for the jerky movement of the inactivation front (Figure 7.2).

Indeed, dosage doesn't seem to matter for most genes. Deliberate deletion of bits of the yeast genome showed that only about 3% of genes are dosage sensitive. These genes have vital functions in transcription and RNA processing, protein transport and folding.

In humans, many genes must work just as well in one copy as two because mutation on one homologue is completely covered up by the normal allele on the other. For instance, cystic fibrosis is a recessive disease affecting the lungs; parents who are heterozygous for mutants have no idea that they each harbour a mutant gene until they have an affected child who is homozygous for the mutant. On the other

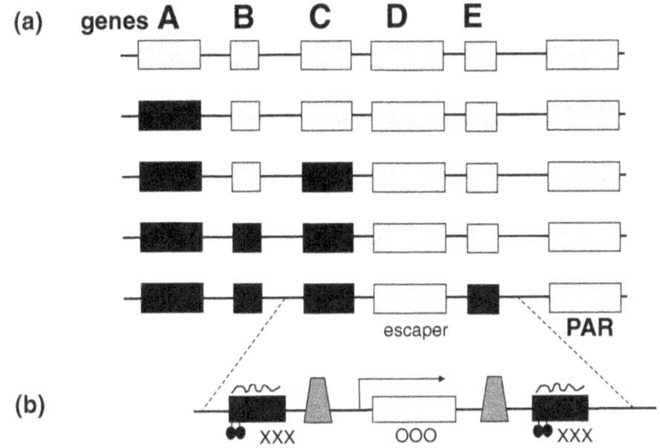

Figure 7.2   Recruitment of X genes into inactive domain, and escape. (a) Lines represents DNA of part of the X chromosome and letters A, B, C, D, E represent different genes. Initially the whole region is paired with the Y and all genes are active (grey). Inactivation accompanies Y degradation and one by one, genes are recruited into the inactive domain (black), not necessarily in order. Gene B initially escapes because its partner on the Y is retained, but when this is lost, it becomes inactivated on the X. Gene D escapes inactivation because it is required in two doses in both sexes, so the Y allele remains active. The PAR is just what's left after this degradation and recruitment. (b) Gene D escapes inactivation by the interposition of boundary elements (dark grey barriers) on either side which protect it from binding XIST RNA (wiggly line), inactive histone modifications (XXX) and DNA methylation (black dots) and permit active histone modifications (OOO) and transcription initiation (arrow).

hand, there are genes for which deletion or mutation on one homologue produces disease; for instance, deletion of one copy of the *SOX9* gene produces drastic bone deformities and sex reversal.

A study of abnormal phenotypes of patients with copy number variants detected 3,000 genome segments that give rise to phenotypes that are sensitive to loss of one copy (haploinsufficiency) and 1,500 that are sensitive to an extra copy (triplosensitivity), so the frequency of dosage sensitive genes may be higher in mammals than in yeast.

The low proportion of dosage-sensitive genes suggests that selection for recruitment into the X inactivation system will be weak – until Y degradation hits one of them.

Genes that retain active partners on the Y chromosome and escape inactivation on the X seem to be a special class of dosage-sensitive genes; either they cannot tolerate even a brief period of dosage inequality, or their dosage must match that of autosomal partners. Indeed, such genes as *UTX* and *KDM5* that have active alleles on both X and Y do have the hallmarks of particularly dosage sensitive genes.

However, many Y partners of X-borne genes are dead genes (pseudogenes); for instance *STSY* and *KALY* near the PAR on the human Y have both been recently truncated and inactivated. However, their X alleles have still not been recruited into the X inactivation system, presumably because dosage differences are not too damaging, so selection is weak.

Are escaper genes also particularly dosage sensitive? Perhaps these genes, particularly those that escape in mice as well as humans, are required to be active in two copies in females. In humans, where Y degradation has left large escape regions, 'innocent bystander' genes that are dosage insensitive escape because of their proximity to dosage sensitive gene.

There are other genes on the Y that are active, but now have a specialized function and can no longer back up their X-borne partner. For instance, *RBMY* has become a spermatogenesis gene on the Y, and no longer complements the vital function in brain development of its partner *RBMX* on the X, so it makes sense that *RBMX* is inactivated. This may explain some of the inconsistencies between species. For instance, *ZFX* escapes inactivation in humans because it has an widely expressed Y partner *ZFY*. But mouse *Zfx* is inactivated, although it has Y partners. However, in mice *Zfy* is expressed only in testis and seems to have acquired a male-specific function, so it no longer complements *Zfx*.

Different genes escape inactivation in different mouse tissues, so it seems likely that selection for their silencing depends on their activity and dosage requirements in different tissues.

Thus the pattern of inactivation of X-borne genes in different mammals makes sense in evolutionary, if not functional, terms. Genes are recruited into the X inactivation system as their Y partners become either inactive or specialized for a male role. The strength of the selective forces will depend on the function of each gene; some are critically dosage sensitive, whereas for others it doesn't matter.

7.3 Evolution of Silencing in Therian Mammals

The inactivation system into which unpartnered X genes are recruited is a marvel of efficiency and stability in eutherian mammals. Picking it apart (Chapter 6) shows that it is built up from several silencing mechanisms; changes in chromatin conformation and nuclear position, differential DNA methylation, histone variants and many types of activating and repressive histone modifications. It is all coordinated by the long ncRNA transcribed from the *XIST* gene and many interacting proteins. How did it start? How were these mechanisms co-opted and integrated?

Details of the molecular mechanism of X chromosome inactivation differ between humans and mice, and between eutherians and marsupials. Can we use these differences to detect how different inactivation mechanisms came together to form one big hyperstable silencing machine in eutherian mammals?

Comparing X chromosome inactivation between humans, bovines, horses, pigs, rabbits – even elephants – shows that inactivation and the molecules that enact it are basically similar in eutherian species. They all have a noncoding *XIST* gene that is expressed from the inactive X as pluripotency genes are turned off when the epiblast differentiates. *XIST* RNA interacts with the same cast of proteins that tether the inactive X to the nuclear membrane, recruit modified histones and silence genes. The spectrum of histone modifications seems to be almost invariant; deacetylation of H3 and H4 is an early event, followed by accumulation of the major repressive mark H3K27me3. Modifications and silencing are controlled by same suite of protein factors, starring polycomb repressive complex PRC2.

The differences between X inactivation in eutherians and marsupials are more substantial, though they are not as extreme as we first thought. The inactivation mechanism at first seemed very different from that in humans, being paternal, incomplete and tissue-specific; however, all these features have since been observed for eutherian X inactivation. It was originally claimed that DNA methylation of X-borne marsupial genes did not differ between the sexes, with the implication that DNA methylation does not lock in marsupial X inactivation as it does in eutherians. However, it now seems that, although methylation of transcription start sites is the same in males and females, sequences either side of transcription start sites are differentially methylated in marsupials as well as eutherians.

There may be some differences in the spectrum of histone modifications. In kangaroos, the major modifications (histone deacetylation, histone H3K27 trimethylation) occur, but not some other changes, suggesting modifications are more like those typical of repetitive sequences. However, a full spectrum was reported in the opossum.

The big differences between the components of the silencing systems in different species lie in the identity and function of the long ncRNAs that control X inactivation. Even among eutherians, there are major differences in the long noncoding RNAs. Humans have a ncRNA (*XACT*) that is transcribed from and binds to the active X. It is shared with gorilla, but is missing from all other eutherians, implying that it evolved in primates only recently.

Rodent X chromosome inactivation uses a suite of ncRNAs within the X inactivation centre to control *Xist* transcription. The *Xite* ncRNAs and two enhancers that control the activity of *Xist* seem to be specific to mice, implying that they evolved in the rodent lineage since they diverged from primates about 70 Mya.

Thus, noncoding genes seem to be much more variable than protein-coding genes, implying that they evolved rapidly. Evidently, once the *Xist/Tsix* interaction was set up, evolution of other ncRNAs that enhance or suppress *XIST* (*Xite* in mice or *XACT* in humans) would be strongly selected for independently in rodent and primate lineages.

The most spectacular difference between X chromosome inactivation in eutherians and marsupials is in the identity of the long noncoding RNA that controls the whole process, *XIST* in eutherians and *RSX* in marsupials. How can we account for this fundamental difference?

7.4 X Inactivation by Rival Noncoding RNAs

XIST is the lynchpin for control of X chromosome inactivation in eutherian mammals. For decades, we expected that this long ncRNA would also control X inactivation in marsupials.

My lab spent 10 years hunting fruitlessly for a marsupial version of *XIST*. It was hard to prove that *XIST* doesn't exist, but this became obvious when we found that genes that lie on either side of *XIST* in humans and mice map to two distant sites on the marsupial X, and neighbouring sequences had no homology to *XIST*. Thus, the region that should harbour *XIST* had been split in two in marsupials, and neither half contained anything that looked like *XIST*. Indeed, the region contained genes that were present in birds and frogs which had been obliterated on human and mouse X chromosomes.

Years later, a noncoding gene was discovered, again by chance, during investigation of the inactivation of the marsupial *HPRT* gene by James Turner. The large insert clone that contained *HPRT* produced, instead of a single point on RNA FISH, a cloud over one of the X chromosomes. *RSX* (for <u>R</u>NA on the <u>S</u>ilenced <u>X</u>) is moderately conserved between marsupial species such as koala and opossum.

RSX has no sequence homology to *XIST* and maps in a completely different part of the conserved region of the X, right next to *HPRT*. However, like *XIST* it is striking in its arrangement of several blocks of repeated sequences. Bioinformatic tricks that compare these repeats in a non-linear context do spot some similarities between *XIST* and *RSX* functional domains. The all-important *XIST* repeat A showed similarity to RSX repeats 2, 3 and 4. And *XIST* repeats B, C and D were highly similar to *RSX* repeat 1.

RSX is transcribed into a very, *very* long (45 kb) RNA and processed into a 27 kb mature RNA with no significant open reading frame. The spliced transcript is stuffed with repeats, in four major blocks (1, 2, 3, 4). In the absence of good marsupial sequence assemblies over this region, it was not clear if *RSX* is accompanied by other noncoding RNAs like *Tsix* or proteins such as RNF12 that regulate its activity.

Does *RSX* serve the same function as *XIST*? *RSX* is transcribed from the inactive (paternal) X and coats it. So far there is no direct evidence that *RSX* has a role in binding to the nuclear membrane, late DNA replication, histone modification or silencing on the inactive marsupial X. Knockouts and transgenesis are still not feasible in marsupials, but expression of an *RSX* transgene inserted into a mouse autosome induced some local gene silencing. This suggests that, like *XIST*, *RSX* induces X inactivation.

RSX RNA also has the ability to bind crucial protein factors, as assessed by the presence of specific binding motifs. The suite of factors overlaps significantly. Repetitive sites of *XIST* that are thought to be required for recruitment of polycomb repressive complex PRC2 are present also in *RSX*, and antibody to PRCs pulled up *RSX* in an immunoprecipitation assay.

So it seems likely that *RSX* acts in the same way as *XIST*, although it is entirely non-homologous. It is hard to make evolutionary sense of this extraordinary convergent evolution. Perhaps both systems grew out of a common ancestor that had two or three – or hundreds – of long noncoding RNAs performing silencing jobs at a local level (Figure 7.3). *XIST* and *RSX* might simply represent selection of two different molecules that perform the same function in the 148 Myr since marsupials and eutherians diverged.

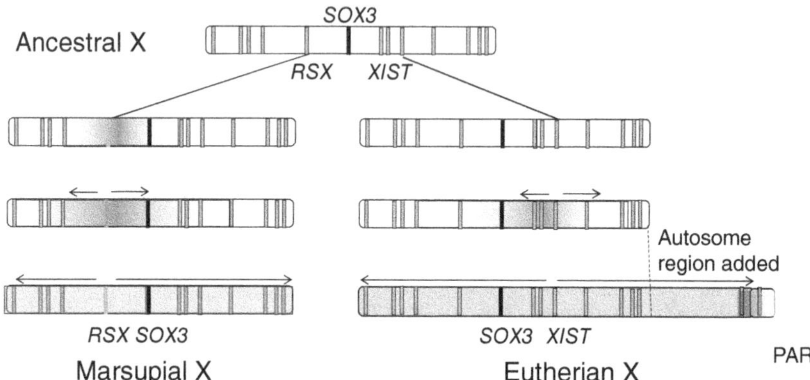

Figure 7.3   Evolution of long ncRNAs that control X chromosome inactivation. The ancestral mammalian X contained many long ncRNAs (grey stripes) that were involved in short-range silencing. The region containing *SOX3* (black bar) was the first to be differentiated from the Y, and inactivation of dosage-sensitive genes in this region was selected for. Noncoding ncRNAs on either side of *SOX3* were recruited to coordinate inactivation of genes: upstream *RSX* in marsupials and downstream *XIST* in eutherians. Inactivation controlled by *RSX* or *XIST* spread as the unpaired region on the Y was extended (arrows). In eutherians, *XIST*-controlled inactivation spread into an autosomal region that was fused with the X and Y, leaving a short pseudoautosomal region (PAR) shared with the Y.

7.5 The Evolution of *XIST*

Where did *XIST* come from and how did it evolve its complex function? Mapping genes around the X inactivation centre can establish the history of the X inactivation centre and provide clues to the origin of *XIST* and other noncoding RNA genes in the X inactivation centre.

The genes that flank the X inactivation centre, *CDX4* and *CHIC1* on one side, and *XPCT* and *RNF12* on the other, have homologues in all vertebrates. These four genes lie together on autosomes in birds and frogs, and on the X in humans, mice and all eutherian mammals, implying that the region has been conserved for more than 360 Mya. But the distance between the flanking markers is much smaller in birds and frogs, suggesting the region has greatly expanded in mammals by insertion of repetitive sequences (Figure 7.4).

So what lies between these flanking sequences in birds and frogs? There is no sign of *XIST*, nor any of the other ncRNAs. Instead, the gap between them is filled with five perfectly good protein-coding genes. These genes all lie in the same order and orientation in chickens and frogs with respect to the flanking markers, so this arrangement must be ancestral.

None of these coding genes survive in eutherians. Whatever became of them? Searching for sequence similarities between the ancestral genes and the ncRNAs

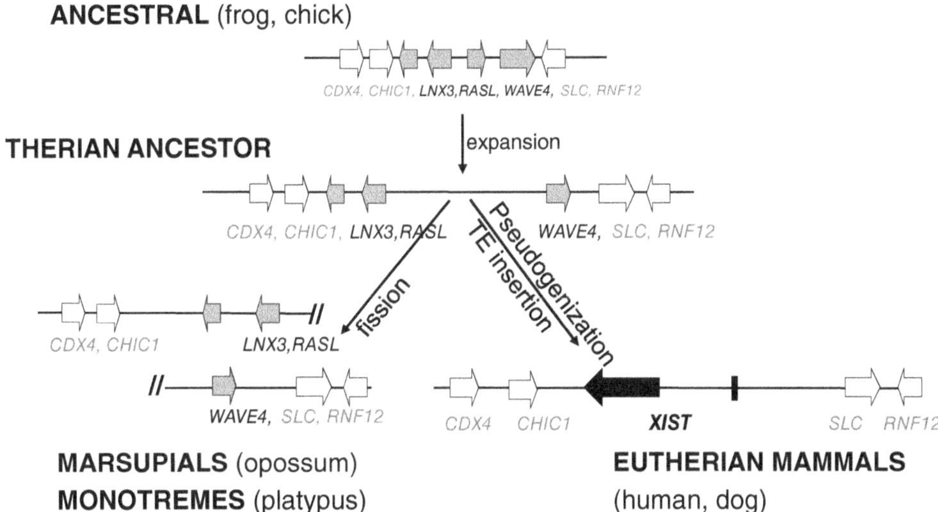

Figure 7.4 Evolution of *XIST*. Comparative gene mapping in reptiles and amphibians shows that the region containing genes (shown in pale grey) that flank *XIST* in eutherian mammals contains three protein-coding genes (dark grey). In the therian ancestor the region expanded with insertions of repetitive sequences. In marsupials and monotremes, this unstable region underwent fission independently, and is now apparent as two gene blocks far apart on the marsupial X and platypus chromosome 6. In eutherian mammals, TE insertions, deletions and mutations occurred to inactivate the three genes and create a long noncoding gene *XIST*; only a tiny portion of *LNX3* is recognizable in the promoter of *XIST*.

of the X inactivation centre came up with 134 short alignments between the ancient genes and regions of eutherian ncRNAs. Of these, the most interesting were some short stretches with homology to *XIST*. One of these ancient genes, *LNX3*, which codes for a ubiquitin ligase, has four exons that aligned with four *XIST* exons, as well as part of one intron and the promoter region. Regions of other ancient genes also aligned to other genes and pseudogenes in and around the eutherian X inactivation centre. This suggests that the promoter and half of the 10 exons of *XIST* evolved from an ancient gene *LNX3*.

What makes up the rest of *XIST*, especially the tandem repeats with crucial functions in binding and tethering? Comparing sequence elements of the tandem repeats with a huge database of mobile elements in chordates revealed that *XIST* Repeats A, C and D have homology to several well-known endogenous retrovirus sequences, left in genomes many millions of years ago after infection. Repeat F has homology to a different class of DNA transposon, repeat H to the very common L1 LINE element, and repeat E to another widely dispersed element, IAP.

These elements now make up most of *XIST*, including 78% of Exon 1. Copies have diverged widely both within and between eutherian species. In addition, mobile elements such as the primate-specific Alu have invaded more recently and are responsible for much of the species diversity.

It was of special interest to examine this region in marsupials, since these mammals lack *XIST*. Remarkably, the opossum sequence was found to contain *LNX3* and another ancient gene, apparently still functional.

However, the rot had already set in, with the loss of the other three coding genes. Mapping flanking markers to two distant locations in monotremes as well as marsupials showed that the region split independently in marsupials and monotremes. Perhaps it had already been invaded by mobile elements and had become structurally unstable in a common mammalian ancestor long before it evolved its function in controlling X chromosome inactivation. The invading repeats found a use in eutherians, but strained the region past its breaking point independently in marsupials and monotremes.

Thus the long noncoding RNA *XIST* was generated when repetitive elements invaded a protein-coding region, hijacking the promoter and repurposing coding sequence (Figure 7.4).

7.6 X Inactivation and Upregulation – Testing Ohno's Hypothesis

X chromosome inactivation does a good job of equalizing the concentration of X-borne gene products between XX females and XY males, because they each have a single active X.

But there is another problem. Most genes act in networks with other genes from all around the genome. Halving the amount of product from genes on the X would be disruptive if they must interact with genes that are present in two copies on autosomes.

A brilliantly satisfying hypothesis was put forward by Susumu Ohno in the 1960s to explain how mammals solved this problem, and how this led to the evolution of X chromosome inactivation. He suggested that the first step was to double the expression of genes on the X in males to compensate for the loss of the Y allele and maintain parity between X-borne and autosomal genes (Figure 7.5).

This would leave males with a balanced complement. But now females would have too much product from X-borne genes. While this is not as damaging as halving the amount in males, two-fold expression of X-borne genes in females would be detrimental. Selection to downregulate X genes in females would restore the balance.

This hypothesis predicts that genes on the active X should be, on average, twice as active as autosomal genes in males and females. Are they? You would think it a simple matter to measure transcripts from genes on the X and compare their concentration with transcripts from autosomal genes, but this has been far from the case.

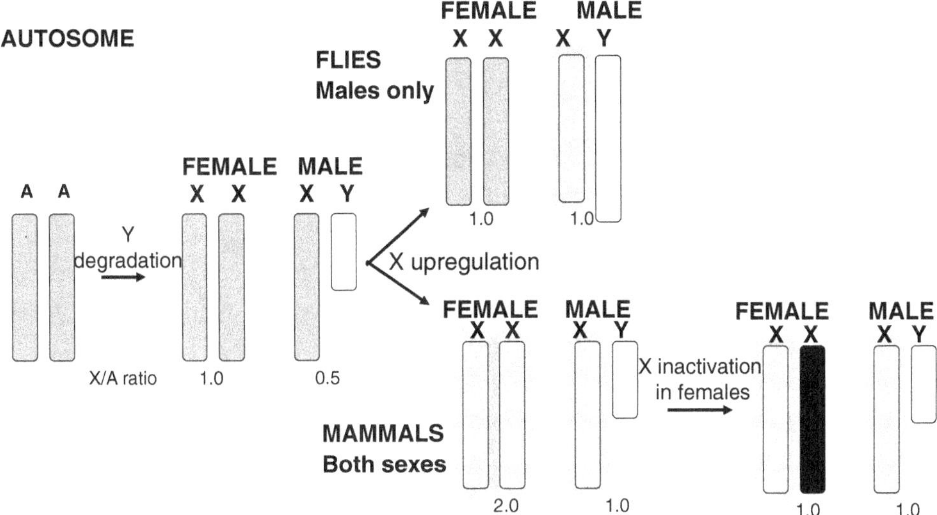

Figure 7.5 Ohno's hypothesis that upregulation of the X drove the evolution of XCI. An autosome pair (grey) evolved into differentiated sex chromosomes (XX in female and XY in male) by degradation of the Y chromosome (white). This did not alter X/A balance (shown below chromosomes) in XX females but did in XY males, and it left the balance between female and male as 2:1. In *Drosophila*, upregulation of activity on the X in males (shown as pale grey, with glow) restored this balance (top images), so that X/A was 1.0 in both sexes and male:female was balanced. However, in mammals (bottom), upregulation of the X occurred in both sexes, producing a 2:1 sex ratio. This male-female and X/A dosage imbalance drove X chromosome inactivation (black) to restore X/A dosage to 1.0 in both sexes and a balanced 1:1 sex ratio (bottom right).

In the first test of this hypothesis, Christine Disteche compared expression of a gene that lies on the X in one mouse species but on an autosome in another. Expression of the single copy of the X gene was equivalent to the two copies of its autosomal homologue, supporting Ohno's hypothesis. Analysis of X-borne and autosomal gene expression on microarrays agreed that genes on the single X were expressed at a higher rate that would balance autosomal genes.

However, using RNA seq to compare transcription of all the genes on the mouse X with transcripts of all the autosomal genes gave inconsistent results. One study of X:AA ratios in three tissues in five primates produced male:female ratios between 1.0 (doubling X expression) and 0.5 (no doubling). But another study found ratios of 0.5, expected if there is no doubling. A further study found no upregulation in eutherians, but upregulation in marsupials, another found upregulation in every tissue in sheep.

The problem with these comparisons is they must assume that X-borne and autosomal gene sets are equivalent and their expression will average out. But there is no 'average gene'; some are expressed at high levels, others at low levels, or are tissue specific.

One of the most rigorous approaches is to compare sets of X-borne genes in eutherians to their orthologues that lie on autosomes in chickens and platypuses, which were used as proxy for the ancestral transcription profiles of mammal X genes. The ratio was barely above 0.5 for any tissue in primates and mouse, contradicting Ohno's hypothesis that expression of X-borne genes was doubled in a mammalian ancestor. One RNA-seq study found no upregulation in eutherians, but upregulation in every tissue in marsupials. However, in another study, genes on the opossum X also showed no upregulation when compared with orthologues on chicken autosomes.

Single-cell RNA seq has been employed to compare X-autosome balance of genes that are moderately transcribed. This study detected an increase in concentration of transcripts of X-borne genes compared to autosomal genes.

It is therefore still not clear to what extent transcription rates support Ohno's hypothesis that upregulation of the mammalian X occurred as the Y degraded, and X inactivation evolved to compensate for this in XX females. These inconsistencies may just mean that highly dose-sensitive genes are upregulated, and others aren't. It is also possible that different lineages have more or less X upregulation.

7.7 Did X Inactivation Evolve from Meiotic Sex Chromosome Inactivation?

The discovery of meiotic sex chromosome inactivation (MCSI) seemed to connect the dots. It was suggested that inactivation of the X (and Y) at male meiosis leads directly to paternal inactivation in the embryo. This fitted in nicely with the idea that paternal inactivation evolved first in a common ancestor and random X inactivation evolved subsequently as an improvement which produced a quasi-heterozygous state.

Paternal inactivation was discovered first in marsupials in the early 1970s, leading to the speculation that paternal inactivation was the primitive state, an attitude encouraged by later discoveries that marsupial X inactivation was incomplete and tissue specific. Later it was observed that the first wave of inactivation in the rodent extra-embryonic membranes is also paternal. This led to the idea that the primitive paternal inactivation was retained by the extraembryonic membranes in eutherians, perhaps because it offered a way to suppress antibodies against embryonic antigens contributed by the male parent that might provoke an immune reaction in the mother.

The molecular mechanisms of paternal and random inactivation in mice share many components, including the roles of *Xist* and its suppressor *Tsix*. Paternal inactivation depends on the early expression of *Xist* from the paternal X. Expression of *Tsix* and its enhancers from the maternal X keeps the maternal *Xist* silent and the maternal X active. (Section 6.12).

The connection was soon made with MCSI, which inactivates both the X and Y chromosomes in sperm. Perhaps, then, the X was inactivated during male meiosis, and simply remains inactive through fertilization and into the first stages of embryonic development.

But this is not so. Heroic attempts were made by Edith Heard in Paris and Jeannie Lee in Boston to observe whether the paternal X was expressed in the fertilized egg, and the two- and four-cell stages, both in mice and in marsupials. The results were unequivocal – there is a brief stage in the earliest mouse embryos in which genes are expressed from both X chromosomes. So the X is inactivated at male meiosis, switched on at fertilization, and switched off again at the two-cell stage, then off again in the blastocyst. The same conclusion was drawn for the marsupial embryo.

Thus, X inactivation during spermatogenesis does not directly carry over into the embryo. Rather, the paternal X must retain some kind of molecular imprint on the paternal X through fertilisation and the first blastomere divisions. Indeed, inactivation of the paternal X in mice seems to be under the control of a subunit of maternal PRC2.

Paternal X inactivation is not seen in the embryo of humans, or any other eutherian except the cow. Does paternal X inactivation in mice, cows – and marsupials – offer any kind of selective advantage in these species? Or is paternal inactivation a side effect of some difference in embryonic development?

Probably the latter. Mouse embryonic development differs from that in rabbits, ungulates, and humans by its early genome activation and early differentiation of extraembryonic and embryonic cells. Early differentiation is also an attribute of the cow embryo. And it is particularly pronounced in marsupial development, in which not only the extraembryonic membranes, but the entire embryo begins to differentiate from a simple two-layered blastocyst within a few days of fertilisation.

This suggests that imprinted paternal inactivation is entirely a product of the timing of the first wave of differentiation. It may be that the paternal imprint, once set, takes a while to be erased. Early differentiation in mouse and cow trophecto-derm, and in the entire marsupial embryo, triggers inactivation before the imprint has worn off.

We conclude, therefore, that paternal X inactivation is a reflection of the timing of differentiation, rather than a response to special requirements. It evolved independently in different mammals that show early differentiation. It is not a pathway from precursor to random X inactivation in differentiation, or necessarily in evolution. Rather, both MCSI and X inactivation are selective responses to dosage inequities, and evolved by using the same silencing mechanisms.

7.8 Dosage Compensation in Monotremes?

To investigate further the conserved molecular silencing mechanisms, we must go outside therian mammals and ask how other vertebrates – and invertebrates – achieve dosage compensation and epigenetic silencing.

I will start with monotremes, extraordinary mammals that retain many reptilian features. Platypus and echidna have bizarre multiple sex chromosomes (Section 4.4). The male platypus has 5X and 5Y chromosomes, which evolved by multiple translocations between an original sex pair and four different autosomes. These sex chromosomes share no homology with the therian X chromosome, but extensive homology to the chicken ZW pair. Given that monotreme and therian sex chromosomes evolved independently from a different autosomes, we must ask if they share an X chromosome inactivation system. Surprisingly, platypus does not seem to subscribe to MCSI, although, like birds, X chromosomes are downregulated during meiosis. This would imply that MCSI, as well as X chromosome inactivation, evolved in therian mammals.

If any species should need dosage compensation, it is the platypus. The five degenerated Y chromosomes are all greatly reduced from their erstwhile partners on the X, leaving nearly 10% of the genome unpaired in males.

In the absence of gene mapping or molecular techniques, early studies focused on cytological hallmarks of inactivation. However, cells from female platypus and echidna had no obvious sex chromatin body, and replication patterns were hard to interpret, given that the identity of sex chromosomes had yet to be sorted out. Levels of DNA methylation on the X chromosomes of females gave no indication of global silencing.

However, we cannot entirely dismiss the concept of global dosage compensation in monotremes because there is a 10 kb noncoding gene on the largest platypus and echidna X (called *PSX* for P̲latypus S̲ex-biased on the X̲) that is expressed only in females. Could it have a role in partial sex chromosome dosage compensation in monotremes?

Comparison of the level of expression of platypus X genes with expression of their orthologues on chick autosomes (proxy for ancestral monotreme sex chromosomes) showed some upregulation – but only in males. This would achieve partial dosage compensation without the necessity of X inactivation.

Recent chromatin immunoprecipitation of active and inactive histone variants along X-specific and paired regions showed the active mark H3K4me1 was much weaker in females in X-specific regions, suggesting this difference might be subject to transcriptional upregulation in males.

When genes were mapped to monotreme X chromosomes, it became possible to monitor their activity directly. SNP analysis showed that each of seven

X-specific genes was transcribed from both alleles, implying no, or random inactivation.

Later RNA-FISH on platypus female cells suggested partial stochastic silencing. Nuclei showed either two transcription signals (2X active) or one (1X active). Each gene had a characteristic and reproducible frequency of 1X active nuclei (25–62%). Single active alleles were located on the same X chromosome, suggesting at least local control in cis. In the absence of platypus families, we could not determine whether inactive alleles were on the paternal or maternal X.

Subsequently, RNA-seq data was used to measure the ratio of expression of many genes on the two largest X chromosomes, in fibroblasts and several tissues from males and females. There was wide variation between genes, but overall, expression in female cells was greater than in male cells (male:female ratios of about 0.6). This implies weak and variable compensation of X-borne genes in platypus, consistent with the RNA FISH results. We had to conclude that sex chromosome dosage compensation is not, after all, critical for life – at least in platypuses.

However, recent research on monotreme sex chromosomes in Paul Waters' Sydney lab completely contradicts this conclusion. Whole genome transcriptomics, coupled with measurement of protein output, shows that though dosage compensation at the transcriptional level is weak, there is complete compensation at the protein level. Whereas X-specific genes showed a male:female ratio of 0.6 at the mRNA level, the male:female ratio was 1:1 at the protein level. This implies that control at the transcriptional level is supplemented by an additional level of post-transcriptional control (Figure 7.6).

Moreover, comparisons of the output of X-specific genes with those of autosomal genes between males and females showed a lower value in males at the transcriptional level, but parity at the protein level. Thus, X-specific genes are only partially compensated at the transcriptional level, but fully compensated at the protein level.

Thus, dosage compensation in platypus is not the half-hearted affair we first thought, but is an extraordinary interplay of variable transcriptional control brought up to parity by different levels of RNA stability or transport, protein translation or stability, at least in monotremes. How this delicate balance is achieved remains quite mysterious at present, but it signifies that dosage control is, after all, critical.

This revolution in appreciation of dosage control demands our rethink of the role and mechanism of transcriptional and post-transcriptional control in birds, which were first demonstrated to have incomplete and variable control at the mRNA level.

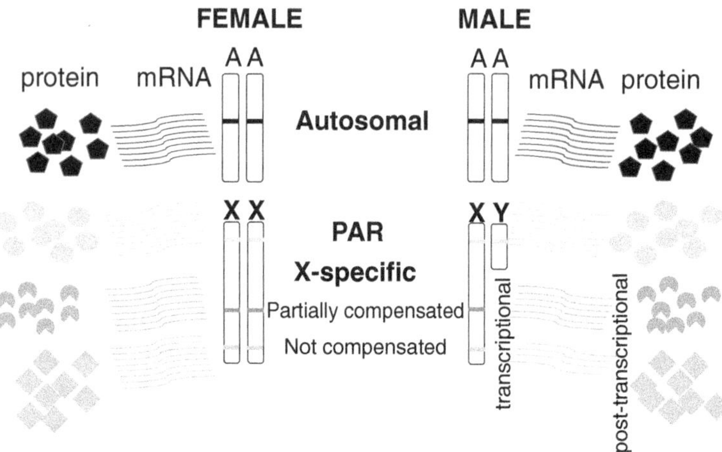

Figure 7.6 Monotreme dosage compensation by transcriptional and post-transcriptional regulation. Autosomal genes (black bars) are present in two doses in both sexes, and are equally transcribed and translated, as are genes (pale grey) in the pseudoautosomal regions (PAR). However, genes in the X-specific regions (darker grey) are present in single dose in males and double dose in females. These genes are variably transcribed in males and females; some are partially compensated by regulation in males, and others are not compensated at all, so mRNA is less abundant in males than it is in females. However, protein abundance is equivalent in both sexes, and equivalent to autosomal genes, implying post-transcriptional upregulation occurs to balance protein abundance in males and females.

7.9 Dosage Compensation in Birds?

Birds, too, seemed to lack global sex chromosome dosage compensation. This was suspected long before techniques were available to look directly at transcription.

Birds also have differentiated sex chromosomes (Section 4.9), but they are the other way around: males are ZZ and females are ZW. The Z chromosome is highly conserved, bearing about 900 protein-coding genes. The W is a small heterochromatic element in fowl and songbirds, bearing only a handful of genes that are partners of Z genes. However, in ratite birds (emus and ostriches), the W is hardly differentiated and still largely pairs with the Z.

It was assumed for a long time that dosage inequity of 900 genes between males and females would have mandated some form of compensation between ZZ males and ZW females. However, no one could discover cytological symptoms of an inactive Z in males. There was no sign of a heterochromatic body or a late-replicating Z. Early isozyme studies of three Z-linked genes showed that both alleles are expressed by heterozygous males, but could not assess dosage.

So is there really no dosage compensation in birds? Microarray studies and more recently RNA-seq all show that there is some compensation at the transcriptional level, but it is incomplete and differs between genes. For a female heterogamety system (ZZ male and ZW female), the expected male:female expression ratio for full compensation is 1.0, and the expectation for no compensation is 2.0 (the other way around from mammals). The male:female ratio for different Z genes was between 1.2 and 1.6 in chickens, implying partial compensation.

RNA in situ hybridization of Z-borne genes in males also showed partial – but stochastic – transcriptional inhibition. Individual genes had reproducible frequencies of 2Z-active (two signals) and 1Z-active fibroblasts (one signal), similar to that seen for the X chromosomes in monotreme cells (as well as partially inactivated marsupial X-borne genes and eutherian escape genes). This stochastic control appeared to be at least regional, for neighbouring genes on the same Z were either both on or both off in a particular cell.

Incomplete and variable compensation was also observed in distantly related birds (zebra finch, crow, and flycatcher). But not in ostrich or emu, because most of these genes are pseudoautosomal in ratites.

Another songbird, the common whitethroat (Section 4.9), presents a unique system to study the evolution of bird dosage compensation because an auto-somal region was fused to the ancestral sex pair about 19 Mya. Both the Z and the W received a large addition, and Z-W sequence divergence showed that many genes on this neo-W have already degenerated. The male-female expression ratio of about 1.36 showed that genes on the new Z region were partially compensated, but not as much as genes on the old part of the Z. This implies an ongoing process of W degradation and a gene-by-gene compensation.

Are Z genes upregulated in ZW females or downregulated in ZZ males? The data suggest both. Comparison with mammalian orthologues suggests Z-borne genes are variably upregulated in females to partially compensate for their presence in only one copy.

In the absence of a global Z compensation mechanism, we would expect no regional differences in silencing. Indeed, male-female ratios of gene transcription bump along uniformly over most of the chicken Z. However, they show a clear patch of more compensated genes lying in 7 Mb of the short arm of the chicken Z chromosome. This coincided with a 250 kb valley of differential DNA methylation, suggesting a region of the Z (dubbed 'Male HyperMethylated', or MHM) is differentially regulated in males and females.

MHM was hypermethylated in males (condensed and transcribed at a low level) and hypomethylated in females (accessible and more highly transcribed); it might therefore control upregulation of the single Z in females. This MHM (and another smaller one on the Z long arm) were highly repetitive and contain – guess what – long ncRNAs. Local and female-specific histone hyperacetylation (H4K16Ac) has also been described on the Z. Thus, long ncRNA might bind to the region, facilitate acetylation and increase gene transcription. Indeed, the MHM loci have interesting parallels to the X inactivation centre in therian mammals, but we don't yet know whether it controls Z upregulation or has some other female-specific role.

Although other birds share this part of the Z chromosome (the oldest stratum), there is no sign of a valley of differential expression or methylation at the orthologous site in zebra finches or the white-throated sparrow. It must therefore have evolved in fowls after they diverged from other birds 90 Mya.

The incomplete dosage compensation at the transcriptional level in birds made us re-examine the need for dosage compensation, and recognize that there may be other ways to achieve it, or to live without it. Indeed, recent experiments measuring mRNA and protein abundance for Z-borne and autosomal genes in males and female chickens show that, as for monotremes, post-transcriptional regulation takes up the slack so that male-female, and Z-autosome outputs are balanced at the protein level (Figure 7.6).

Given that the platypus sex chromosome chain evolved by multiple translocations from a bird-like sex chromosome (Section 4.9), I consider it likely that this multilevel partial transcription/complementary translation system represents an ancestral dosage compensation mechanism. It is therefore interesting to consider dosage compensation in reptiles, amphibians and fish.

7.10 Sex Chromosome Dosage Compensation in Other Vertebrates

Little is known about dosage compensation in fish, amphibians or other reptiles (birds are, of course, a branch of reptiles).

Reptiles display a great variety of XY and ZW systems and their variants, and a great variety of degrees of differentiation (Section 4.12). Frogs generally have monomorphic sex chromosomes, so you would not expect dosage compensation to be needed. And fish have every imaginable type of sex chromosomes.

It is interesting to look for dosage compensation in species with different degrees of sex chromosome differentiation. Snakes offer such a system. Vipers and colubrids have the same ZW pair, in which the W is greatly degraded and full of repetitive sequences, so most Z genes are present singly in ZW females. RNA-seq showed the male:female ratio of expression of Z-borne genes in vipers and rattlesnakes to be much higher than 1.0 (complete compensation) but lower than 2.0 (no compensation). Comparing expression of genes on the Z in rattlesnakes with orthologous genes that lie on autosomes in chickens and lizards showed no evidence for upregulation of Z genes at the transcriptional level that is predicted by Ohno's hypothesis.

There was variation in male:female expression ratio of genes along the Z chromosome that suggested regions with higher or lower compensation. However, there was no sign of any fully compensated region that could mark some kind of control centre. Thus snakes appear to have no Z upregulation in females, no global dosage compensation, but do partially control gene activity on a gene-by-gene, or regional basis.

However, very different results were found for the green anole lizard. Like mammals, the anole has an XX female:XY male system, and also a recent autosome addition which, as for eutherian mammals, generated a new region on the X and Y. RNA-seq data reveals almost perfect transcriptional dosage compensation for genes on the oldest part of the X with male:female ratios close to 1.0. Genes in the newer parts of the X show incomplete compensation, with ratios below 1 but way above the 0.5 expected for no compensation.

Moreover, expression of X genes in males is equivalent to expression of autosomal genes, implying that X-borne genes are upregulated in XY males. This is accomplished by acetylation of histone H4K16 by an acetylase that is twice as active in males. Evidently, upregulation of X genes occurred to match autosomal levels, but this occurred only in males, so X inactivation in females was unnecessary (Figure 7.5).

In several fish with differentiated sex chromosomes, gene expression studies imply an incomplete and variable gene-by-gene dosage compensation. This includes the stickleback, which has a young XY system (Section 4.12), and the tongue sole with a ZW system.

A study of several closely related species of livebearing fish, including the guppy, revealed an XY system with different levels of differentiation, from undetectable to nearly complete degradation. Expression of X-borne genes in XY males and XX females were equivalent even in the species with largely degraded Y chromosomes, implying complete dosage compensation. As yet, we don't know how this works.

Thus, most fish show an incomplete, gene-by-gene dosage compensation at the transcriptional level, but at least one group has evolved complete X dosage compensation.

The picture of sex chromosome dosage compensation in vertebrates is therefore one of incredible variety. XY and ZW systems of male and female heterogamety, with more or less differentiated Y or W chromosomes, may be compensated at the transcriptional level by a whole chromosome, regional, or gene-by-gene upregulation in the heterogametic sex and/or downregulation in the homogametic sex. At least in birds and monotremes, partial compensation at the transcriptional level is bolstered by regulation at the post-transcriptional level (translation? protein stability?), implying a critical role for sex chromosome dosage compensation in vertebrates.

7.11 The Big Picture of Dosage Compensation in Animals

The most detailed studies of sex chromosome dosage compensation came, not from mammals or other vertebrates, but from invertebrates, particularly the fruit fly *Drosophila melanogaster*, and the nematode *Caenorhabditis elegans*. These two models are highly manipulable, and many mutants are available.

Drosophila was by far the earliest model for sex chromosome function, and is the best understood at a molecular level. As in mammals, males are XY and females are XX, and the Y chromosome bears few active genes. Based just on its appearance, it was recognized in the 1950s that the single X in males works twice as hard as each X in females. Indeed, subsequent studies showed that complete dosage compensation is achieved by the twofold upregulation of transcription of X genes. X upregulation occurs only in males, so there is no need to downregulate the X in females (Figure 7.5).

Two-fold upregulation of X genes in males is achieved by an enzyme that acetylates histone H4K16 (sound familiar?) (Figure 7.7). Its action is complicated by the need to make it male-specific. This is achieved by the formation of a protein complex initiated by the male-specific lethal locus *msl*. MSL protein is expressed only in male embryos. In females, it is inhibited by a gene *sex lethal (sxl)* that is active only in XX embryos in which the X/autosome ratio is 1.0. In male embryos *msl2* binds to several other epigenetic modifying proteins, including an acetyltransferase, as well as long ncRNAs transcribed from two X-borne genes (*rox1* and *rox2*),

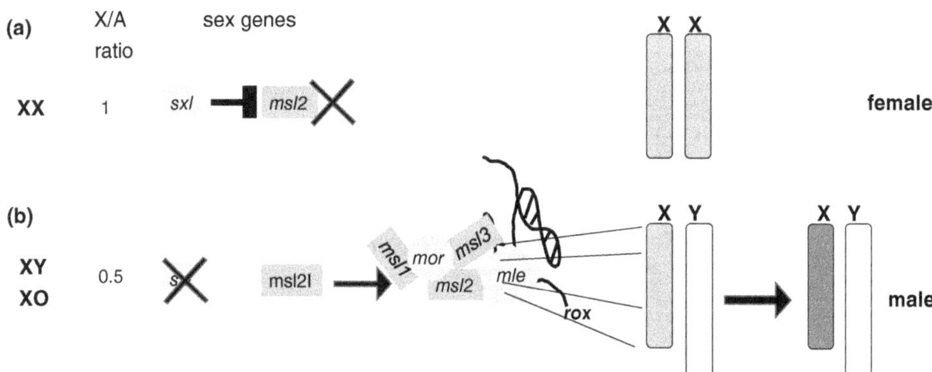

Figure 7.7   Dosage compensation in *Drosophila*. The X/autosome ratio X/A determines the expression of gene *sex lethal, sxl*. (a) *Sxl* is expressed when X/A = 1. It inhibits *male specific lethal msl2* in XX embryos. (b) *Sxl* is not expressed when X/A is 0.5 in XY or XO embryos, so *msl2* (a ubiquitin ligase) is expressed. It forms a complex with *mof* (H4K16 acetylase) and other epigenetic modifiers as well as a long ncRNA (*rox*), which binds to sequences on the X chromosome and spreads, causing changes to chromatin organization that result in hyperactivation of all genes on the single X.

and is recruited to entry sites on the X marked by a repetitive DNA motif. The MSL complex boosts H4K16 acetylation and changes chromatin configuration all along the male X chromosome to loosen chromatin and increase transcription of the X in males.

Some *Drosophila* species have autosomal additions to the conserved X, allowing us to see how Y degeneration and X dosage compensation spreads into new regions. Y degeneration is very rapid, and hyperacetylation of the unpaired portion of the X follows it closely, as DNA sequences mutate to form new MSL entry sites.

The nematode *C. elegans* has XO male: XX hermaphrodite sex determination. Again, X chromosome dosage compensation between XO and XX is achieved by doubling transcription of the X to match that of an autosome pair. However, upregulation is not male-specific. Overexpression of X genes in XX hermaphrodites is avoided, not by inactivating one of the two X chromosomes, but by downregulating transcription from both.

Again, a protein complex is key. The <u>D</u>osage <u>C</u>ompensation <u>C</u>omplex (DCC) comprises condensin (the protein that condenses and segregates chromosomes at meiosis) and five other proteins, one of which is expressed only in XX hermaphrodites. DCC binds to recruitment sites on the X defined by a 12 bp motif, and spreads along both X chromosomes (and also into fused autosomal regions). DCC regulates X chromosome structure by boosting the repressive H4K20me1 along both X chromosomes and depleting activating H4K16ac. This downregulates transcription from both X chromosomes in XX hermaphrodites.

Recent studies on other insects and invertebrates reveal many different ways of achieving dosage compensation. In mosquitoes, with an unrelated XX female:XY male system, male-specific upregulation of X-borne genes also occurs, but at the behest of a Y-borne gene. Most bugs (Hemiptera) also have an XY system, with a highly conserved X. All species examined had some kind of dosage compensation, showing male:female ratios of X-borne gene expression of 0.8 that denotes partial compensation at the transcriptional level.

Butterflies and moths excited a lot of interest because they all have a ZZ male:ZW female mode of sex determination and could be used to test the hypothesis that only XY systems have global dosage compensation, which is absent in ZW species such as birds (Section 7.9). Early results suggested that ZW females had lower expression of Z-borne genes than ZZ males, but when sex-biased genes were removed, gene expression in both sexes was equal for all Z-borne genes in several tissues in four species, in contradiction to this hypothesis.

The Monarch butterfly has a compound Z produced by fusion with an autosome. Remarkably, the old part of the Z had nematode-like compensation but the new part had *Drosophila*-like compensation. Evidently organisms can select for and maintain two different global systems of dosage compensation simultaneously.

Dosage compensation systems might also be different at different stages of development. Schistosomes (parasitic liver flukes), also with a ZW system, have complete compensation in the free larval stage, effected by Z upregulation in females. However, as parasites in vertebrate hosts, they default to incomplete compensation. Z gene upregulation depends on differences in chromatin structure which accompany differences in histone modification, including enrichment with the active H3K4me3 and depletion of repressive H3K27me3 and H4K20me1.

Remarkably, dosage compensation even occurs in some plants. Campions of the genus *Silene* have evolved – very recently (4 Mya) and very rapidly – an XY system of sex determination. Upregulation of the single X was seen in males, and upregulation, of only a single X, in females. In both sexes, the upregulated X was inherited from the female parent, so it constitutes a different example of paternally imprinted dosage compensation. This young system is remarkably efficient; it immediately incorporated unpaired X sequences (constructed by targeted Y deletions) into the X upregulation system. This might account for the record rate at which the *Silene* Y chromosome has degenerated.

The big picture of dosage compensation is thus one of great variety of mechanisms resulting in different degrees of transcriptional regulation. There is no great divide between XY and ZW systems, as was originally proposed. Upregulation of the single X or Z occurs in most species but is not ubiquitous. Male-specificity occurs in some XY systems (and female specificity in some ZW systems) but not in others, and is offset by X or Z downregulation of different types and completeness (inactivation or hyperactivation of a single chromosome, or downregulation of both). The role of post-transcriptional control is yet to be investigated, and it remains possible that all the systems with partial control at the transcriptional level will turn out to be fully compensated at the protein level.

7.12 Other Ways to Compensate – or Avoid – Gene Dosage Differences

Sex chromosomes are not the only generators of gene dosage problems. Many genetic accidents result in unbalanced complements of autosomes (Section 1.9) or parts of autosomes. For instance, patients with Down's Syndrome have three rather than two copies of chromosome 21. But the activity of genes on chromosome 21 is not 50% higher than normal as you'd expect. There must be some system that damps down the activity on chromosomes that are present in abnormal numbers in aneuploids.

This dosage compensation system, general to the whole genome, also seems to apply to the mouse X chromosome. When *Xist* was knocked out specifically in the epiblast which forms the embryo, the trophectoderm that forms extra embryonic membranes underwent paternal X inactivation as usual, but XX female embryos could not inactivate either X. A quarter survived to term, although they were severely abnormal and died at weaning. The really surprising thing, though, was that expression of X genes in these 2X-active females was not double, but only 14–36% higher. Thus, the two X chromosomes in female embryos are partially compensated, even in the absence of normal X inactivation.

Dosage compensation can also be achieved in XY males, not only by upregulating genes on the single X, but by downregulating autosomal genes with which they interact. There is at least some evidence that this occurs in mammals.

Recent experiments in David Page's laboratory deliver the surprising finding that extra Y chromosomes and inactive Xs both have major effects on gene expression, both from the active X, and many genes (21%) on autosomes. These effects are positive for some genes and negative for others, and are strikingly different in different tissues, so it is hard to see how this would smooth out dosage compensation, but it does underscore the complexity of dosage regulation.

There are also ways of avoiding dosage compensation by the simple expedient of expressing X genes only in one sex. For instance, all the cancer-testis antigen genes on the human and mouse X are expressed only at male spermatogenesis, and several genes on the bird Z are expressed only in the egg. There is a strong drive for genes on sex chromosomes to evolve sex-specific and tissue-specific expression. In fact, very many genes are sex-biased, in that they are expressed much more strongly in one sex than the other. Indeed, a recent comparison of gene expression in men and women found that fully one third of our 20,000+ genes are expressed more in one sex than the other.

Thus, there is even more variation between animals in the way that gene dosage differences are compensated – or avoided.

7.13 The Even Bigger Picture – Other Epigenetic Silencing Systems

Chromosome inactivation and parent-specific gene activity were described long before Mary Lyon formulated her X chromosome inactivation hypothesis.

Cytological observations on weird sex chromosome behaviour of some peculiar flies and scale insects first showed that one X chromosome may become condensed, inactivated, and even eliminated at meiosis or from the developing embryo. The process, then called 'facultative heterochromatinization', involved the condensation of the whole X chromosome and cessation of RNA transcription, observed by the crude methods of the time as dark-staining blobs that failed to incorporate radioactive RNA precursors. In some mealybugs the entire paternal set of chromosomes becomes heterochromatic and is eliminated.

Classic genetic crosses showed that the disposable X, referred to as 'imprinted', always came from the male parent. Heterochromatinization appeared to be under the control of a gene on the other X (referred to as the 'controlling element') which acts in cis. The molecular changes underlying imprinting and inactivation in these systems are not yet defined, but differences in DNA methylation and histone modification between maternal and paternal chromosomes have been reported.

Digging further into transcriptional silencing also reveals even deeper roots. It now is clear that many autosomal genes are expressed from only one of the two homologous loci in a range of species from yeast to human. Some instances of monoallelic expression are merely the consequence of a snapshot of a very jerky transcriptional process, in which bursts of RNA synthesis are separated by long pauses. But some of these genes are members of large families (such as antibodies and odour receptors), and share mechanisms to ensure that only one allele, of only one gene copy, is expressed in any clone of cells. For instance, individual human immune cells and their descendants make only one type of antibody by a process of 'allelic exclusion', and some parasites, such as Trypanosomes use the same process to switch their surface proteins to evade the immune system.

But studies of transcription in yeast showed that expression from one allele ('monoallelic expression') is far more widespread. Fully 13% of genes were expressed predominantly from one allele in diploid cells, and there was genetic evidence to suggest that there was control in cis of this allelic silencing. The ability now to look at RNA transcripts from a single mammalian cell reveals that many genes – perhaps 12% of the genome – are expressed from only one of the two alleles. This expression pattern is stable and mitotically heritable, giving rise to a clone of cells expressing one or the other of the two alleles. Different clones express different alleles, showing that this is not parental imprinting, nor a DNA variant.

The developing animal is therefore a genetic mosaic, much like the blotchy mice described by Mary Lyon. The molecular basis of random monoallelic expression shares several characteristics with the randomly inactive X chromosome, including proximity and concentration of LINE sequences, nearby or overlapping long ncRNA, different positions in the nucleus, DNA methylation differences, late DNA replication and differential histone modifications. Transcription repression seems to be stochastic, as are the partially silenced genes on chicken, monotreme and marsupial sex chromosomes, and the escaper genes in human and mouse X chromosomes.

Thus, a wide variety of other systems of epigenetic silencing share molecular mechanisms with X chromosome inactivation.

7.14 X Chromosome Inactivation and Genomic Imprinting

Even more striking similarities are seen with the parental silencing of a group of 'imprinted' autosomal genes in mammals.

Discovery of paternal X chromosome inactivation in kangaroos in 1971 was the first indication that parent-specific gene expression occurred in mammals as well as mealybugs. It was another decade before parental control of gene silencing was discovered in autosomes of eutherian mammals.

Nuclear transplantation experiments in 1984 established that a mouse embryo required genomes from both a female (egg) and a male (sperm) to develop normally. Different combinations of pronuclei produced androgenetic eggs with two paternal nuclei and gynogenetic eggs with two maternal nuclei. Neither got very far in development; the androgenetic eggs were good at developing extra-embryonic tissue but bad at embryonic development, and the gynogenetic eggs were the reverse.

The failure of both turned out to be due to genes that are expressed solely from one or the other parental genome. Parent-specific silencing of autosomal genes was first detected because deletion or mutation of the expressed allele can produce genetic disease. For instance, Prader–Willi disease (compulsive eating and mild mental retardation) was found to be caused by the deletion of a region from the paternally derived chromosome 15, and quite a different disease, Angelman Syndrome (ataxia and severe mental retardation) by the deletion of same region from the maternal chromosome 15. It turns out two genes whose deficiencies result in these diseases lie close together in this region and are oppositely imprinted.

Several other groups of imprinted genes are known from human genetic diseases, and unbiased studies of transcripts now identify more than 200 imprinted genes, about half of which are silenced if they come from the father, and half if they come from the mother. Some regions contain several genes that are imprinted one way or the other – or both ways. Imprinting may be tissue-specific, and is particularly rife in the placenta.

Genomic imprinting also occurs in other mammals. In mice, an overlapping set of more than 200 genes are paternally, or maternally imprinted to be inactive. Many genes are imprinted widely across eutherians, and imprinted regions have the same structure even between humans and elephants. Only a few of these were found to be imprinted in marsupials, but unbiased molecular methods found ten more and predicted a total of 60, largely specific to marsupials.

The imprinting marks that distinguish maternal and paternal alleles are established during male and female gametogenesis, passed to the zygote through fertilization, maintained throughout development and adult life, and erased in primordial germ cells before new imprints are set.

Imprinted genes are usually clustered in large regions (hundreds or thousands of kb) that are replete with repetitive DNA sequences. Clusters often contain both maternally and paternally imprinted genes.

Differential DNA methylation was the first mechanism to be discovered. CpG rich regions called Differentially Methylated Regions show different levels of methylation on the maternal and paternal alleles, and also in male and female germ cells. There are also changes in chromatin conformation, DNA replication timing and nuclear position within imprinted regions that are correlated with gene silencing.

Each imprinted cluster has an Imprinting Centre (IC) that is differentially methylated on the paternal and maternal chromosomes and acts in cis to control transcription of up to 3 Mb of DNA. The IC has receptor sites for binding the insulator CTCF, and this seems to operate the switch that controls which allele is expressed.

Every imprinted cluster harbours long noncoding RNAs, usually originating at the IC. For instance, the Prader–Willi –region contains an enormous (~1 Mb) ncRNA, whose truncation results in reactivation of the silenced alleles. It is processed into coding sequences, as well as a noncoding RNA that is antisense to coding genes and overlaps with the differentially methylated site. Long ncRNAs form a 'cloud' in the nucleus like the *XIST* cloud, and they may act in cis by coating the chromosome region to form a local silencing compartment. Histone modification must play at least a minor role, since mutation of histone modifying enzymes (including PRC) affects imprinting (Figure 7.8).

Transcriptional silencing may be due to antisense interference by the long ncRNA, DNA methylation directly or by facilitating histone modification, and action of CTCF; maybe combinations of all four, which are not necessarily the same for all imprinted regions.

When did genomic imprinting evolve? Its presence in marsupials but absence from egg-laying monotremes and other vertebrates implies that it arose in a therian ancestor 146–105 Mya, along with viviparity. However, some regions were imprinted only recently. For instance, the genes within the Prader Willi–Angelman region lie on two different chromosomes in chickens, frogs, monotremes and even marsupials, so they must have come together only recently in the eutherian lineage. Gene duplication and invasion of repetitive sequences and mobile elements occurred within the fused region. At another imprinted locus, differential methylation is lacking in marsupials, so it must have evolved in eutherians.

Why did genomic imprinting evolve, when it forsakes the advantages of diploidy and exposes a gene to Ohno's 'peril of hemizygosity'? Many imprinted genes code for proteins that regulate embryonic growth, and overall maternally imprinted/paternally expressed genes favour growth of the embryo or neonate (at the expense of the mother), whereas paternally imprinted/maternally expressed genes limit growth. The conflict-kinship theory proposes that the different interests

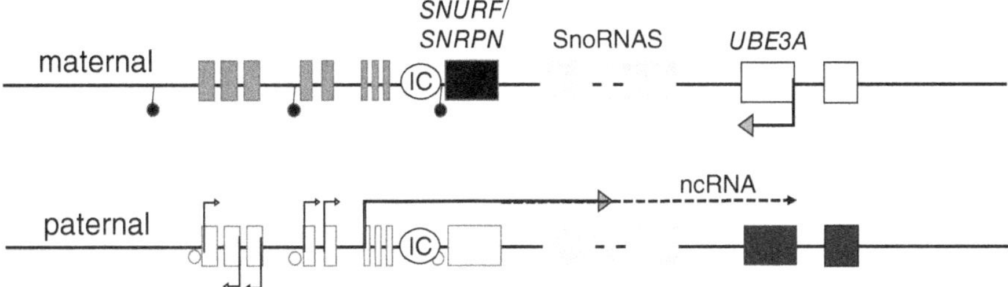

Figure 7.8   Genomic imprinting. Imprinted regions share many characteristics with the inactive X, including an imprinting centre IC, DNA methylation and histone modifications, generation of ncRNA. Two genes close together on human chromosome 15 are oppositely imprinted; *SNURF/SNRPN* is expressed only from the paternal and *UBE3A* from the maternal chromosome 15. On the maternal chromosome 15, the *SNURF/SNRPN* cistron (two genes regulated together by the imprinting centre IC) is repressed by DNA methylation (black circle), as are other genes upstream, while *UBE3A* and a neighbouring gene downstream are expressed. On the paternal chromosome 15 the *SNURF/SNRPN* cistron is transcribed from 5'exons into a very long RNA that is spliced into several coding transcripts (including *SNURF* and *SNRPN*), small nucleolar RNAs (snoRNAs) and a ncRNA antisense to *UBE3A*, which represses *UBE3A* transcription and inactivates the gene. Deletions in this region produce different diseases. Deletions (or uniparental disomy – both copies of chromosome 15 from the same parent) that remove paternal *SNRPN* cause Prader–Willi syndrome, and deletions or mutations that affect maternal *UB3A* cause Angelman syndrome.

of the parental genomes in species with internal gestation drive the evolution of imprinting; the interests of the paternal genome are to exploit female partners to have as many offspring as possible, whereas the interests of the maternal genome are to limit expending resources on one foetus in order to have many more.

Alternatively, genomic imprinting may have evolved to reinforce the necessity for sexual reproduction between male and female, since genomic imprinting makes parthenogenesis impossible in mammals.

There are many obvious parallels between genomic imprinting and X chromosome inactivation. Both result in monoallelic expression. Both involve transcriptional silencing across a region of the genome, controlled in cis by a long noncoding RNA. Both involve DNA methylation and histone modification, and are confined by insulator sequences.

But the two systems appear to have evolved independently, at different times and in response to different drivers. Imprinting was a response to the sexual antagonism engendered by internal nutrition of the young, whereas X chromosome inactivation was a response to dosage imbalance on the degradation of a novel Y chromosome. It seems more likely, then, that the two silencing mechanisms simply re-used ancient components.

7.15 Ancient Mechanisms for Epigenetic Silencing?

As we learn more about epigenetic silencing, it becomes obvious that X chromosome inactivation shares many functional and molecular characteristics with other dosage compensation systems, and more generally with a wide variety of epigenetic silencing systems in plants and fungi as well as invertebrates and vertebrates.

These silencing systems have in common many or all of:

- Transcriptional silencing
- Delayed DNA synthesis
- Altered chromosome conformation and position in the nucleus
- A landscape of repetitive DNAs, including insulator element CTCF
- Long noncoding RNA molecules nearby or antisense
- DNA methylation differences
- Post-translational modification of histones
- A mechanism for marking parental differences for future silencing (imprinting)

It is stunning to find that the molecular changes that underpin these silencing events reuse the same strategies, the same DNA and histone modifications and even the same enzymes to achieve them. The polycomb repressor complex PRC is a case in point, popping up throughout the phylogeny in all sorts of situations (we will meet it again in considering sex determination in Part III).

It is also clear that many components of the X inactivation system have other functions in gene regulation. For instance, histone acetylation controls neural development as well as X inactivity in females or hyperactivity in males. Many of the conserved histone modifications evolved to silence repetitive elements concentrated at centromeres, and several proteins controlling the conformation and nuclear location of the inactive X (e.g., SMCHD1, lamin B, cohesin) are general architectural factors involved in nuclear structure. A bevy of RNA-binding proteins, including SPEN, a major transcription factor that delivers the final *coup de grace* to the inactive X, belong to families that are ubiquitous in eukaryotes.

These silencing systems are usually multilayered. Again and again we find that if one component is removed or interfered with, the effects are muted, attesting to a fine backup system. Mammalian X chromosome inactivation is a good example; it appears to use almost all of these mechanisms to produce a hyperstable and somatically heritable silencing, which is robust to interference.

These systems go back way before the evolution of vertebrates, and it seems sensible to envisage a molecular toolbox of ancient silencing mechanisms that were selected in different lineages in response to different selective forces. Some silencing systems evolved simply to regulate the supply of mRNA, others as weapons in a war of sexual antagonism, still others to smooth out problems of differential gene dosage.

7.16 Conclusions – Ad Hoc Control of Sex Chromosome Activity

The ubiquity of epigenetic silencing mechanisms makes it difficult to deduce a pathway by which these elements were put together into the complex hyperstable X chromosome inactivation of eutherian mammals.

Ohno's hypothesis that it all started with selection for hyperexpression of genes on the single X of males is a very attractive starting point. And this does, indeed, seem to have been the start of dosage compensation systems in *Drosophila*, nematodes, and anole lizards. In mammals there is evidence that overexpression of genes on the mammalian X occurred, but perhaps not for all genes, and all species.

Upregulation of the X only in XY males would be the most parsimonious way to dosage compensate the X, both smoothing out sex differences, as well as restoring X:autosome parity. Indeed, male-specific upregulation occurs in flies and the anole lizard. But not in mammals or nematodes; they upregulate both X chromosomes in females, so they had to evolve some way to downregulate X expression. Different solutions were selected in different lineages, including lowering both as in nematodes, or inactivating one X as in mammals.

Across animals, many solutions have been cobbled together. Though they share ancient molecular silencing mechanisms, there is surprising diversity in the players (especially the ncRNAs) that accomplish this. Particularly extraordinary is role of *XIST* in eutherians, which is mirrored by a different ncRNA (*RSX*) in marsupials.

Given the variation between species and the ubiquity of the molecular strategies underpinning epigenetic silencing, it is hard to perceive a step by step buildup of the inactivation process. For instance, the idea that paternal marsupial X chromosome inactivation was a leaky, imperfect version was undermined by the variation within eutherians, and the idea that X inactivation arose from meiotic sex chromosome inactivation was scotched by the demonstration that the X is reactivated on fertilization. Shared characteristics do not necessarily mean identity by descent because virtually all the elements of the silencing machines are also seen all over the animal family tree, in fish, flies and mealy bugs.

So there is no obvious pathway by which X inactivation evolved. It looks more like a playground in which pieces of play equipment abound, and you can take your pick. The grand conclusion from studying the evolution of X chromosome inactivation in mammals, and epigenetic silencing systems in other organisms is that there is a huge range of differentiation to be compensated, big differences in the dosage sensitivity of genes that require compensation and lots of ways to regulate gene dosage.

But what they have in common are a raft of molecular mechanisms for upregulating and silencing genes, as well as the repurposing of genes that have functions in other organs or systems. These are shared to a remarkable degree by mammals, other vertebrates and invertebrates. They even reappear in Part III because they are critical for environmental sex determination.

FURTHER READING

Classic Papers

Chandra HS, Brown SW, 1975. Chromosome imprinting and the mammalian X chromosome. *Nature* 253: 165–168

Charlesworth B, 1998. Sex Chromosomes. *Current Biology*, 8: R931–R933 http://biomednet.com/elecref/09609822008R0931

Cooper DW, 1971. Directed genetic change model for X chromosome inactivation in eutherian mammals. *Nature* 230: 292–294

Reviews and Research Articles

Arnold AP, Itoh Y, Melamed E, 2008. A bird's-eye view of sex chromosome dosage compensation. *Annual Review of Genomics and Human Genetics* 9: 109–127

Chandler CH, 2017. When and why does sex chromosome dosage compensation evolve? *Annals of the New York Academy of Sciences* 1389: 37–53

Chen J, Wang M, He X, et al., 2020. The evolution of sex chromosome dosage compensation in animals. *Journal of Genetics and Genomics* 47: 681–693

Duret L, Chureau C, Samain S, Weissenbach J, Avner P, 2006. The Xist RNA gene evolved in eutherians by pseudogenization of a protein-coding gene. *Science* 312: 1653–1655

Gendrel AV, Marion-Poll L, Katoh K, Heard E, 2016. Random monoallelic expression of genes on autosomes: Parallels with X-chromosome inactivation. *Seminars in Cell & Developmental Biology* 56: 100–110

Grant J, Mahadevaiah SK, Khil P, et al., 2012. Rsx is a metatherian RNA with Xist-like properties in X-chromosome inactivation. *Nature* 487: 254–258

Graves JAM, 2016. Evolution of vertebrate sex chromosomes and dosage compensation. *Nat Rev Genet* 17: 33–46

Gu L, Walters JR, 2017. Evolution of sex chromosome dosage compensation in animals: A beautiful theory, undermined by facts and bedeviled by details. *Genome Biology and Evolution*, 9: 2461–2476, https://doi.org/10.1093/gbe/evx154

Huynh KD, Lee JT, 2003. Inheritance of a pre-inactivated paternal X chromosome in early mouse embryos. *Nature* 426: 857–862

Kuroda MI, Hilfiker A, Lucchesi JC, 2016. Dosage compensation in *Drosophila* – a model for the coordinate regulation of transcription. *Genetics* 204: 435–450

Lee JR, Bartolomei MS, 2013. X inactivation, imprinting and long noncoding RNAs in health and disease. *Cell* 152: 1303–1323

Lister NC, Milton AM, Patel H, et al., 2024. Incomplete transcriptional dosage compensation of vertebrate sex chromosomes is balanced by post-transcriptional compensation. *Proceedings of the National Academy of Sciences of the United States of America* 121(32): e2322360121, www.pnas.org/doi/10.1073/pnas.2322360121

Renfree MB, Hore TA, Shaw G, et al., 2009. Evolution of Genomic Imprinting: insights from marsupials and monotremes. *Annual Review of Genomics and Human Genetics* 10: 241–262

Part III

Sex Determination

8

Sex and the Single Gene

The whole point of sex chromosomes, the reason they have evolved and differentiated, is that they specify sex.

Biological differences between men and women are profound; it has been said, half seriously, that men are genetically more similar to male chimpanzees than they are to women. Yet all these differences come down to the action of a single gene. This 'Testis Determining Factor', TDF, kickstarts the development of a testis in the embryo. The testis makes hormones, and these male hormones make the baby a boy. In the absence of TDF, an ovary forms, and female hormones promote female development.

In this final section, I will review the differences – phenotypic and genetic – between the sexes, leading to the concept of the 'Testis Determining Factor'. I will describe the successful search for this gene on the human Y chromosome and our gathering understanding of other genes that interact with it to specify males and females. I will also examine alternative ways to specify sex in other vertebrates and what this tells us about how sex determination systems evolved.

In this chapter I will summarize the differences between males and females, and lay out the evidence for a single gene that triggers these alternative pathways. I will tell the story of how TDF was located and then identified in one of the first, and perhaps the most spectacularly successful, approach to the positional cloning of a human gene.

8.1 Sex Differences

When I ask audiences to write down, in one minute, the differences between men and women, boys and girls (Table 8.1), the emphasis is on the most visible sexual characteristics, and usually nobody even mentions eggs and sperm. Penises and breasts win outright, body type and behaviour come a close second. This was true of learned adacemies as well as high school classes.

Table 8.1 *Differences between men and women, as seen by high school students.*

Trait	Men	Women
Gametes	Sperm	Eggs
Reproductive equipment	Testicles Semen Penis Prostate Male hormones	Womb Vagina Labia Breasts Female hormones
Secondary sexual characteristics	Bigger, Taller More Muscly Beard Baldness Voice lower Mature later Shorter Lifespan	Smaller, shorter Less muscly Fat distribution Don't go bald Voice higher Mature earlier Live longer
Behaviour	Aggressive, disruptive Attracted to females Adventurous Loud, silly Better at math, diagrams Wear daggy clothes	Calm, submissive Attracted to males Bitchy Caring Better at languages, art Obsessed with appearance

Before giving a talk/discussion on sex determination to several classes across many Australian schools, I handed out cards and asked students to write down as many differences between men and women, boys and girls as they could think of in one minute. This is the response from a year 9 class (average age 15) in Victoria, but it was absolutely typical of many high school classes, as well as learned Academies in Australia and New Zealand. Disheartening stereotypes they may be, but this is what people note.

Although it may be popular to minimize morphological and physiological (and especially behavioural) differences in this gender-aware age, average sex differences are truly profound. Sex is hailed as the most striking normal polymorphism in humans.

Differences in size, morphology and physiology between men and women are the result of gene activity that differs between males and females. And it is not just genes on sex chromosomes that are differentially active. Recent data on gene activity in 44 different tissues suggest that every tissue and organ in the human body is directly or indirectly sexualized by differential gene action. At least a third of our genome is differentially expressed in male and female cells of one or many tissues, and every tissue shows sex differences of many genes, some major and some subtle. Not just gonads and reproductive tissues, but also hearts, kidneys and the brain show differences in gene action. Some regulatory pathways show sex differences, even if the outcome is the same.

What could this mean for male and female kidney function, or male and female behaviour? And for male and female responses to diseases and therapies? This insight reinforces the observations that men and women may have quite different vulnerabilities and responses to many diseases, and to therapies. As has been said, women should not be regarded just as 'small men' from a medical viewpoint.

Although some sex differences result from different presence or dosage of genes on sex chromosomes, most of the obvious sex differences are the downstream result of the sex determination decision, and are mediated by sex hormones.

8.2 Sex Hormones

Hormones are chemical messengers made in one part of the body that have effects in other parts of the body. Some are protein chains (polypeptide hormones), and others are fat-soluble steroids. The major sex hormones are steroids (androgens and oestrogens and their derivatives), but some protein hormones also play important roles in sexual development and function.

Major sex hormones are testosterone and oestrogen, and their derivatives (Figure 8.1). In general, androgens have masculinizing effects and oestrogens feminising effects. But although testosterone is thought of as the quintessentially male hormone and oestrogens as female, it is really a matter of levels because both these hormones are produced in both sexes. As well as their function in reproduction, sex hormones have major effects on the immune system, neural function, cognition and the vascular system.

Testosterone and oestrogen belong to the same steroid hormone pathway, which starts with cholesterol. An enzyme cleaves off a sidechain to produce a precursor of progesterone, as well as cortisone (the stress hormone). Other enzymes then convert this precursor to androgens, which are converted to testosterone. Another enzyme (5-α reductase) converts testosterone into its more active derivative dihydrotestosterone (DHT).

The oestrogens (oestradiol and its relatives) derive from testosterone and a precursor by the action of an important enzyme (aromatase, which comes into its own in Chapter 10).

Steroid hormones turn on particular genes in the nucleus. They enter a cell and bind to a steroid hormone receptor, which may be in the nucleus, or in the cytoplasm temporarily protected by a protein chaperone (heat shock protein), which then dissociates. These receptors act as transcription factors, having a hormone-binding domain that ensures transport into the nucleus, and a DNA-binding domain to lock onto target genes and initiate transcription.

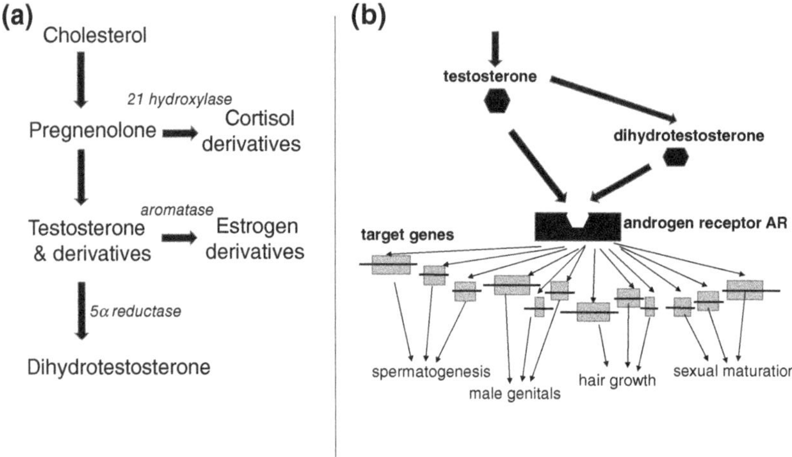

Figure 8.1 Sex hormones and their functions. (a) Simplified pathway for biosynthesis of steroid sex hormones. Androgens (testosterone and its relatives) and oestrogen and its relatives are synthesized in the same pathway, starting from cholesterol, and have very similar structures. Cholesterol is converted to pregnenolone, which is converted to both androgens and to cortisol and its relatives (via the enzyme 21 hydroxylase). Testosterone is converted both to dihydrotestosterone, the more active form (via enzyme 5α reductase), and to oestrogens (via enzyme aromatase). (b) Action of testosterone and dihydrotestosterone on male characteristics. Both hormones bind to a common receptor (androgen receptor, AR) in the nucleus. The AR-hormone complex then activates batteries of genes scattered around the genome, which contribute to aspects of male phenotype such as spermatogenesis, male genitals, hair growth and sexual maturation.

Testosterone is made mostly in the testes, with a little made in the adrenal glands atop the kidneys. The production of testosterone differs at different life stages. It triggers the changes at puberty (growth of penis and testicles, facial and body hair, muscle development), and is also important for fat distribution, bone growth and mood. It is the most important androgen for gonad function, being needed for making sperm, as well as for sex drive. After puberty testosterone is produced in a constant daily amount but, like most other hormones, declines with age.

Testosterone synthesis is controlled by another hormone (gonadotropin) made in the hypothalamus at the base of the brain. This controls release of two other polypeptide hormones (luteinizing hormone LH and follicle stimulating hormone) from the pituitary gland, which in turn stimulates the testes to make testosterone.

Testosterone in the blood is converted into dihydrotestosterone (DHT), which is active in other parts of the body, as well as oestradiol, which contributes to bone strength in men as well as women. DHT is vital for development of the penis and scrotum; in its absence, a clitoris, vagina and labia form.

Female sex hormones are made mostly in the ovaries, in a cyclical pattern. Oestrogens play important roles in sexual development and reproduction, and are responsible for many of the biological changes that occur at puberty and pregnancy; levels drop sharply at menopause.

Another steroid derivative, progesterone, prepares for and supports pregnancy. It is produced in the ovaries after an egg is ovulated, and the placenta produces some after pregnancy is established.

In women, testosterone is made by the ovary and adrenal glands. Although present at low concentration, testosterone is important for regulating the menstrual cycle, imparting muscle and bone strength and regulating sexual desire.

The roles of sex hormones in health and disease have been studied intensively. Like every genetic pathway in humans, there is considerable variation in enzyme activity and hormone levels, leading to a great range of morphological and physiological differences. Most of these are trivial, but some cause differences in sexual development in otherwise healthy children, and a few cause serious diseases. These <u>D</u>ifferences/Disorders of <u>S</u>exual <u>D</u>evelopment are collectively known by the umbrella term DSD.

Blocks in different reactions of androgen biosynthesis by mutation of genes that encode enzymes in the pathway cause several DSD conditions. At the most severe end is congenital adrenal hyperplasia (CAH), caused by a block in the conversion of the androgen precursor to cortisone. This causes a lack of cortisol and derivatives, which affects blood pressure and sugar balance and may be life-threatening because of uncontrolled salt loss. The block also produces a build-up of androgens so that XX babies may be born with genitals that look masculine.

Defects in the conversion of testosterone to DHT results in the opposite: incomplete genital virilization. Mutations in the enzyme that catalyses this step (5α reductase 2) disrupt the testosterone pathway so that newborns look externally female, but affected children develop male characteristics after testosterone and other factors are upregulated at puberty. This recessive condition is relatively common in some populations; in islands north of Australia, a third sex is referred to as 'Makim man'.

One of the first conditions to be identified was 'testicular feminization', which is sex-linked in humans and mice. The cause of this XY female sex reversal was traced to mutations of the androgen receptor gene *AR* on the X chromosome. XY embryos have functional (but undescended) testes, which make typical androgens. However, the inactive hormone receptor means that the androgens have no effect, and the baby is born apparently a girl. Among girls with this Androgen Insensitivity Syndrome (AIS) are athletes and models; they can be taller (because of Y-borne growth genes) and the absence of androgens avoids teenage skin problems. Partial AIS is variable, depending on the degree of tissue responsiveness to androgens.

The variable phenotypes resulting from this genetic variation lead to difficulty diagnosing and responding appropriately (discussed in Chapter 11).

Thus, the most obvious differences between men and women are the immediate results of their different hormone levels, which lie downstream from primary sex determination, which directs undifferentiated gonads to become either testes or ovaries.

8.3 The Testis and Sex Determination

The sex hormones that direct male and female development are produced by developing gonads and other organs they direct in the embryo. So we must ask, what controls gonad development?

It was suggested as long ago as the 1930s that a single sex-determining gene could control a common step in the complex sexual differentiation pathway. This 'single genetic switch' theory was given credence by the discovery that the critical factor in mammalian sex determination and differentiation is the development of the embryonic gonad. If a testis develops, the embryo assumes male characteristics. If no testis develops, the embryo develops as a female.

In a series of classic experiments on embryonic gonads in the 1940s to 1960s, Alfred Jost demonstrated the central role of the testis in male development (Figure 8.2). When he removed the gonads from XY embryonic rabbits in utero, he found that subsequent development was female. Removal of the gonads from XX embryos had no effect; development was still female. These and other experiments

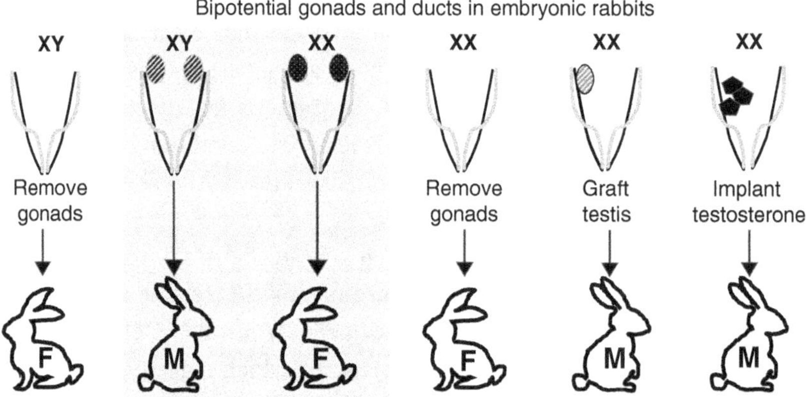

Figure 8.2   Jost's experiments showed that a testis, or testosterone, was required for male development. If the bipotential gonad was removed from XY foetuses, they developed as females. If the gonad (ovary, spotted) was removed from XX embryos, they still developed as females. Grafting a testis (striped), or implanting testosterone into XX foetuses promoted male development.

and observations suggested that it is the absence of a testis, rather than the presence of an ovary, that determines female development. In agreement with this hypothesis, he found that grafting a testis into XX embryos produced male development. These results suggested that the basic mammalian body plan is female, but the presence of testes switches development to a male pathway.

With the discovery of androgens, Jost was able to show that implanting a crystal of testosterone into XX embryos also produced male development, demonstrating the central importance of this hormone.

Thus sex determination can be reduced to its essence: the choice between forming testis or not forming a testis in the embryo.

We therefore need to understand how the gonads develop. The embryonic gonad undergoes two clear phases of development. The first phase is the initial development of a bipotential ('indifferent') gonad that is identical between male and female embryos, but has the ability to become either a male gonad (testis) or female gonad (ovary). The second phase is the specific differentiation of that bipotential structure into either a testis or an ovary.

8.4 Development of the Bipotential Gonad and Genital Tracts

The bipotential gonad is absolutely unique. No other mammalian tissue has the potential to develop into two completely different organs.

In mammalian embryos of both sexes, the gonads develop from an embryonic cell layer (mesoderm) that fills most of the space on either side of the embryo between the limb buds. The epithelial cell layer under the embryonic kidney thickens, and three regions develop (from head end) into adrenal glands, gonads, and kidneys. Long, thin paired structures arise within the middle region; these 'genital ridges' will form testis or ovary, but at this stage are indistinguishable between the XX (female) and XY (male) embryos.

This undifferentiated embryonic gonad contains precursors of two major somatic cell types – supporting cells that nourish the germ cells, and steroid-producing cells. Supporting cells will differentiate into Sertoli cells in XY embryos, and granulosa cells in XX embryos. Steroidogenic cells will develop into Leydig cells (producing androgen) in XY embryos, and theca cells (making oestrogen) in XX embryos. Other cell types are muscle like ('myoid') cells that will perform a structural role.

An early idea was that the bipotential gonad is a mixture of presumptive testis and presumptive ovary cells and the fate of the bipotential gonad simply depends on which population triumphs. But it is now clear that the same supporting cell primordia differentiate into Sertoli or granulosa cells, and a single steroidogenic lineage into Leydig or theca cells making androgens or oestrogens. This was confirmed by labelling experiments showing that marked progenitor cells can end up as either, implying that sex determination is a real switch, thrown by the presence or absence of a Y chromosome. However, there is some regional specification so that the outer layers of the bipotential gonad (medulla) grow preferentially in testis and the inner layer (cortex) in the ovary.

As well as gonad differentiation, different plumbing is required for male and female function. The early mammalian embryo develops two sets of tracts (called the Müllerian and Wolffian ducts after their discoverers). In XX embryos, the Müllerian ducts differentiate into female structures (oviduct, uterus, cervix and the top half of the vagina) while the Wolffian ducts degenerate. In XY embryos the opposite happens: the Wolffian ducts develop into male structures (epididymis and vas deferens that transport sperm, seminal vesicle and prostate that add semen constituents) while the Müllerian ducts degenerate (Figure 8.3).

Jost's experiments manipulating the gonads of embryonic rabbits revealed that a testis is required to stabilize the male Wolffian ducts, and the implantation of testosterone showed that this hormone is solely responsible. However,

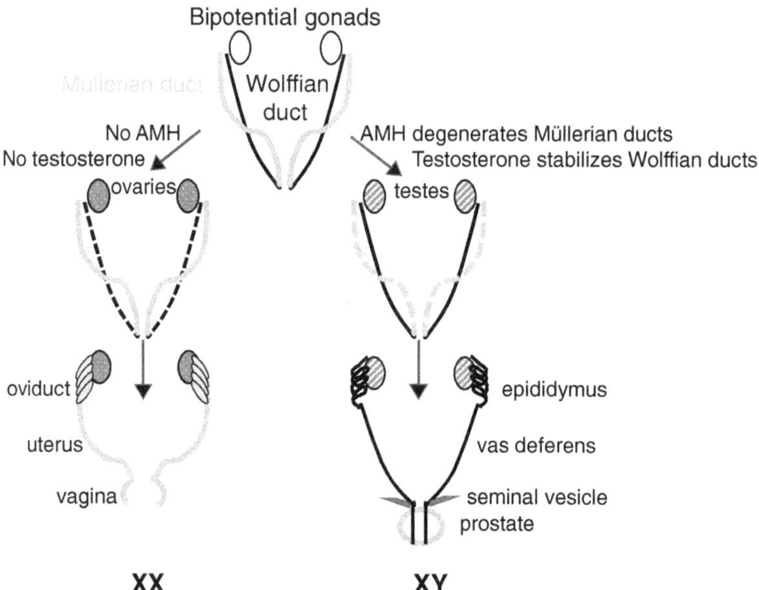

Figure 8.3   Development of genital ducts. The bipotential gonads of the embryo have two sets of ducts; the Müllerian ducts and Wolffian ducts. In XY embryos testosterone stabilizes the Wolffian ducts, which differentiate into the epididymis and vas deferens (tubes that transport sperm from the testis) and the seminal vesicle and prostate that produce the constituents of semen. Another hormone AMH (anti-Müllerian hormone) degrades the Müllerian ducts. In XX embryos, there is no testosterone, so Wolffian ducts degenerate. In the absence of AMH the Müllerian ducts differentiate into oviduct, uterus, and the upper part of the vagina.

implantation of testosterone did not cause degeneration of the female Müllerian ducts (Figure 8.3). This was the first evidence of another hormone produced in male but not female embryos, called the anti-Müllerian hormone (AMH, sometimes called the Müllerian inhibiting substance MIS) which directs the degeneration of the Müllerian ducts.

In XX embryos, the absence of AMH permits retention of Müllerian ducts that become the female reproductive tract, while the low level of testosterone causes involution of Wolffian ducts.

8.5 Gonad Differentiation

The differentiation of the bipotential gonad into either of two quite different organs requires remarkable switches in cell fate.

How can bipotential gonads differentiate into either of two entirely different organs; testis in XY embryos and ovary in XX embryos? Three main types of cells constitute the genital ridge; two somatic lineages (supporting cells, hormone-producing cells) and germ cells. Each of these follows different fates in XY and XX embryos.

In XY male embryos the supporting cells of the bipotential gonad are the first to differentiate. They surround primordial germ cells (PGC) that have already migrated into the bipotential gonad. These clusters are surrounded by a flattened layer of myoid cells that migrate from the embryonic kidney to form 'testis cords', visible by 13.5 days after fertilisation (days post coitum, dpc). Sertoli cell development marks the transition from a bipotential gonad toward testis.

Steroidogenic progenitor cells migrate from the embryonic kidney into the developing gonad, where they occupy the regions outside the testis cords. Here they differentiate into Leydig cells and start producing the androgens that drive male development.

While the Sertoli cells are differentiating, the gonad continues to enlarge as cells migrate from the adjacent embryonic kidney, then proliferate. They are critical in forming the male-specific vasculature, with a prominent blood vessel on the upper surface of the developing gonad and smaller side branches between the testis cords. These vessels facilitate crosstalk between Sertoli cells and other cells in the developing gonad and send hormones throughout the foetus.

In XX embryos, the somatic cells reorganize differently around germ cells. Supporting cells differentiate into granulosa cells that form clumps with germ cells. Shortly after birth the ovary undergoes major reorganisation. The somatic cells break down the clumps of germ cells and enclose individual oocytes, leading to the formation of primordial follicles that are surrounded by pre-granulosa cells which form connections that support them. These cyst-like structures develop into follicles within which the germ cells mature. Steroidogenic cells differentiate into theca that surround the follicles. They produce androgens that are converted to oestrogens by the granulosa cells.

Gonad development has been extensively studied in mice, and most of our current insights are based on experiments using this model mammal, in which development is very rapid during a 19–21 day pregnancy. In XY embryos the first signs of gonad differentiation are not obvious until day 10.5 dpc. The ovary differentiates later than the testis; no visible sign of differentiation occurs until 16 dpc. Thus, gonad differentiation is quite a late event.

In humans, the equivalent stages take much longer. In XY male embryos the first signs of differentiation are visible by 7 weeks and development of testis cords is obvious by 12 weeks. Ovary development in XX human embryos occurs later, and follicles are obvious by 16 weeks.

Thus the alternate pathways of gonad differentiation in males and females are very extensive and complex. Not surprisingly, making a functioning ovary and completing female development is every bit as complex as making a functioning testis and developing as a male, so a conclusion from Jost's experiments that

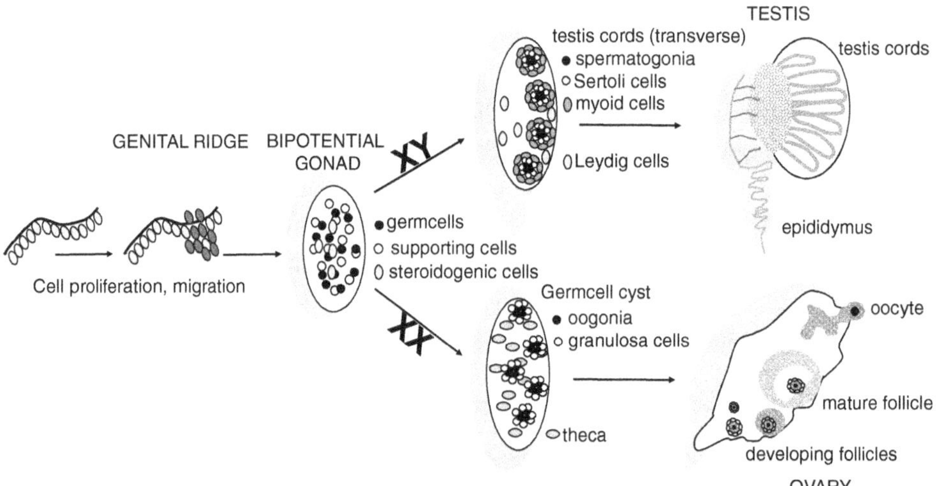

Figure 8.4   Gonad differentiation in mice. The bipotential gonad at E12.5 develops into a testis in XY embryos and into an ovary in XX embryos. It contains two somatic cell types, supporting cells and steroidogenic cells, as well as the germ cells that migrate in. In XY embryos, the supporting cells form Sertoli cells and the steroidogenic cells differentiate into androgen-producing Leydig cells. Sertoli cells surround germ cells and are encased in myoid cells, forming looped cords, within which germ cells differentiate into spermatogonia, which undergo meiosis to form mature sperm. In XX embryos, the supporting cells form granulosa cells, which surround the germ cells in cysts, which grow and mature as follicles in the ovary, eventually bursting from the surface as an ovum.

female development is somehow a passive 'default state' must be rejected. It is obvious that many genes must be differentially expressed in males and females, and somehow these must be coordinated (Chapter 9).

The differentiation of somatic cells (supporting cells and steroid hormone producing cells) into different structures in testis and ovary are all about supporting the growth and maturation of eggs and sperm (Figure 8.4).

8.6 Development of Germ Cells

Surprisingly, the primordial germ cells (PGC) that will develop into eggs or sperm have a quite different origin, arising from the yolk sac outside and at the tail end of the developing embryo. They proliferate and make a remarkable journey right through the developing gut and into the bipotential gonad (Figure 8.5).

Cell labelling experiments in mice show that initially a few (about 45) PGCs arise from the epiblast. They multiply and move into the embryo, then migrate into the developing genital ridges. They move in an amoeboid fashion, possibly guided by nerve fibres and a chemotactic gradient. As they migrate, they continue to multiply until there are about 3,000 cells. They arrive at the genital ridge by 8.5 dpc, and must wait a few days; their further differentiation into sperm or eggs does not occur until the differentiation of the gonad into either testis or ovary (Figure 8.6).

In XY embryos PGC ('prospermatogonia') stop dividing. Clusters of germ cells are then surrounded by Sertoli cells and myoid cells to form cordlike structures, within which they differentiate into spermatogonia and resume proliferation a few days after birth.

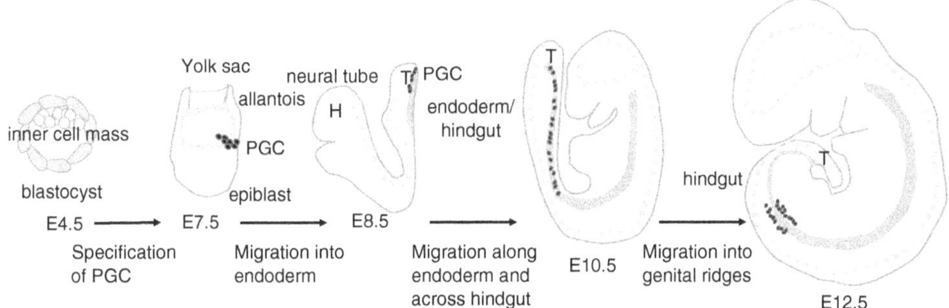

Figure 8.5   Germ cell origins. Differentiation and migration of primordial germ cells (PGC). PGCs (black dots) arise in the yolk sac above the tail-end (T) of the primitive streak by day 7–8 after fertilization (E7.5). They migrate along the endoderm and right across the hindgut (dark grey), spreading out into the two genital ridges on either side by E12.5. As the neural tube (dotted grey line) develops, the head end (H) bends around. PGCs develop into either eggs or sperm, depending on whether the genital ridge develops into a testis or an ovary.

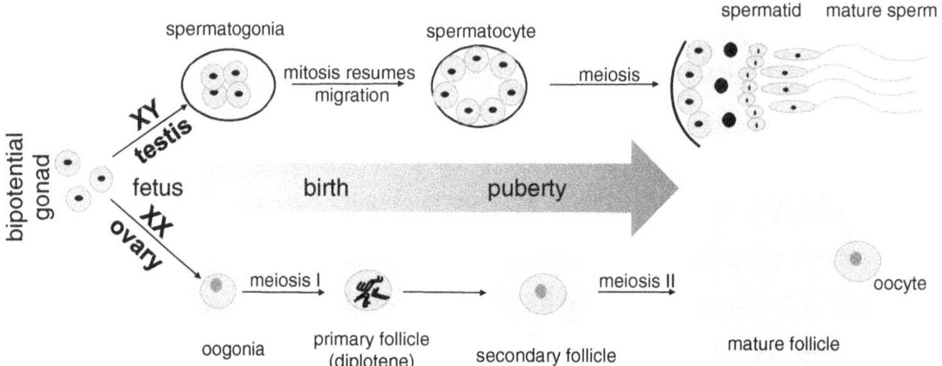

Figure 8.6   Germ cell development. Differentiation of germ cells into sperm and eggs. In XY embryos, PGCs develop into spermatogonia in the seminiferous tubules of the testis. Mitosis expands their number and meiosis is inhibited; cells move to the periphery of cords. At puberty meiosis initiates in testis cords, forming haploid spermatids (small cells), which elongate, eject cytoplasm and complete their differentiation into motile sperm. Spermatogenesis continues throughout life. In XX embryos, PGCs in the ovary divide a few times to become oogonia, then undergo the first round of meiosis, arresting at diplotene. They become surrounded with supporting cells (theca, pattern background). After puberty, hormones stimulate ovulation monthly, during which one of the primary follicles becomes greatly expanded and bursts from the periphery of the ovary. It enters the second meiotic division, which is completed on fertilisation to form a haploid oocyte.

Importantly, they do not enter meiosis until puberty. At this stage they set up a production line within the cords, undergoing meiosis to produce haploid cells (called 'spermatids') which migrate inwards. Spermatids elongate and reorganize, losing their cytoplasm and acquiring motile tails. Mature sperm are produced on the inner surface of the cords, and migrate into the ducts, where they receive their final constituents, and semen is produced. Sperm production starts at puberty and after this time meiosis is continuous.

In XX embryos the PGC in the genital ridge undergo limited proliferation to become oogonia. Oogonia stop proliferating and enter the first stage of meiosis, then arrest at prophase I of the first division around the time of birth. They do not resume meiosis until ovulation, which does not occur until females reach sexual maturity. At ovulation, oocytes complete the first meiotic division, enter the second meiotic division, then arrest again. The second meiotic division is not completed until after fertilisation.

In the adult ovary, germ cell clusters occupy the outer edge (cortex). These ovarian follicles are composed of oocytes surrounded by somatic cells that include granulosa and theca cells. Blood vessels develop between them, demarcating strings of germ cells.

All mammals show a spectacular difference between the output of sperm and eggs. Males produce billions of sperm continuously from puberty to death. Females have a limited supply of eggs, which are released sparingly (one a month in humans), and stop ovulating at menopause.

This is because oocytes undergo apoptosis (programmed cell death), limiting the number of primordial follicles. Human females have an initial store of 5–7 million germ cell progenitors that have grown by 20 weeks of embryonic development, but by birth only one or two million remain, nested in follicles. By puberty there are 300,000–400,000 left and the number continues to decline. Even if the store of oocytes can be replenished from germline stem cells in bone marrow, as has been suggested, it seems that females have a smaller pool of primordial follicles, limiting the female reproductive lifespan.

Thus germ cells in the ovary follow a very different schedule from those in the testis. The number of progenitors, the timing of meiosis and the duration of maturation mean that in mammals the production of sperm is profligate, whereas eggs are in short supply.

8.7 Sertoli Cells and the Testis Determining Factor (TDF)

As Jost showed, testis differentiation is the key event in mammalian sex determination. The induction of a testis in XY embryos was early ascribed to a signal that initiated testis development; this was called the Testis Determining Factor (TDF).

Many observations pinpoint Sertoli cells as crucial in receiving the testis determining signal. Sertoli cells are critical for orchestrating the organization of other cells within the testis; in their absence the ovary pathway is pursued.

The crucial role played by Sertoli cells in testis development was elegantly demonstrated (Figure 8.7) by making embryos that were a mixture of XY and XX cells by squishing XX and XY blastocysts together to develop as a single embryo. These XX-XY chimaeras had equal numbers of XX and XY cells in all embryonic lineages.

When the cells in the differentiated testis were separated and examined, however, marked differences were found. Surprisingly, the sex chromosome constitution did not seem to matter in Leydig and myoid cells in the testis, which were about half XX and half XY, implying that both XY and XX cells can develop along the male pathway, taking their cues from the cell environment.

This was true even of PGC development. XX germ cells became sperm in an XY gonad, and XY germ cells became eggs in an ovary. This implies that PGC fate, too, is determined by the immediate gonadal environment rather than any intrinsic signal.

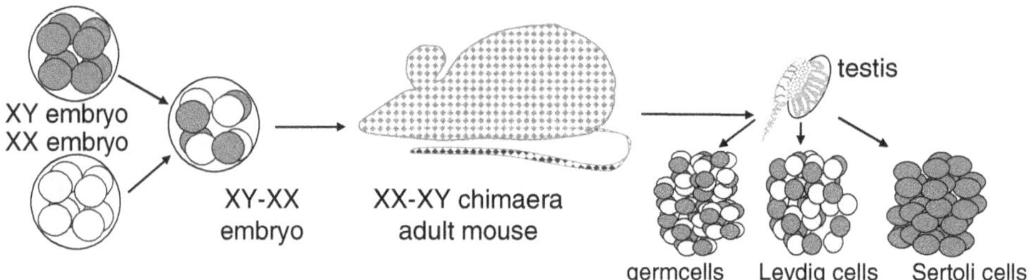

Figure 8.7   Experiment showing that only Sertoli cells receive the testis determining signal from the Y chromosome. Cells from an XY embryo (dark grey) were combined with cells from an XX embryo (pale grey) to make embryos that were XX-XY chimaeras, and grew up with a mixture of XX and XY cells. Testes were removed from these mice, and germ cells, Leydig cells and Sertoli cells were separated and tested for their possession of a Y chromosome, by in situ hybridisation of radioactive Y-specific sequences. Germ cells and Leydig cells were a 1:1 mixture of XX and XY cells. Only Sertoli cells were composed entirely of XY cells.

But Sertoli cells were found to be overwhelmingly XY, showing that a Y chromosome is needed for Sertoli cell differentiation. This implies that the sex-determining trigger in XY cells acts cell-autonomously to initiate Sertoli cell differentiation. Pre-Sertoli cells then signal other cell lineages in the gonad to begin testis-specific differentiation.

Thus, of the cells in the developing bipotential gonad, Sertoli cells were the only ones for which an XY sex chromosome constitution was crucial. Sertoli cells are therefore the one type that responds to the testis determining signal in XY embryos. Hormone-producing cells and germ cells pursued a male or female path according to messages from the Sertoli cells. Leydig cells have nothing to do with determining sex, but make androgens that powerfully influence male secondary sexual development.

So the first and most important decision that the bipotential gonad makes is whether the supporting cells should become Sertoli or granulosa cells. The function of the Testis Determining Factor is therefore likely to act by initiating Sertoli cell differentiation. The frenzied search for this factor is one of the most compelling genetics stories of the twentieth century.

8.8 Chromosomal Basis of Mammalian Sex Determination

Evidence has been available for decades that in mammals a single gene on the Y chromosome triggers the pathway for differentiation of a bipotential gonad into a testis, and gonad differentiation triggers all the other changes in sexual differentiation.

This was not always obvious. Century old studies on *Drosophila* sex chromosomes showed that sex in the fruit fly was determined not by the Y, but by the number of X chromosomes because flies with a single X chromosome were male even if they had no Y, and XXY flies were female. It was surmised that the same mechanism might hold true in humans.

However, it doesn't. Observation of people with unusual sex chromosome constitutions (aneuploidy) showed that the Y, not the X, was responsible for sex determination in humans (Figure 8.8a). Sex chromosome aneuploidy is extraordinarily benign (Chapter 5) because dosage compensation inactivates all but one X chromosome, and the Y contains few genes anyway. In 1959 it was shown that women with Turner's syndrome had a single X chromosome (XO), and men with Klinefelter's syndrome had a Y chromosome as well as two X chromosomes (XXY). The number of X chromosomes seemed not to be the determining factor; in fact, people with XXXY and even XXXXY, are also male.

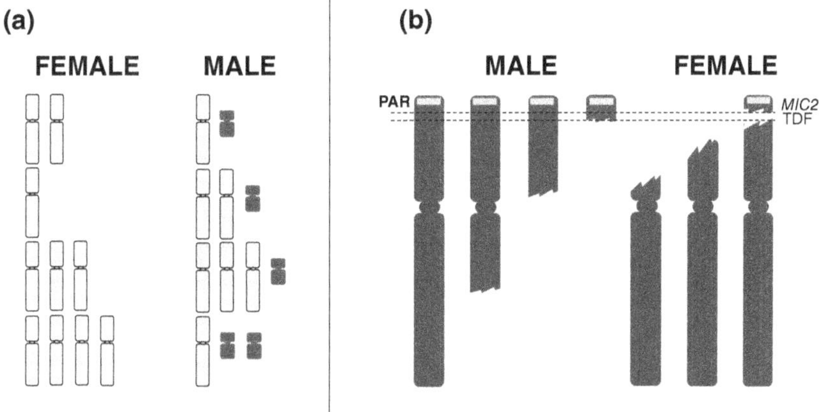

Figure 8.8   Y localization of the testis determining factor. (a) The location of the male determining gene (testis determining factor TDF) was inferred by the sex of people with unusual numbers of X chromosomes (pale grey) or Y chromosomes (black). Individuals with two X chromosomes are typically female. Individuals with a single X (XO) are female with Turner's Syndrome and individuals with three or more X chromosomes are female. Humans with XY are typically male. Individuals with XXY are males with Kleinfelter's Syndrome, as are individuals with multiple X chromosomes and a Y. Individuals with two or more Y chromosomes are male. Thus the Y chromosome(s), rather than the number of X chromosomes, determines maleness, and must contain TDF. (b) The location of TDF on the Y was narrowed down by observing the sex of people with parts of a Y chromosome (black), due to deletion or translocation of part of the Y to X or autosome. Individuals with deletions of the long arm of the Y were male but made no sperm. Individuals missing most of the short arm were male as long as they had the region near to the PAR (lighter grey), implying that this region contained TDF. Individuals with a Y chromosome that lacked this region were female.

This led to the conclusion that the Y chromosome carries the genetic instructions for testis development. Female development does not depend on a dominant gene on either of the sex chromosomes, or on the number of X chromosomes. Consequently, a sex chromosome constitution of XY gives rise to a male and XX to a female.

It was proposed that, as a minimum, a single genetic entity on the Y chromosome could be responsible for testis formation. In humans, this putative gene was called the Testis Determining Factor (TDF) because of the uncertain nature of the genetic signal. Mouse geneticists boldly called it the testis determining Y gene (*Tdy*).

TDF would be expected to be active in the somatic cells of the bipotential gonad just before the first signs of testis differentiation. It should activate other genes in a testis determining pathway.

Identifying TDF became something of a holy grail. Testis determination is likely to be typical of other pathways of organogenesis, and it was even hoped that the testis determining gene may belong to a family of genes that have similar functions in controlling the development of other organs. It therefore presented an ideal opportunity to examine regulation of developmental pathways.

A very great advantage of studying the sex-determining pathway is that, although TDF has a dramatic phenotypic effect, it is irrelevant for survival. Individuals with mutant TDF are not dead, but merely female. Many human variants of sexual differentiation are known and have been well studied.

Over a period of 30 years, many attempts were made to identify this pivotal gene, and to study its function. The hunt for the testis determining gene began in earnest in the 1960s, gathered momentum during the 1970s and reached a frenzy in the 1980s. It involved a huge amount of work on what ultimately turned out to be false trails. However, the pursuit of each has turned up surprises, and enriched our understanding of sex chromosome organisation, function and evolution, so the major candidates are worth following (Figure 8.9).

8.9 Searching for a Male-Specific Protein

A difficulty in identifying and isolating the testis determining gene is that we had no idea what its protein product was or how it worked.

An early approach to identifying *TDF* in the 1970s was to look for proteins that are expressed in males but absent in females. An effective way to identify critical male-specific products was to search for immunological differences between male and female mice within inbred strains that had few other genetic differences.

Before even the role of the Y chromosome in sex determination became clear, a male-specific antigen was described in 1955. Female mice grafted with skin from

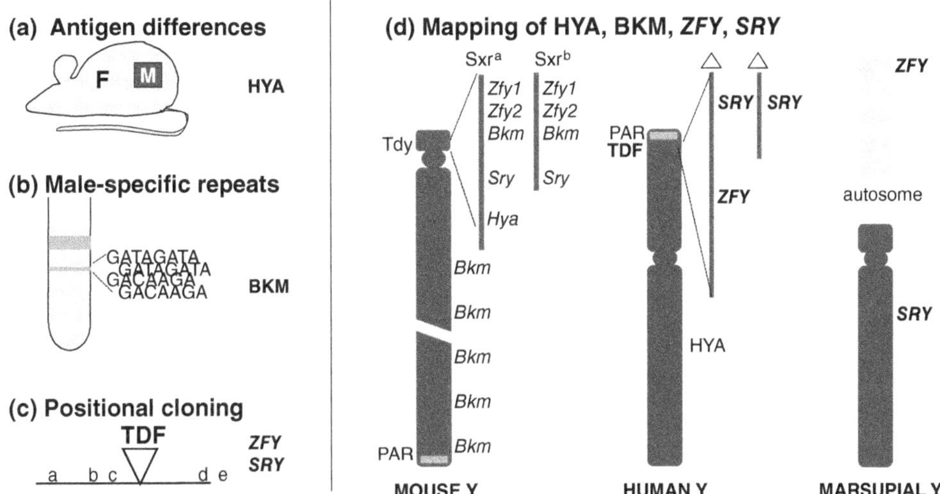

Figure 8.9   Rise and fall of candidate sex-determining genes. (a) Male-specific antigen HYA was identified by grafting skin from a male mouse (black square) onto a female mouse. (b) Density gradient separation of DNA sequences from an Indian snake (the banded krait). A thin minor band beneath the major band containing most of the DNA contained male-specific repeats (GATA, GACA). (c) Positional cloning by sequencing regions of the human Y that were present in XX males and absent in XY females. Deletions (delta symbol) located by marker genes a–e in XY females yielded a gene *ZFY*. (d) Mapping of HYA, BKM, *ZFY* and *SRY* to regions of the Y chromosome in mouse, human and marsupial. In mouse Hya mapped to the short arm of the Y. It was also present on the small male-determining Sxr^a region of the Y that had been translocated to X, and which contained the testis determining factor Tdy. However, it was excluded from the smaller male-determining region Sxr^b. In humans, HYA mapped to the long arm of the Y, rather to the short arm that contains TDF. Thus, HYA was excluded from a role in sex determination. Bkm sequences are spread all over the mouse Y chromosome, but are not present on the human or kangaroo Y. Thus, Bkm was excluded from a male-determining role in mammals. *ZFY* mapped to the short arm of the human Y. It was present also in the mouse sex-determining regions Sxr^a and Sxr^b, However, it was autosomal in marsupials so it was excluded from a male-determining role. *SRY* was cloned from the smaller interval of the human Y, and proved to map to the smallest mouse interval Sxr^b, and also to the marsupial Y chromosome, so it was accepted as the best candidate TDF.

male animals of the same inbred strains were found to mount a weak rejection response, not seen with male-male or female-female grafts. This suggested the presence of an antigen on the surface of male, but not female cells (Figure 8.9a).

Experiments with tumour cells with and without a Y chromosome correlated graft rejection with the presence of a Y chromosome. The antigen detected by graft rejection was therefore called the H-Y antigen, and the gene on the Y chromosome that encoded it was called *HYA* in humans (*Hya* in mice).

The discovery of a male-specific cell surface factor that was apparently encoded by the Y chromosome generated great excitement. It was suggested in 1975 that the H-Y antigen was encoded by the testis determining gene on the mammalian Y chromosome. This hypothesis was debated in thousands of papers over a decade, during which the H-Y antigen was widely supposed to be the product of the sex-determining gene. The great Susumu Ohno was a big fan of *HYA*; others were not so sure.

A sex-specific antigen was also detected in other vertebrates. Graft rejection was found to occur between male and female birds. However, it was ZZ male birds which rejected ZW female tissue, suggesting that a female-specific avian H-W antigen is encoded by a gene on the W chromosome. Mammalian H-Y antibodies were even reported to cross-react with cells from the heterogametic sex in reptile and amphibian species. This was puzzling, since the mammal XY, bird ZW and reptile sex chromosome pairs are not homologous. This work was much disputed among claims and counterclaims of having more or less specific antibodies.

The HYA hypothesis began to look more and more shaky. Any worthy testis determining gene should be expressed only in the developing genital ridge of males, but the H-Y antigen was found to be ubiquitously expressed in adult males as well as foetal and embryonic tissue. Its expression did not correlate in any way with testis development.

It was gene mapping that finally dispatched the HYA hypothesis. One important piece of circumstantial evidence in favour of *Hya*'s testis determining role in mice was the observation that HYA was expressed by males with only a tiny region of the Y chromosome translocated to the X (called Sxr[a]). However, male mice with an even smaller region of the Y(Sxr[b]) later proved to lack *Hya*.

In humans even fairly crude deletion mapping of the Y chromosome was sufficient to debunk *HYA*. In males with deletions of parts of the Y chromosome, HYA expression was concordant with the retention of the long arm of the human Y chromosome, whereas the development of testes, and hence maleness, depends on the presence of the short arm (Figure 8.8b). The position of human *HYA* has since been refined to the middle of the long arm.

We now understand why HYA has been so difficult to track down. The antibody is invoked in females in response to the products of whatever male-specific genes are ubiquitously expressed. These are a mix of several proteins, and are not even the same in different species.

In mice, it seems to be the Y-linked *Kdmd* gene, which encodes a male-specific epitope not present in the product of its X-linked gene partner *Kdmc*. In humans, several genes on the Y chromosome that are ubiquitously expressed (e.g., *UTY, DBY, DFFRY*) also raise male-specific minor histocompatibility antigens that can be a problem for tissue transplants.

Thus, *HYA* is not the mammalian testis determining factor.

8.10 Searching for Male-Specific DNA

Is there a DNA sequence on the Y chromosome that can be shown to determine sex?

The Y chromosome was known to be composed largely of repeated sequences, which seemed unlikely candidates for a sex-determining role. Nevertheless, schemes were considered by which repetitive sequences could control sex determination, and these seemed more credible when a minor 'satellite' sequence was found that separated from the main band of DNA during gradient density centrifugation. DNA from the Banded Krait (an Indian snake) was used to produce the *Bkm* (for Banded Krait Minor) sequence, which was claimed to be conserved on the sex chromosomes of all vertebrates (Figure 8.9b).

Notably, *Bkm* sequences were concentrated on the W chromosome of snakes, but were also detected throughout the vertebrates. The conservation of this 'Garden of Eden' sequence suggested an important function. Of particular interest was the observation that *Bkm* sequences were concentrated on the mouse Y chromosome and were contained within the translocated Sxr region (Section 8.8), then the smallest known region of Y that carries *Tdy*. The concentration of *Bkm* on sex chromosomes in snakes and mammals led to the hypothesis that *Bkm* might play a role in sex chromosome function and may even be *TDF*.

Excitement was somewhat dimmed when sequence analysis showed the *Bkm* satellite to be largely composed of simple tetranucleotides GACA and GATA. Obviously this was incompatible with a coding function. An inventive explanation was advanced for how *Bkm* could determine sex by acting as a sink to bind a testis inhibitor produced elsewhere on the genome. When a Y chromosome, loaded with *Bkm* sequences, was present to absorb inhibitor, testis determination was permitted. In vertebrates with ZZ males and ZW females, the *Bkm* sequences on the W chromosome would have to absorb an ovary-inhibiting substance.

However, studies over several years produced no direct evidence that *Bkm* acted to determine sex in any organism, and *Bkm* sequences were finally eliminated as a candidate for *TDF*, again by gene mapping evidence. Although *Bkm* sequences were present on the Y chromosome in many mammals, they were not concentrated on the human Y chromosome. More specifically, it was found that although the Y region present in Sxr[a] male mice contained Bkm sequences, they were deleted in male mice with the smaller male-determining region Sxr[b].

There are still occasional claims that *Bkm* sequences play an important role in differential condensation and activity of sex chromosomes in mammals and other vertebrates. However, there remains no convincing evidence that repeated sequences on sex chromosomes play a specific role in sex chromosome function. Repeated sequences on the Y are more usually species-specific, and are more likely to have arisen by random amplification of different sequence classes (Chapter 4).

The elimination of *HYA* and *Bkm* from consideration as the testis determining factor left no candidate genes or functions on the Y. How then could *TDF* be found on the Y chromosome? The only clue to finding TDF was its location on the Y chromosome.

8.11 Regional Mapping and Positional Cloning to Identify TDF

Advances in molecular biology offered a new way to clone unknown mammalian genes, by narrowing down the position of the gene until there is only a small region to search for candidates. This 'positional cloning' strategy had just been applied to identifying and cloning the gene mutated in cystic fibrosis. Any gene having a distinct phenotype can be mapped and cloned in this way, even if its characteristics and product are quite unknown. The testis determining factor *TDF* certainly caused a distinct phenotype – maleness.

The human Y is small, but there were still 60 million base pairs to search, a tall order in the days when DNA sequencing was practicable for only a few hundred or a thousand base pairs. The region could be somewhat refined to the 30 Mb euchromatic region because TDF is unlikely to lie in the pseudoautosomal region shared by the X, or in the simple-sequence heterochromatin.

Positional cloning involves constructing a map, identifying markers that flank the gene of interest, cloning and characterizing all the DNA that spans the interval between the flanking markers, and searching this for candidate genes (Figure 8.9c).

TDF obviously cannot be mapped by recombination since the Y does not pair and recombine with the X outside the PAR. Mapping therefore relied on constructing deletion maps and long-range restriction maps (Section 2.16).

Deletion mapping relies on the molecular analysis of DNA from patients with deletions of parts of the Y. The first clue came from the observation of male patients who lacked all or most of the long arm of the Y. Conversely, some patients who were phenotypically female were found who had derivatives of the long arm of the Y, but lacked the short arm. This meant that TDF must lie on the small short arm of the human Y (Figure 8.8b), greatly narrowing down the search.

TDF could be mapped more precisely by examining the chromosomes of so-called XX sex reversed males. Patients who were externally male but had poorly developed testes were described who apparently lacked a Y chromosome. However, screening with Y-specific DNA markers revealed some Y chromosome material in the DNA of many of these XX males. It turned out that they had the top part of the Y chromosome, which had evidently been swapped with the top of the X outside the pseudoautosomal region. This had transferred *TDF* to the X chromosome.

Screening for Y-specific molecular markers allowed the extent of the Y material to be assessed in each patient (Figure 8.9c). Some XX males had a large portion of the Y, whereas others contained only the very tip of the short arm. Lining up the regions present in different patients determined the order of markers on the map of the short arm of the Y. Since all these patients were phenotypically male, they must all have *TDF*, so it follows that *TDF* must be present even in the smallest region of Y present. Lining up the pieces identified the DNA marker that was closest to TDF on the centromere side.

The other important group of patients were so-called XY females, who turned out to have a Y that was missing bits of the short arm due to deletion or swapping with the X. The missing bit must contain TDF. Lining up these deletions pinpointed a critical interval that must contain TDF.

Mapping allowed *TDF* to be mapped to the most distal part of the Y-specific region, near to the pseudoautosomal boundary and flanked by *MIC2* (now called *CD99*, the closest gene within the PAR) on the telomere side.

This critical interval was further subdivided by making a long-range restriction map of the human Y chromosome by lining up overlapping large fragments generated by rare cutters. This map began at the end of the *MIC2* gene in the pseudoautosomal region and crossed the boundary into Y-specific sequences. Groups in England, France and the US then set out to clone and search this entire region to locate the human *TDF* gene.

On the basis of the human Y chromosome maps, Peter Goodfellow's group in London started a Y chromosome walk from *MIC2* across the pseudoautosomal boundary into the Y-specific region, using DNA sequences from one fragment to identify overlapping fragments. Since DNA cloning and sequencing was in its infancy in the 1980s, this was slow going.

Concurrently, a search for the mouse testis determining gene, *Tdy*, was undertaken by Robin Lovell-Badge's group in London, using deletion mapping to localize the gene to the short arm of the Y. Particularly valuable were Sxr (sex reversed) males which had a tiny region of the Y added to the X, which must contain *Tdy*. Mapping Y-specific DNA provided a detailed map of the sex-determining region of the mouse Y chromosome.

Meanwhile, David Page's group in Boston, working with material from other XX men and XY women, also deduced that *TDF* should reside within the top portion of the human Y chromosome. His group began walking along the Y chromosome from a breakpoint on the centromere side toward the PAR. Thus, groups on either side of the Atlantic approached the potential *TDF* gene from opposite ends of the Y chromosome.

A problem was what to look for, given that the characteristics of TDF – size, intron-exon structure, base sequence – were quite unknown. These groups therefore combed the newly defined critical regions for a sequence that contained an open reading frame that would code for a sensible protein, and recognisable signal sequences. In addition, they looked for homologous sequences in other mammals, since it seemed likely that such a critical gene would be highly conserved.

The stage was set for a race to clone TDF from the mammal Y chromosome.

8.12 Cloning the First Gene from the Human Y – the *ZFY* Saga

Page's group used information from two patients to dramatically narrow down the region which must contain *TDF*. One XX male contained only 280 kb of the Y-specific region, which must include *TDF*. One XY female had most of a Y chromosome that was lacking only a small region, which must contain TDF. Overlapping these regions narrowed their search to a 150 kb sex-determining region in which *TDF* should lie.

Page's group then used hundreds of clones from a genomic library to walk across the entire region. They found one clone which detected male-specific fragments in human DNA. It also detected male-specific fragments on a 'Noah's Ark' blot of DNA from males and females of a variety of eutherian mammals.

Sequencing revealed a large gene with identifiable promoter, splice and poly A addition sites, as well as translation initiation and termination codons. The largest reading frame (1.2 kb) encoded a predicted protein with 13 'zinc fingers', a nuclear localization signal and an activation domain. This denoted a gene which was given the name *ZFY* (for Zinc Finger Y).

ZFY looked very convincing. Its protein product, especially the polypeptide fingers coordinated with zinc, resembled many known transcription factors that bind DNA. A DNA-binding function would be a respectable role for a male-determining gene, whose presence kick-starts the testis differentiation pathway.

Thus, the structure and putative function of *ZFY*, its location, and conservation on the mammalian Y chromosome, made it an excellent candidate for *TDF*.

Surprisingly, Noah's Ark blots showed that the *ZFY* gene probe also hybridized to bands shared by male and female DNA. Sex differences in dosage suggested that these represented an X-linked copy of *ZFY*, called *ZFX*. This suggested that ZFY might work in a more complicated way, in conjunction with ZFX (interacting to form a male-determining protein) or antagonistically (if both proteins bind the same target with different efficiencies).

The first indications that *ZFY* was the wrong gene came from mapping homologous sequences in marsupials. For *ZFY* to be a credible mammalian *TDF* gene, it should detect male-specific homologous sequences on the marsupial Y chromosome, since the Y is testis determining in marsupials as well as eutherians (Section 10.3).

Two students in my lab in Australia used probes that detected *ZFY* to screen marsupial Noah's Ark blots. They detected bands in marsupial DNA, but none were male-specific. In situ hybridization showed that homologous sequences mapped to chromosome 5 in the kangaroo, a strange place for a sex-determining gene.

This autosomal region also contained several genes that lie on the human X but are autosomal in marsupials, implying that ZFX/ZFY was part of the region that was added to the eutherian sex chromosomes only recently (Section 4.1). The location of human *ZFX* among other XAR genes confirmed that it lay within the region recently added to the eutherian sex chromosomes, and the similarity of its sequence to that of *ZFY* (97% within DNA-binding domains) implied that the two genes diverged recently from an autosomal ancestor.

These results engendered great speculation. The immediate reaction among human and mouse geneticists was that marsupials are simply weird, having developed a variant form of sex determination early in marsupial evolution. However, our finding that *ZFY* is autosomal also in monotremes implied an autosomal ancestral state.

An alternative and more radical suggestion in our *Nature* cover story, was that *ZFY* was not, after all, the mammalian testis determining gene. This hypothesis (unpopular, at least in Boston) began to look more appealing as details accumulated about the position, structure and expression of *ZFY* and its X-borne homologues in humans and mice.

Homologues (*Zfy*) were found on the mouse Y within the sex-determining *Sxr* region, and there was a *Zfx* homologue on the X within the recently added region. But strangely, *Zfy* was present in duplicate, and even more strangely, other rodent species had multiple copies (up to hundreds).

Another puzzling feature of *ZFY/ZFX* was their expression profiles. Expression of the testis determining gene might be expected to be limited to embryonic genital ridge tissue. However, in humans *ZFY* and *ZFX* were both expressed at all stages in all foetal and adult human tissues tested.

Mouse *Zfy* and *Zfx* expression patterns were different, but no less puzzling. Like its human homologue, mouse *Zfx* is ubiquitously expressed. However, mouse *Zfy-1* and *Zfy-2* show a very restricted expression profile. *Zfy-1* is expressed in male germ cells at all stages of development in both foetal and adult testes. *Zfy-2* is expressed at a low level in male germ cells in the foetal testis, but produces abundant transcripts in the testes postnatally. Neither is expressed in a mutant strain of mice deficient in germ cells. Little or no *Zfy* was expressed in Sertoli cells, which we know to be critical for receiving the male-determining message. This suggests that *Zfy* is expressed mostly in the germ cells rather than the somatic component of the testes, a profile more compatible with a role in spermatogenesis than in sex determination.

Also inconsistent with a role in sex determination was the expression of normal *Zfy* in a strain of XY female mice with a mutant *Tdy* gene, suggesting that sex reversal was not the result of *Zfy* mutation. These observations disqualified *ZFY* as a candidate for *TDF*. More recently it was established that a double knockout of the two *Zfy* genes does not reverse sex.

For decades, the real function of ZFY was unknown; however, *ZFY* has rebounded with the discovery of a critical function in sex chromosome inactivation at meiosis (MSCI), revealed by its expression at interphase following the first stage of meiosis, and the derangement of meiotic sex chromosome inactivation in knockout mutants (Section 5.14). It also has been claimed to play a role in the regulation of many autosomal genes by the Y and inactive X.

But it was back to the drawing board in the hunt for TDF; the critical region of the Y had to be searched in even greater detail to reveal the real sex-determining gene.

8.13 Discovery of the Testis Determining Gene *SRY*

It was critical to re-examine the data which defined the sex-determining region of the human Y. Page's deletion map was based on two patients with different portions of the Y chromosome. This defined the sex-determining region as the 150 kb interval which was present in an XX male and absent from an XY female. The data from the XX male that marked the proximal end of the sex-determining region was unequivocal, so the real TDF should therefore lie somewhere between *ZFY* and the pseudoautosomal boundary.

Goodfellow's group (which by now included Andrew Sinclair, who had mapped marsupial *ZFY* in my lab) found several XX male patients whose DNA lacked *ZFY*. Some had even smaller regions of the Y chromosome, which included the pseudoautosomal boundary but not *ZFY*, confirming the evidence from comparative mapping and expression studies that *ZFY* was the wrong gene. This narrowed the sex-determining region to a stretch of 35 kb, adjacent to the pseudoautosomal boundary. The discrepancy between this newly defined sex-determining region and Page's previously defined region was resolved when a second deletion extending 45 kb into the Y-specific region was found in the XY female, on whom the critical region depended.

Goodfellow's group turned their attention to combing the newly defined 35 kb critical region for TDF. This region was encompassed within their chromosome walk, so it had already been cloned several times over. Logically, one of these segments must contain TDF. Yet initial analysis had detected no genes.

An exhaustive (and exhausting) search of the target 35 kb of Y chromosomal DNA was undertaken for a sequence which could be the testis determining gene.

Problems were posed by the high content of repeated sequence DNA, which meant that clones from this region cross-hybridized confusingly to many other sequences.

The region was divided into very small pieces, and each tested for male-specific sequences by probing Noah's Ark blots of DNA from male and female humans, cattle and mice, as well as to somatic cell hybrids containing the human X or Y chromosome. It was frustrating work because most probes just hybridized to repetitive elements distributed throughout the human and bovine genome.

Just one probe looked more promising because it also detected male-specific bands in the mouse and bovine genomes. The putative gene defined by this probe was given the noncommittal name *SRY* (for Sex-determining Region Y, since the inference that it was the long sought TDF was based solely on its location.

Finding TDF was a pioneering use of a positional cloning strategy based on discovering the exact chromosomal position of the gene and physically isolating the target stretch of DNA. This strategy has since been spectacularly successful in the hunt for many genes associated with genetic diseases.

The long search for TDF was over. Now it was crucial to demonstrate that *SRY* was the right gene.

8.14 Characteristics of *SRY*

At first sight, *SRY* looked singularly unimpressive. It was a small intronless gene, whose sequence didn't match anything much in databases. The only recognizable motif was a short sequence that codes for a 79 amino acid stretch that is shared by High Mobility Group proteins, and was therefore called the 'HMG box'.

A sequence homologous to human *SRY* was cloned from the mouse. *Sry* was located on the short arm of the mouse Y, and was present in Sxr[a] mice, and even Sxr[b] mice that carry the smallest part of the mouse Y chromosome known to be sex-determining. The location of *Sry* to this minimal sex-determining region of the mouse Y is consistent with *Sry* being the mouse testis determining factor *Tdy*.

In line with a conserved role in sex determination, it was found that sequences homologous to *SRY* were male-specific in all eutherian mammals tested.

Given the impact of our discovery that *ZFY* was autosomal in marsupials, it was important to discover whether an *SRY* homologue was located on the marsupial Y chromosome. Probing male and female marsupial genomic DNA with the human *SRY* gene revealed many homologous sequences which were shared between males and females, but only a weak male-specific band. Eventually a tammar wallaby genomic *SRY* clone and a dunnart testis cDNA *SRY* clone were isolated. These marsupial *SRY* clones both mapped to the Y chromosome, confirming *SRY* as a good candidate for the testis determining factor in marsupials as well as eutherians. Phew!

Given the importance of sex determination, you might expect the testis determining gene to be highly conserved between mammalian species. It was a shock to discover that *SRY* is quite poorly conserved. On a Southern blot, human *SRY* barely detected the mouse, let alone the marsupial *SRY* homologue. Homology between human *SRY* and mouse *Sry* is about 80% at the nucleotide level (90% at the amino acid level) in the HMG box region but rapidly drops off outside this region. When the putative *SRY* amino acid sequence of humans, mice, rabbits and the two marsupial species are compared, they show an overall similarity of about 85% within the HMG box, but poor homology outside the box. This suggests that the 79 amino acid HMG box motif is the functionally important region of *SRY* protein.

Detection of male-female shared bands by *SRY* probes on Southern blots implied that there are many *SRY*-related genes on autosomes in every species. It turns out that the HMG box is shared by many genes in all mammals, other vertebrates and even invertebrates.

Thus, *SRY* is evidently a member of a large gene family. The autosomal fragments detected by human *SRY* on Noah's Ark blots were found to share the HMG box motif. Those genes which share more than 60% amino acid sequence similarity to *SRY* within the HMG box region became known as *SOX* genes (for <u>*S*</u>RY related, HMG b<u>ox</u>-containing).

In humans and mice, twenty *SOX* genes fall into ten different groups based on sequence and structural differences. Group A contains only *SRY*, and group B are highly homologous to *SRY*. Most other groups are intronless, like *SRY*, but some contain introns and presumably represent the ancestral *SOX* gene from which processed copies were made. Autosomal *SOX* genes are extremely highly conserved, having orthologues in other vertebrates and even insects. Many members have important functions in development, such as *SOX2*, which is a pluripotency marker that can induce stem cells. Others play important roles in mammalian development; some in sex determination (Section 8.8), some in neuronal or other organ development.

My group in Australia, trying to clone the marsupial homologue of *SRY*, accidentally stumbled upon a particularly interesting *SOX* gene, which showed dosage differences between marsupial males and females on Southern blots. Our somatic cell hybrids confirmed that it lay on the marsupial X chromosome, which has homology to the ancient part of the X of eutherian mammals. Sequence analysis showed that it shared the closest homology to *SRY* in humans as well as marsupials. We proposed that this gene (named *SOX3* in line with a mouse homologue) was the ancestor of *SRY*, an idea subsequently confirmed by sex reversal in mice and humans with altered *SOX3* expression (Section 10.4).

Thus, *SRY* is a member of the large *SOX* gene family which share a conserved HMG motif. *SRY* maps to the Y in all therian mammals, but its sequence is poorly conserved. Direct proof that *SRY* is the long-sought TDF came from two sources; mutational analysis of sex reversed human patients, and *Sry* transgenesis in mice (Figure 8.10).

8.15 Mutation of Human *SRY* Produces Sex Reversal

If *SRY* is the testis determining gene, mutations in *SRY* should prevent testis formation and result in a female rather than a male phenotype. At least some XY females with complete gonadal dysgenesis would therefore be expected to carry mutations in the *SRY* gene. They do.

SRY sequences were amplified by PCR from XY females, and from XY male controls. Fragments of these products were tested for sequence differences using a technique that identifies mismatches between the hybridizing single DNA strands.

Most of the DNA from XY females contained no *SRY* sequences, and they were presumed to have *SRY* deleted from the Y. Some had a normal *SRY* gene and were presumed to have mutants in other genes. But one XY female showed an altered *SRY*, and sequencing revealed a single base change which would encode a different amino acid within the HMG box (Figure 8.10a). Another XY female had a frameshift mutation in the conserved HMG box, which would wipe out any normal protein.

Their fathers had a normal *SRY* sequence, indicating that these two XY females had received a *de novo SRY* mutation.

Subsequent studies demonstrated that 15% of XY females tested carry mutations in *SRY*. These mutations cluster within the HMG box, as you would expect if this is the business end of *SRY* (Figure 9.2). Some XY females proved to be mosaic for the mutation, and could have gonads with both testicular and ovarian development (ovotesticular DSD). For instance, a mutant *SRY* gene was found in only part of the ovotestis in one patient, and presumably resulted from a somatic mutation of a single base.

Since *SRY* mutation produces a sterile female phenotype, you would expect that these *SRY* mutations could hardly be inherited. But surprisingly, one of the first eleven XY females studied shared an *SRY* mutation with her father (obviously a normal fertile male). Both showed a G to C base change within the HMG box, causing a valine to leucine amino acid substitution. Interestingly, the father had an XY female sister, implying that the variant is conditionally sex reversing, depending on other genes.

No natural single base substitutions have been detected in mouse *Sry*. However, a sex reversing mutation was produced by inserting sequences into the DNA of mouse embryonal carcinoma lines using a virus vector. In XY female mice of this strain, the Y chromosome was no longer sex-determining and molecular analysis

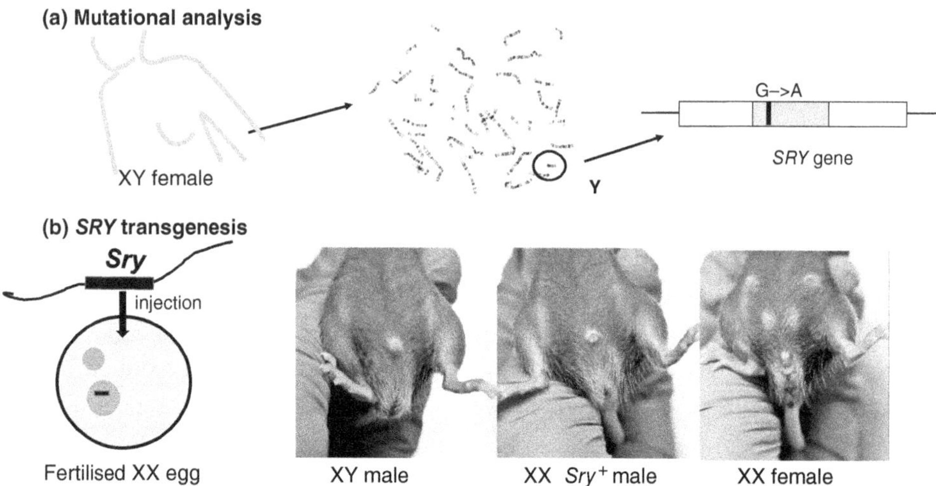

Figure 8.10   Demonstrations that *SRY* is the mammal sex-determining gene. (a) Human *SRY* codes for a polypeptide containing 204 amino acids. The sequence of the middle portion has similarities to the High Mobility Group proteins (the HMG box, grey). Of hundreds of XY females who have *SRY*, many have single base mutations (represented as a dark line) that alter the amino acid, confirming that the SRY protein is necessary for male determination. Photograph supplied by Victorian Clinical Genetics Services, Murdoch Children's Research Institute, Melbourne. (b) A 14 kb fragment of DNA containing the mouse *Sry* gene and flanking sequences was microinjected into the pronucleus of a fertilized mouse egg. Of the XX embryos resulting, several contained the *Sry* transgene, and these showed testicular differentiation. Two were allowed to go to term, and their phenotype and behaviour was male. The photos show the genitals of an Sry+ XX mouse compared to normal male and female. Photograph supplied by Haley Hrncir and Art Arnold, University of California, Los Angeles.

showed that the *Sry* gene had been deleted. There was no adverse effect on somatic tissues, suggesting that *Sry* plays no role outside gonadal development.

Thus, mutations in the SRY gene cause a failure of testis development and phenotypic sex reversal in humans and mice. This demonstrated that *SRY* is necessary for testis determination and for male development.

8.16 *Sry* Transgenesis Produces Male Mice

Is *SRY* all that is needed to kick-start testis differentiation? The de novo mutations in XY females show that *SRY* is necessary for sex determination but do not tell us whether it is sufficient.

This question was addressed by Peter Koopman in Lovell-Badge's lab by injecting a large fragment of the mouse Y chromosome, containing the *Sry* sequence, into fertilized XX mouse eggs (Figure 8.10b). A 14 kb genomic fragment that encompassed all the identified *Sry* sequence and substantial amounts of flanking sequence was used in order to include all the regulatory sequences necessary for the precise expression of *Sry*. An examination of the XX embryos which carried the *Sry* transgene revealed that one third had started to develop testes by 14 dpc.

Some of the mice in this experiment were allowed to develop to term and of these, three XX mice were found to lack a Y chromosome but retain the *Sry* transgene. One of these XX *Sry*⁺ transgenic mice (fetchingly named 'Randy'), had normal male internal and external genitalia including normal, though small testes. Randy also showed normal male sexual behaviour, repeatedly mating with female mice.

Histological examination of the testis in XX *Sry*⁺ transgenic males revealed normal tubules, Leydig cells, peritubular myoid cells and Sertoli cells. But spermatogenesis was absent and germ cells were degenerating, accounting for the small testes. Sterility was not surprising, since this mouse lacked Y chromosome gene(s) such as *Rbmy*, and *Spy*, which are required for spermatogenesis (Section 2.17).

Detailed sequence analysis of the inserted 14 kb genomic fragment revealed no genes other than *Sry*, implying that *Sry* is the only gene on the Y chromosome required for initiating testis development.

It was not surprising that many other XX transgenic mice showed no sex reversal. We know from other systems that the site of integration of a transgene can suppress its expression and block its action.

The ubiquity of sex-determining function of SRY was demonstrated by transgenesis of XX mouse eggs with human or goat *SRY*, which also produced some XX *SRY*⁺ transgenic males.

More recently, the reverse experiment has been done; knocking out the mouse *Sry* gene resulted in mouse pups that look, and act, like females.

The demonstration that XX mice transgenic for a 14 kb segment of the Y carrying only the *Sry* gene constitutes incontrovertible evidence that *Sry* is both necessary and sufficient for testis determination. Thus, *SRY/Sry* is the mammalian testis determining factor, TDF.

8.17 Conclusions

I once volunteered to present a series of lectures on the biology of sex differences for a new course on women's studies, stressing the role of the newly discovered *SRY* gene. My offer was summarily rejected on the grounds that there are *no* biological differences between men and women. 'It's only one gene, so it can't be very important', sniffed the organizer.

However, it is now clear that a single gene, *SRY*, is responsible directly and indirectly, for the very profound differences between male and female mammals.

The fundamental differences between males and females are defined by their differential reproductive roles. In mammals, these differences are most obviously manifest in gonads, germ cells and reproductive tracts. Males have testes and produce sperm, and females have ovaries and produce eggs.

Most other, more visible, secondary sexual dimorphism – for instance, size, musculature, fat and hair distribution – are the result of sex hormone differences.

Most remarkably, these sex differences, direct and indirect, are ultimately the product of a single gene, the Testis Determining Factor TDF that triggers a genetic pathway that leads to the development of a testis in the embryo. The embryonic testis makes male hormones, which are responsible for the secondary sexual characteristics; the baby is a boy. In the absence of TDF the gonad develops as an ovary and the baby is a girl.

The search for TDF went on for three decades, using new technologies as they came to hand. Screening for male-specific proteins and male-specific DNA sequences turned up interesting male-specific factors that increased our knowledge of the Y chromosome, although they proved not to reveal TDF.

The search for the sex-determining gene by mapping and cloning regions of the Y, and the discovery of *SRY* is a classic demonstration of the power of genetics to define and locate an unknown gene, and the power of molecular biology to isolate it, verify it and discover how it acts. *SRY* is conserved on the Y chromosome in other mammals; its testis determining function was proven by analysis of mutant *SRY* in XY female patients, and by *Sry* transgenesis in mice.

But although SRY triggers sex determination, it acts through many genes in a huge array of developmental networks. In Chapter 9 I will explore how a single gene on the Y chromosome could possibly control all the sexual dimorphisms in humans and other mammals.

FURTHER READING

Book

Mills A, 2018. *Biology of Sex*. University of Toronto Press

Classic Papers

Jost A, Vigier B, Pre´pin J, Perchellet JP, 1973. Studies on sex differentiation in mammals. *Recent Progress in Hormone Research* 29: 1–41

Koopman P, Gubbay J, Vivian N, et al., 1991. Male development of chromosomally female mice transgenic for *Sry*. *Nature* 351: 117–121

Page DC, Mosher R, Simpson EM, et al., 1987. The sex-determining region of the human Y chromosome encodes a finger protein. *Cell* 51: 1091–1104

Sinclair AH, Berta P, Palmer MS, et al., 1990. A gene from the human sex determining region encodes a protein with homology to a conserved DNA-binding motif. *Nature* 346: 240–244

Reviews and Research Papers

Arnold AP, 2019. Rethinking sex determination of non-gonadal tissues. *Current Topics in Developmental Biology* 134: 289–315

Eggers S, Sinclair AH, 2012. Mammalian sex determination – insights from humans and mice. *Chromosome Research* 20: 215–238

Graves JAM, 2015. In retrospect: Twenty five years of the sex determining gene. *Nature* 528: 343–344

Kiefer JC, 2007. Back to basics: *Sox* genes. *Developmental Dynamics* 236: 2356–2366

Migale R, Neumann M, Lovell-Badge R, 2021. Long-range regulation of key sex determining genes. *Sexual Development* 15: 360–380

Nagahama Y, Chakraborty T, Paul-Prasanth B, 2021. Sex determination, gonadal sex differentiation, and plasticity in vertebrate species. *Physiological Reviews* 101: 1237–1308. https://doi.org/10.1152/physrev.00044.2019

Oliva M, Muñoz-Aguirre M, Kim-Hellmuth S, et al., 2020. The impact of sex on gene expression across human tissues. *Science* 369: 1331–1339

Simpson E, Scott D, Chandler P, 1997. The male-specific histocompatibility antigen, H-Y: a history of transplantation, immune response genes, sex determination and expression cloning. *Annual Review of Immunology* 15: 39–61. The HYA story

Singh L, Purdom IF, Jones KW, 1976. Satellite DNA and evolution of sex chromosomes. *Chromosoma* 59: 43–62. The Bkm story

Spiller C, Koopman P, Bowles J, 2017. Sex determination in the mammalian germline. *Annual Review of Genetics* 51: 265–285

Wilhelm D, Palmer S, Koopman P, 2007. Sex determination and gonadal development in mammals. *Physiological Reviews* 87: 1–28

Zhao F, Yao HH-C, 2019. A tale of two tracts: history, current advances, and future directions of research on sexual differentiation of reproductive tracts. *Biology of Reproduction* 101: 602–616

9

Molecular Biology of Mammal Sex Determination

Famously, Peter Goodfellow authored an article in 1986 entitled 'Sex is simple'. But the sex-determining pathway is not a simple A–>B–>C–>male/a–>b–>c–>female, toward one fate or the other, nor is there a simple on-off switch that determines whether the male or the female pathway will be followed. Rather, it is a network, full of checks and balances between antagonistic genes. Gonad differentiation is a real pushmi-pullyu.

Primary sex differences occur in gonads, germ cells and reproductive tracts that are under the control of the *SRY* gene. The more visible secondary sexual dimorphisms, evident as differences in genitalia, physique and behaviour, are largely controlled by hormones. In this chapter, I will explore how a single gene could possibly control all these differences.

The discovery that the *SRY* gene represents the long-sought testis determining factor (TDF) excited expectations that a few months' study of its transcription and a year or two's biochemical work on its protein product would completely resolve the age-old mystery of sex determination. With *SRY* in hand, the whole mammalian testis determination pathway should fall like a stack of dominoes.

But it took more than 20 years to discover the targets of *SRY* and how they interact with other pathways to effect differentiation of the testis in XY embryos and an ovary in XX embryos. It required a mixture of biochemical studies of the transcript, the protein and its binding partners, and of genes identified from mutant mice and patients with atypical sexual differentiation.

Here, I will summarize the work that has led us, at last, to an understanding of the network of pathways that control sexual differentiation in humans and other mammals. It is not as simple as we thought.

9.1 Structure and Sequence of *SRY*

The newly discovered human *SRY* genomic clone was analyzed minutely, in the hopes that the base sequence of the gene, or the presumptive amino acid sequence of its product, would provide clues to its function in sex determination.

Human *SRY* was seen to be a small gene with a single exon (Figure 9.1a). Its nucleotide sequence contained two open reading frames, not interrupted by chain-terminating codons, that could code for proteins of 99 and 204 amino acids. Neither sequence looked familiar, nor did a search of DNA and protein databases reveal related sequences.

However, the product of the longer open reading frame of *SRY* contained a small tract similar to a 79 amino acid motif in several proteins in the high mobility group (HMG) and contained within several transcription factors, so it was called the 'HMG box'. The HMG box is associated with DNA binding in these proteins, suggesting that SRY protein may also act as a transcription factor. This made sense, since the ability to bind DNA and regulate the transcription of other genes in the sex-determining pathway was a respectable function for a master regulatory switch.

To define the coding unit of *SRY* and to deduce its structure and function, it was important to identify the RNA transcript. Where does it start and stop?

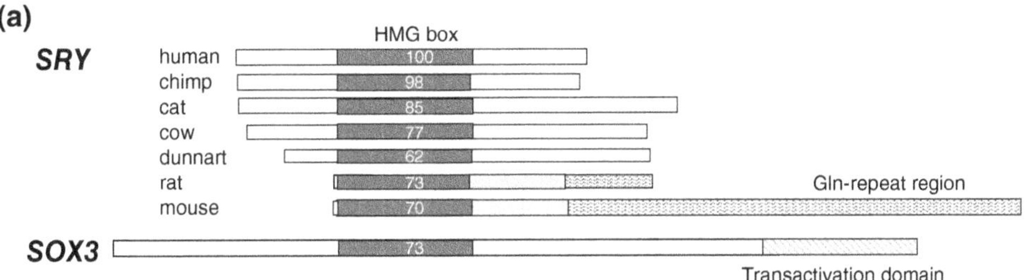

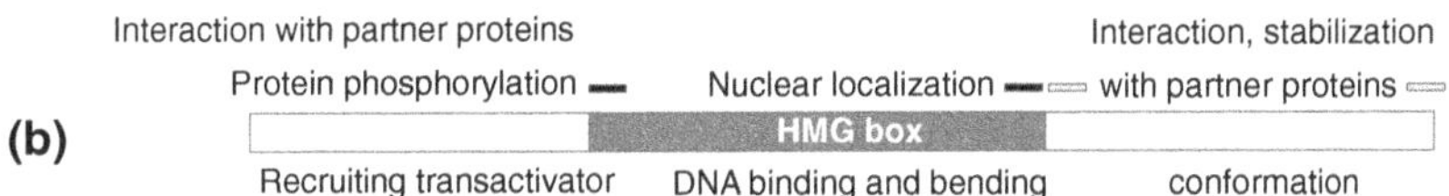

Figure 9.1 *SRY* structure and conservation. (a) Human *SRY* codes for a polypeptide containing 204 amino acids. The sequence of the middle portion has similarities to the High Mobility Group proteins (the HMG box). *SRY* genes of other mammals share a conserved HMG box (dark grey; figures represent % conservation of amino acids). All non-rodent eutherians share N- and C-terminal domains, but in mice and rats there is a long Q-rich region (patterned) separated by a bridge (light grey). *SRY* is a member of the *SOX* gene family that shares the HMG box, other members of which (e.g., *SOX3)* also contain a transactivation domain (striped). (b) Sequence signals that suggest putative functions of regions of the human SRY protein. The HMG box binds to DNA and bends it through an acute angle. At either end is a nuclear location signal (black bars) that allows the protein to move into the nucleus. The N-terminal domain may bind to other proteins which activate target genes. The C-terminal domain contains signals (grey bars) that would stabilize interaction with partner proteins and protein conformation.

Is *SRY* transcript processed? Is it conserved between species? Answering these questions was a lot harder than anyone expected.

Messenger RNA was prepared from adult human testis and analyzed by separating molecules by size on a gel and probing with the genomic *SRY* clone ('Northern blot'). A band was detected at 0.9 kb, indicating that *SRY* is transcribed into a 900 base RNA molecule. With some difficulty, a complete DNA copy (cDNA) of the transcript was isolated and the beginning and end of the gene were established. The structure of the *SRY* gene could now be deduced by lining up the sequence of the genomic and cDNA clones. They were identical, confirming the absence of introns. An exhaustive search upstream of *SRY* revealed no other exons. This must be the whole human gene. Not surprisingly, embryonic human *SRY* proved hard to isolate, but a transcript of the same size was eventually detected in the genital ridge. *SRY* was also expressed by some human stem cells, and had the same structure as the adult testis *SRY*.

The Open Reading Frame (ORF) of human *SRY* expressed in the adult testis was predicted to encode 204 amino acids, containing the 79 amino acid HMG box in the middle. Upstream (toward the 5' end) sequence encoded a 57-amino acid N-terminal domain, and downstream sequence (toward the 3' end) encoded a 68-amino acid C-terminal domain of the SRY protein.

Surprisingly, mouse *Sry* proved to have a very different structure. It was much longer (encoding 395 amino acids). This included the 79 amino acid HMG box – but its position in the protein was radically different, occurring right near the start. Even weirder was a long C-terminal domain, which was largely (73%) composed of trinucleotide (CAG) repeats that would code for a glutamine-rich (Q-rich) protein tail.

RNA was isolated from mouse genital ridges, reverse transcribed into complementary DNA (cDNA) and lined up with the genomic clone. This indicated a linear *Sry* transcript derived from a single exon. However, a second downstream exon was discovered more recently. The single exon *Sry* that was originally described is transcribed into mRNA with only exon I sequences, but this proves to harbour sequences that promote RNA degradation ('degron') so the mRNA is unstable. An alternative transcript splices out this region and substitutes Exon II sequences that code for the last 15 amino acids of mouse *Sry*.

SRY has been cloned and transcripts have been characterized from many mammals, showing that eutherian mammals, even marsupials, subscribe to the human protein structure, having an HMG box flanked by an N-terminal domain of 30–60 amino acids and a C-terminal domain of 70–100 amino acids. The HMG box is relatively well conserved but the N- and C-terminal domains are not. Rodents are bewilderingly different, having variable numbers of CAG repeats in other mouse and rat species. Thus, rodent – and specifically mouse – SRY protein differs in structure, suggesting it could differ in its mechanism of action from that of other mammals.

9.2 Function of SRY Protein

How does SRY protein work? If SRY were a typical transcription factor that turns on testis differentiating genes, we would expect it to contain DNA-binding and transactivating domains. However, it is far from typical and decades after its discovery, we still lack a complete picture of how it activates the testis determining pathway. It does not help that there may be major differences between species in how it packs its punch.

Direct studies of SRY protein were excruciatingly difficult. Human SRY protein was present at far too low a concentration to purify it directly, so instead the gene was inserted into an expression vector and cloned in *Escherichia coli*. Extracts were purified and the products were run on a Western blot that separates proteins by size and detects SRY by probing with anti-SRY antibody raised in rabbits. SRY appeared as a single band of 23.9 KD, indicating a small protein.

The properties of SRY were originally inferred from the amino acid sequence deduced from the DNA sequence, informed by analogy with similar proteins. Many lines of evidence show that the HMG box is critical for *SRY* function. For a start, this is the only region that shows evolutionary conservation, always a good clue to its importance. Mutation analysis of human XY female patients showed that nearly all the sex reversing mutations (66/80) occurred within the box (Figure 9.2).

The sex-determining potential of the HMG box could also be tested by making transgenic mice with constructs containing normal, or deleted or mutated *SRY* sequence, and *SRY* from various species. Refinement of the technique showed that *SRY* from human or goat, as well as mouse, could sex-reverse XX mouse embryos. Deletions or stop codons within the HMG box of the transgene, but not outside it, completely wiped out male development.

The reverse was also true. Knocking down the *Sry* gene in XY mouse embryos produced XY females, and specifically targeting the HMG box region was just as effective. Specific CRISPR knockouts of the HMG box region of pig *SRY* produced XY piglets with female internal organs, genitalia and behaviour.

What does the HMG box do? By analogy with other HMG box-containing proteins, the major function of the HMG box was predicted to be binding DNA and bending it sharply. Some HMG proteins bind DNA with little specificity, but others, including SRY, bind specific DNA sequences with high affinity.

To identify the SRY binding site, recombinant SRY protein was incubated with a random pool of labelled primer DNA sequences, and those that were bound were pulled out by the effects of binding on slowing their mobility in a gel ('gel retardation'). Multiple rounds of selection and enrichment identified several short DNA molecules that all shared A/TAACAAT, suggesting that this is a high-affinity binding site for SRY protein. However, SRY can also bind DNA that forms specialized crosses and loops without sequence specificity.

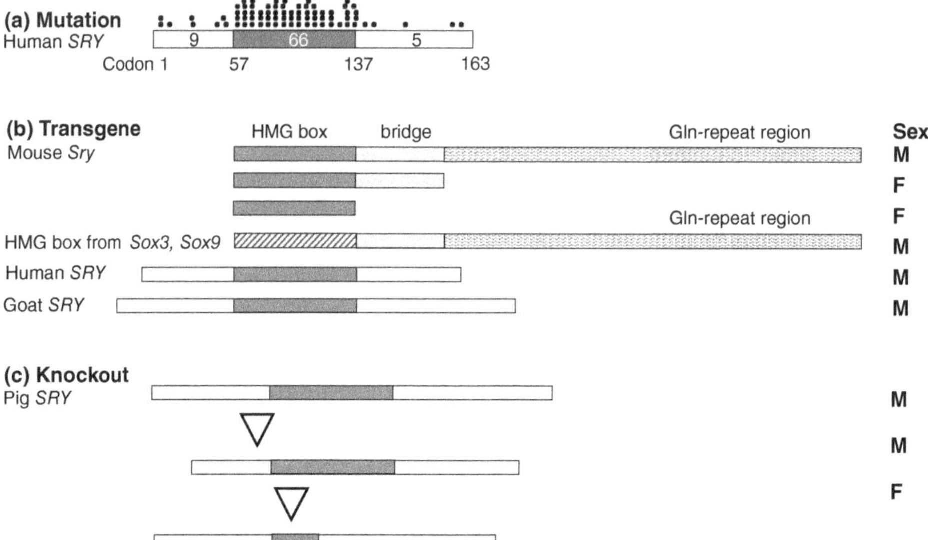

Figure 9.2 *SRY* mutation and manipulation. (a) Sex-reversing mutations in
different regions of the human *SRY* gene (dots, numbers within boxes). Nearly all
(66) are within the HMG box (dark grey). (b) Transgenesis: *Sry* transgenes were
transferred into XX mouse eggs. Transgenesis with whole *Sry* produced male
pups (M), but constructs lacking the glutamine-rich tail produced females (F).
Constructs in which the HMG box of mouse *Sry* had been replaced by the HMG
box of human or goat *SRY*, or of other *Sox* genes (*Sox3* and *Sox9*) produced male
pups. (c) Knockout of *SRY* in pigs by CRISPR-Cas targeted deletion. Removal
(depicted by delta) of 72 bps outside the box in XY pigs did not cause sex reversal,
but removal of about 300 bps within the box caused female development.

What happens when SRY binds to DNA? A three-dimensional structure of
SRY-DNA was resolved by nuclear magnetic resonance (NMR) spectroscopy.
This revealed a twisted boomerang shape, in which three helices make up a con-
cave surface that interacts with the minor groove of DNA, bending it by 65°.

Altered interaction of mutant SRY protein with DNA confirmed that binding and bending are critical to SRY activity. Recombinant SRY protein from
XY females with mutations in the HMG box was examined by gel retardation.
Whereas normal SRY protein bound to the A/TAACAAT target sequence, producing a band shift, no shift occurred with the three mutant proteins. This implies
that disrupting DNA binding disrupted the critical function of SRY. Some other
mutant SRY proteins from XY females proved to bind to DNA, but to bend it
through an abnormal angle.

Mouse SRY protein differs from human SRY in its binding dynamics. It has a
higher degree of sequence specificity than human SRY, contacts DNA predominantly through the major groove, and induces an 85° bend in the DNA.

Is DNA binding and bending through the HMG box all there is to SRY function? This would be surprising because most transcription factors have two separate domains, one that binds DNA and the other that regulates transcription of the target sequence. The poorly conserved N- and C-terminal domains do not point to a conserved function.

At one end of the HMG box lies a conserved nuclear localization signal that binds a calcium-binding protein, and at the other a nuclear importation motif (Figure 9.1b). These sequences allow the SRY protein to access DNA in the nucleus. We know they are essential because mutations of these codons cause XY sex reversal in humans.

The discovery of sex-reversing human *SRY* mutations outside the box suggests these regions have at least a modifying function. Some of these mutants are conditional, in that they were inherited from the normal father, or manifest by a brother, implying that the function of this region is critical only in combination with other genetic changes.

Secondly, transgenesis by constructs with deletions outside the box does have an effect on the efficiency of testis formation in transgenic XX mice by weakening its DNA binding and perhaps interfering with protein modification (phosphorylation). In particular, the ablation of the rodent-specific glutamine-rich tail of the *Sry* gene removed its male-determining potential, implying that it is necessary for male determination in mice.

Thus, SRY protein functions by binding and bending DNA at particular target sites defined both by DNA sequence and conformation (Figure 9.3). The

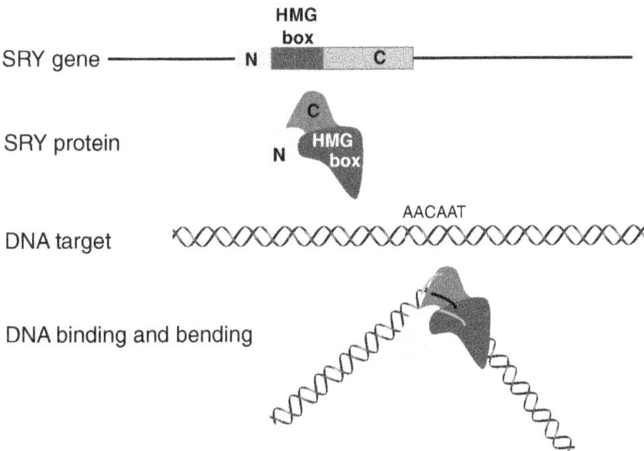

Figure 9.3 Function of SRY protein. The three domains of the *SRY* gene encode an N-terminal region (light grey), the HMG box (black) and a C-terminal region (dark grey), giving the SRY protein its shape. The SRY protein binds DNA in the minor groove at a recognition sequence A/TACAAT/A and bends it through an angle of 85°.

interaction may bring otherwise distant regions of the genome into juxtaposition, allowing bound transcription factors to interact and activate transcription of target genes.

9.3 Expression of *SRY* in the Developing Gonad

We know that the testis determining factor must be expressed in the supporting cell lineage of the bipotential gonad before testis differentiates (Section 8.4), so we would expect to see *SRY* expressed in the genital ridge before the testis differentiation becomes visible.

Not surprisingly, details of *Sry* expression have relied on a mouse model. In the foetal mouse, the genital ridge develops between 9.5 and 10.5 dpc (days post coitum) and by 12.5 dpc, the male genital ridge develops testis cords which form as Sertoli cells differentiate and align. Any worthy candidate testis-determining gene should be expressed in the genital ridge before 12.5 dpc.

The tiny amount of tissue available from the embryonic genital ridge ruled out Northern analysis of mouse *Sry*, so the more sensitive reverse-transcribed PCR (RT-PCR) was used. This involves making a DNA copy of the mRNA from the genital ridge using a poly T primer (which binds to the Poly A tail of the mRNA). *Sry* specific primers are then used to PCR amplify the *Sry* transcript. This technique revealed linear *Sry* transcripts only in the embryonic genital ridge, and nowhere else in the embryo.

As predicted, *Sry* transcripts in the genital ridge were detectable only within a narrow window, between 10.5 and 12.5 dpc, with a peak at 11.5 dpc. *Sry* is expressed in a wave from the centre, encompassing the whole gonad by 11.5 dpc, then it recedes from the anterior (head end) and is gone by 12.5 dpc. In situ hybridization patterns were consistent with this tissue-specific and stage-specific transcription of *Sry* RNA.

What cell type in the genital ridge expressed *SRY*? Experiments with XX:XY chimaeras (Section 8.4) had fingered Sertoli cells because they must be XY, whereas Leydig cells and germ cells could be either. The resolution of radioactive in situ hybridization was too rough to pinpoint a specific cell type, but a Sertoli cell location was indicated by the expression of *Sry* in the testes of mouse embryos with a mutation that prevents germ cell proliferation and migration, but still allows development of the testis. This confirmed that *Sry* is expressed only from the somatic cells of the developing testis.

Thus, the time and tissue in which *Sry* is expressed in the mouse gonad is consistent with predictions made about the testis determining factor.

Both the timing and level of *Sry* expression are absolutely critical to proper testis development in mice. Earlier work showed that when a Y chromosome from another mouse strain (the pocket mouse, *Mus poschivinus)* was bred into the lab mouse strain, XY sex-reversed females and intersexes were produced. This turned out to be because the *M. poschiavinus Sry* is expressed later than the lab mouse gene, so a threshold concentration of SRY protein is reached too late to interact with other genes responsible for testis differentiation.

The critical timing was confirmed by constructing a mouse strain transgenic for *Sry* under the control of a promoter sequence that could be induced. Turning on *Sry* at different times during the development of *XX Sry* transgenic mice showed that there was only 6 hours, between 11.0 and 11.25 dpc during which the *Sry* transgene would determine testis. Thus *Sry* action occurs as a brief flash in mouse development.

The idea that *Sry* expression must reach a threshold level to turn on testis also explains why only a few XX mice transgenic for *Sry* develop testes, because in many transgenic *Sry* is expressed below this threshold level.

The situation in human embryonic gonads is dramatically different. A 1.1 kb linear *SRY* transcript is detected from 6 weeks after conception, peaking at the time that testis cords become visible. However, surprisingly, *SRY* was also transcribed in gonads at earlier stages, and continued to be expressed at low levels throughout embryonic development. Even more surprisingly, *SRY* was expressed in many human tissues at every stage. *SRY* was found to be widely expressed also in many other placental mammals (horses, pigs, sheep and dogs) and even marsupials (tammar wallaby). So, again, mouse is the outlier.

Thus, expression patterns of *SRY* in the gonad are remarkably different in rodents and other therian mammals. Mouse *Sry* is expressed briefly during two days of testis differentiation, but in humans and other therians *SRY* is almost ubiquitously expressed.

It is just as well that this was not known at the time *SRY* was discovered, or this gene might have been abandoned along with *ZFY*!

9.4 *SRY* Expression in Other Tissues – What Does It Mean?

The expression of *SRY* outside the developing gonad in most mammals was a big surprise. Is this transcript functional? What does it do?

Unlike human *SRY*, mouse *Sry* is also copied into quite different RNA transcripts in adult testis. Transcripts were 1.3 kb long, but they were weirdly jumbled up, with 3′ sequences from the end of the gene stuck onto the 5′ beginnings. These circular permutations arise because, extraordinarily, the transcript is circular, the product of transcribing RNA within a stem-loop structure.

Sry transcripts in adult mouse testis were deemed not to be functional because they were circular and not polyadenylated or associated with ribosomes, so probably not translated. But transcription occurs at about the same time as the first wave of spermatogenesis, initially suggesting a role in sperm development. However, mice chimeric for normal XY cells and XY cells lacking *Sry* can pass on this mutation to the next generation, indicating that germ cells do not need to express *Sry* to produce functional sperm. The circular transcript was then dismissed as junk. However, it has now been found to bind to inhibitory small RNA molecules (Section 8.14), so it may function as a sink that removes these inhibitors.

In humans, *SRY* is expressed in several adult tissues, including adrenal glands, kidneys, heart and lungs. *SRY* transcripts in the kidney are polyadenylated and presumably translated. However, it is hard to see how they could serve a vital function in non-gonad tissue, since XY females with a mutant *SRY* have normal kidneys.

Particularly interesting are reports, over many years, that *SRY* is expressed in parts of the brain in mice and humans. In mice these transcripts are initially circular, but after gonad differentiation they become linear and may be translated and functional. *Sry* is active in cells that express an enzyme that affects dopaminergic neurons in regions concerned with sexual differentiation and voluntary movement. This enzyme was downregulated by inhibiting *Sry* using short antisense DNA molecules that bind to *Sry* and stop it being translated. *Sry* also mediates expression of several other genes in the brain. *SRY* might therefore have a role in the regulation of sexual differentiation of the nervous system.

Could *SRY* expression in the brain directly affect behaviour? Differences in male and female behaviour had always been ascribed to sex hormones produced by the differentiated gonads. But studies with the four core mouse model show that the presence of an *Sry* gene (in normal XY mice and XX mice with an *Sry* transgene) influenced some behaviours (activity, food consumption, anxiety, pup retrieval) compared to normal XX mice and XY females lacking *Sry*.

This may be relevant to understanding and treating Parkinson's disease, which affects 1/300 people and is much more common in men than women. It usually appears in later life as problems in starting and maintaining voluntary movement, and may be accompanied by severe tremor. Of great medical interest is the finding that *Sry* is overexpressed in the brains of mice and rats with symptoms of Parkinson's disease due to mutation or chemical treatment. Inhibiting *Sry* in the brains of these rodents using antisense DNA ameliorates the severity of symptoms in males, but not females. Perhaps suppressing the activity of *SRY* in neurons of men with Parkinson's disease could mitigate this distressing disease.

Thus *SRY* may, after all, have some direct, non-hormone-induced effects in tissue other than the gonad, and these might be important in development, ageing and disease.

9.5 Pathway Genes Revealed by Atypical Sexual Development

What genes does SRY target? Even with *SRY* in hand, it proved extremely difficult to identify genuine *in vivo* targets, for instance, by comparing genes turned on in XX and XY mouse embryos. Progress depended on the identification of other genes whose mutation or loss caused disruption of sex determination, and much information was provided by studies of human patients with variations of sexual development (Table 9.1).

Not surprisingly, given the number of genes involved in sex determination, there is a lot of variation in sexual development. Many of these variants do not affect sexual function, but some are severe and a few are lethal. Where these variants caused severe effects on function, they were called disorders of sexual development (DSD) by medicos, otherwise they are called differences in sexual development.

DSD is defined as congenital variation of chromosomal, gonadal or anatomical sex. DSDs occur in up to 2% of live births and comprise 7.5% of all birth defects. Most are common and mild conditions (e.g., cryptorchidism in which testes fail to descend into the scrotum, and hypospadias in which the urethra is displaced from the tip of the penis). Several conditions are caused by mutations in sex hormone synthesis (Section 8.1). But a few conditions produce severe gonadal abnormalities and complete sex reversal.

Gonadal abnormalities are rare (1/4,500) but they have received much attention because they are very distressing for new parents, and have major implications for the welfare of the affected child (Section 11.5). Great efforts have therefore been made to discover the molecular basis of different DSD conditions, and to develop diagnoses.

DSDs include atypical complements of sex chromosomes (XO females with Turner's syndrome and XXY males with Klinefelter's syndrome), rearrangements of sex chromosomes, such as the XX males with small Y translocations bearing *SRY* (Section 8.9) and XY females who have swapped a piece of X for a piece of Y and lack *SRY*. Several sex-reversed mouse mutants also have *Sry* on an X or a truncated Y, or a Y chromosome with an *Sry* deletion.

Of greatest interest here are conditions in which another gene in the sex-determining pathway is mutated so that sexual development is incomplete or even reversed. As well as providing clinical advice to DSD patients, studies of the molecular basis of some of these rare conditions have yielded invaluable information on one or other step of what turns out to be a complex biochemical network. Many XY female conditions turn out to have mutations in one of the genes encoding factors or cofactors downstream of *SRY* that are required to form a testis. Several XX male conditions reveal genes that encode factors involved in making an ovary.

Table 9.1   *Genes identified in human DSD patients.*

GENE	Sex chromosomes	Phenotypic effect
Genes involved in development of a bipotential gonad		
SF1	XY	GD, female development, adrenal failure
SF1	XX	Premature ovarian insufficiency
WT1	XY	GD, genital abnormalities, kidney failure (Denys-Drash, Frasier syndromes)
CBX2	XY	Female development, DSD
GATA4	XY	DSD
MAP3K4	XY	GD
Genes involved in development of testis		
SRY	XY	GD, female development
SRY (+)	XX	DSD, male development
SOX9	XY	GD, female development, skeletal malformation (campomelic dysplasia)
SOX9 (+)	XX	DSD, male development
SOX3 (+)	XX	Male development
AMH	XY	persistent Müllerian glands
DMRT1	XY	DSD
DHH	XY	GD
FGF9 (+)	XX	Testicles, male development
FGFR2	XY	female, DSD
Genes involved in development of ovary		
WNT4	XX	Ovo-testicular DSD, male development
WNT4 (+)	XY	GD
RSPO1	XX	Testicular DSD, male development
RSPO1(+)	XY	Ovo-testicular DSD, female development
FOXL2	XX	Premature ovarian insufficiency, eyelid malformation
DAX1 (+)	XY	GD
DAX1	XX	Low sex hormones, adrenal hypoplasia (CAH)
ATRX	XY	DSD, incomplete female development, alpha thalassaemia, mental retardation

Most mutations involve loss of gene function. Gain of function (usually by duplication of translocation) is depicted as (+). DSD denotes disorders (differences) of sexual differentiation and GD to gonadal dysgenesis (atypical development of gonads, replaced by fibroid tissue 'streak gonads').

Until recently, the molecular basis of most DSD cases was mysterious. It took many years to track down the gene that was mutated, and diagnosis was possible for only a few conditions. However, the advent of cheap whole-genome sequencing has identified SNPs, rearrangements or repetitive sequences in many other genes, and diagnosis of XY female patients has increased from 15% to 43%. A targeted gene panel to look at all 64 known DSD genes has been developed to streamline diagnosis.

Most of the genes identified in DSDs are not on sex chromosomes. This is not surprising; only the gene that triggers testis determination needs to be on the Y.

As well as human mutants, there are observations of sex reversal in mammal species that have also been valuable in identifying steps of sex determination. Sex-reversed goats and beagles add to the compendium of genes involved in making a testis or an ovary.

In the next sections, I describe how the major pathway genes were discovered and how their function in making gonads was deduced.

9.6 Genes Involved in Development of the Bipotential Gonad

The first step in sex determination is the development of the bipotential gonad, which will differentiate into either a testis or an ovary, according to whether or not *SRY* is expressed. Three major pathways shape the bipotential gonad and regulate *SRY* expression (Figure 9.4).

A critical gene, Steroidogenic Factor 1 (*SF1*), was discovered through its action in binding to and activating genes encoding enzymes in steroid pathways. *SF1* is now known to belong to the <u>N</u>uclear <u>R</u>eceptor family, so it is also called *NR5A1*. *SF1* encodes a protein with zinc fingers that recognize DNA, and sequences that stabilize DNA binding.

Mutations in *SF1* were detected in XY female patients with streak gonads (gonadal dysgenesis, describing undifferentiated fibrous material), female genitalia and Müllerian ducts. This implies that *SF1* acts far upstream before the choice of testis or ovary is made, or even before gonad and adrenal gland are differentiated. Consistent with this idea, *SF1 is* expressed early in human embryos. *SF1* expression in the endothelium seems to define the progenitors of the somatic lineages of the genital ridge.

Confirming an early role in differentiation of the genital ridge in mice, *Sf1* is also transcribed in the genital ridge before *Sry* expression in mice. Knocking out this gene produced complete failure of gonad development and absence of adrenal glands in both sexes. Failure of testis differentiation produced XY female animals.

Hooking *Sf1* to *a* reporter gene showed that it up-regulates *Sry*. After gonad differentiation, *Sf1* is downregulated in the ovary, but continues in Sertoli and Leydig cells of the developing testis, where it functions in steroid pathways.

Another crucial gene, <u>W</u>ilms' <u>T</u>umour suppressor 1 (*WT1*), was first identified by mutations in paediatric kidney tumours. *WT1* was found to be mutated in patients with syndromes (Denys-Drash, WAGR and Frasier) that include gonadal, kidney or urogenital abnormalities.

Knocking out the homologous *Wt1* gene in mice stopped kidney and gonad development at the bipotential stage. This suggested an early action, and indeed, *Wt1* was expressed early in the genital ridge and embryonic kidney.

WT1 encodes a nuclear zinc finger protein that can act both as a transcription activator and repressor, depending on the cell and promoter context. Alternative splicing produces different isoforms differing by the presence or absence of three amino acids (Lysine, Threonine and Serine, abbreviated as KTS) with distinct roles in gonad differentiation. The −KTS isoform can bind to sequences in the *SF1* promoter, and the +KTS form seems to regulate *SRY* expression, since knocking it out reduced *SRY* transcription.

Another early expressed gene essential for gonad development is *GATA4*, a member of a transcription factor family that binds the nucleotide sequence GATA via two zinc finger domains for DNA binding, stabilization and protein-protein interaction. Human *GATA4* is expressed in the somatic cells of the bipotential gonad, where it interacts with *SF1* to regulate several genes, including *SRY* during sexual development.

Mutations in *GATA4* cause gonadal dysgenesis. In mice, *Gata4* is expressed in somatic cells of the undifferentiated genital ridge in both sexes at 11.5 dpc, but it becomes sex-specific by 13.5 dpc, being up-regulated in pre-Sertoli cells of XY embryos, but down regulated throughout the developing gonad in XX embryos. Knocking out *Gata4* in mouse embryos before E10.5, but not later, produces severe testis abnormalities. *Gata4* seems to be the earliest gene expressed. It controls the thickening of the epidermis to form the genital ridge and may initiate *Sf1* transcription. It later interacts with other proteins, and a phosphorylated form upregulates *Sry*.

Each of these three genes are themselves regulated by other genes, and loss of any of them disrupts gonad development and may affect kidneys, ureters, gonads, genital tracts or adrenal glands. Some of these genes encode transcription factors or co-factors, regulating *SF1* and other early genes, and some downstream genes. Others, such as <u>C</u>hromo<u>box</u> Homolog 2 (*CBX2*) are chromatin remodelling factors that help access promoters.

The bipotential gonad is unique in that it may follow one of two quite different developmental pathways to become either a testis or an ovary. The point of decision is whether or not *SRY* is expressed.

9.7 Turning on *SRY*

The sex switch comes down to turning on *SRY* transcription in the pre-Sertoli cells of the bipotential gonad (Figure 9.4). In mice, this switch is on for a brief flash – only 6 hours.

It requires several transcription factors and cofactors, some shared with previous steps in gonad development. Genes in three major pathways ultimately regulate *SRY* transcription. *SF1* is everywhere, acting as a transcription factor at multiple points in both male and female gonad and adrenal development. Sequences upstream of *SRY* include binding sites for *GATA4* and *WT1*, and there are hints of other binding sites further upstream from the transcriptional start site. *SF1, WT1* and *GATA4* and their regulators are crucial for expression of *SRY*, and mutations in any of them disrupts male development.

Other factors, including genes that encode protein kinases and insulin signalling may regulate *SRY* by phosphorylating other proteins. Mutations that disrupt phosphorylation of any of several proteins that ultimately regulate *Sry* transcription in mouse embryos, and lower *Sry* concentration, cause loss of Sertoli cells and female development.

Surprisingly, epigenetic pathways are also involved in *SRY* initiation. Some sex reversed XY patients show *SRY* hypermethylation, and *SRY* hypermethylation causes partial sex reversal in dogs, suggesting that DNA methylation suppresses *SRY* transcription. Indeed, methyl groups are lost from the human *SRY* promoter just before it is activated.

Histone modification is also involved in *SRY* regulation. Several histone methyltransferases are testis-specific and appear to control *Sry* expression, as well as

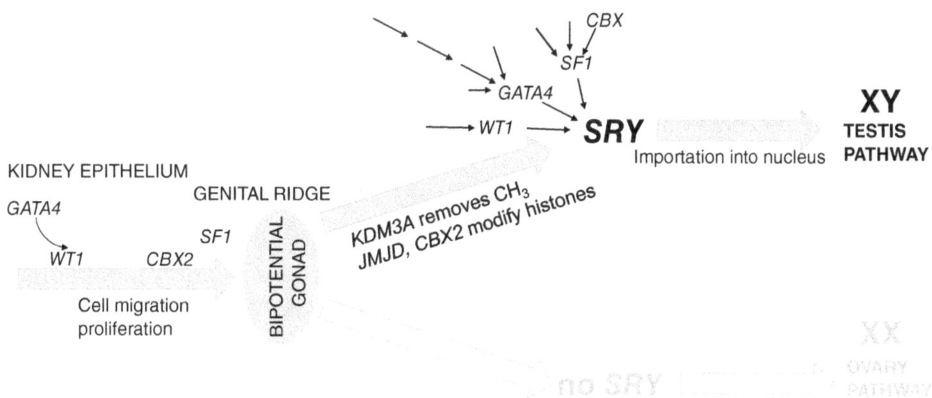

Figure 9.4   Regulation of *SRY*. Several genes are involved in the differentiation of the bipotential gonad. Other genes are required to turn on *SRY*, and several genes modify their action. Epigenetic modification of histones upstream of *SRY*, as well as removal of DNA methylation, is also required.

being essential for spermatogenesis. The Jumonji gene *Jmjd1a* (also known as *Kdm3a*) encodes an enzyme that removes methylation from the repressive histone H3K9me2. XY mice deficient for *Jmjd1a* had low *Sry* expression leading to XY female development. Adding in a histone methylase restored the function. This suggests that the balance between 'writing' and 'erasing' histone methylation marks (another loop in the battle of the sexes) is also finely balanced in activating *SRY* by catalyzing H3K9 demethylation.

Other genes that activate or repress testis formation (e.g., *ATRX* and *CBX2*) are involved in chromatin modification and remodelling. *CBX2* (related to Polycomb Repressor Complex) is mutated in some XY females, suggesting that epigenetic modifications contribute to male development.

To activate genes in the testis pathway, SRY protein must get into the nucleus. This is accomplished by the nuclear localization signal at the end of the HMG box; mutation of which can cause sex reversal.

What happens next in XY and XX embryos? In the next sections, I discuss the targets of *SRY* action in testis differentiation, and the genes that are active in the ovary in the absence of *SRY*.

9.8 Discovering *SOX9*, the Key Target of *SRY*

The discovery of *SRY* did not, as had been expected, immediately reveal the target gene(s) to which it binds. *SRY* acts in an unhelpfully convoluted way by binding DNA at poorly defined sites, with other proteins, and bending DNA to bring unknown sequences into contact and activate unknown genes.

The immediate target of *SRY* was discovered, instead, by studies of a rare sex-reversing syndrome. Campomelic dysplasia (CD) is a horrible autosomal disease. Affected babies are born with severe bowing of the long bones and survive for only a short time before succumbing to respiratory failure. Three quarters of affected XY newborns have female morphology.

Two groups independently pursued the CD gene responsible for bone deformity and sex reversal. Many cases were sheeted home to translocations involving the same region of chromosome 17, pointing to an unknown gene that was disrupted by the rearrangement. Lining up the translocation breakpoints defined a small critical region that contained only one gene. To everyone's surprise (including the authors'), this gene turned out to be a relative of *SRY*, a member of the <u>S</u>RY-related HMG b<u>ox</u> (SOX) gene family (Section 8.12) called *SOX9*.

SOX9 is a member of the group E of intron-containing *SOX* genes, mapping to human chromosome 17. SOX9 protein is much longer than SRY and much better conserved between mammals, even other vertebrates. It has two trans-activation domains and two nuclear localization signals, suggesting that it acts as a transcription factor.

It was proposed that *SOX9* is directly up-regulated by *SRY* as the first step in testis determination. *SOX9* could then trigger testis determination in the undifferentiated XY gonad (Figure 9.5).

Consistent with this role, *Sox9* is expressed at 10.5 dpc in the genital ridge of mouse embryos at low levels in both sexes, but is then dramatically ramped up in pre-Sertoli cells of XY embryos just a few hours after *Sry* expression. *Sry* shuts down, but *Sox9* maintains its expression in Sertoli cells. In addition, *Sox9* is also expressed at sites of cartilage formation in the mouse embryo, explaining the bone defects of CD patients.

Knocking out *Sox9* confirmed that it is critical for testis determination. Conditional loss of *Sox9* in the gonads of XY mouse embryos resulted in ovary development. Expressing *Sox9* in the gonad of XX embryos, and XY embryos lacking *Sry*, induced testis formation; this is the strongest evidence that the only role of *Sry* is to turn on *Sox9*. Consistent with this role were observations that *Sox9* over-expression in XX mice resulted in male development, and in humans XX male babies were found with duplications involving *SOX9*.

The evidence is therefore consistent with the idea that *SOX9* is the direct target of *SRY* in mammals, and holds a pivotal role in Sertoli cell development, which in turn sets up testis differentiation. Unlike *SRY*, *SOX9* is conserved across all vertebrates, suggesting it is an important sex-determining gene. Thus *SRY* transcriptional regulation of *SOX9* is *the* key step in testis determination.

9.9 Regulating *SOX9*

How is *SOX9* upregulated by *SRY*? Much effort has been put into understanding this crucial first step of testis determination; it is surprisingly complex, and still not entirely clear.

Does *SRY* upregulate *SOX9* directly? The dominant male determining action of TDF predicted that *SRY* was a transcriptional activator that binds to and turns on other testis differentiating genes. However, if *SRY* acted as a repressor of a repressor, it would have just the same effect, and there were some data (including its poor conservation) suggesting that *SRY* action on *SOX9* was indirect.

Several lines of evidence now show that the *SRY* protein acts as a transcriptional activator, albeit a rather weak one that requires several cofactors. It is translated in the cytoplasm, translocates to the nucleus and binds directly to a testis-specific enhancer of *SOX9*.

Identifying the *SRY* binding site upstream of *SOX9* was vital. However, scanning sequence in the megabase upstream found nothing that looked like an enhancer sequence; indeed, this region seemed to be something of a 'gene desert', although it was highly conserved among mammals.

More successful were studies of mouse mutants with aberrant *Sox9* expression. A sex-reversing mutation in mice called *Odsex* was found to be due to a failure to downregulate *Sox9* in XX embryos, which develop as XX sex-reversed males. *Odsex* mapped to a 150 bp deletion far upstream (1.3 Mb) of *Sox9*, suggesting that a repressor was deleted in the *Odsex* mutants. However, the mutation was hard to pin down because deleting these regions did not reverse sex.

Later, a region was identified in mice only 13 kb upstream of the *Sox9* promoter that seemed to contain a testis-specific enhancer of *Sox9* (*TES* for Testis Enhancer Sequence). TES contained binding sites for *Sry* and *Sf1*. Further refinement defined a 1.4 kb TES Core Element (TESCO). Functional analysis of this region using transgenic mice and reporter gene assays showed that Sry and Sf1 proteins can synergistically bind this TESCO enhancer to initiate *Sox9* expression.

When TESCO was knocked out, *Sox9* expression was about halved – however, XY mice were still male. This implied that there were other enhancers that regulate *Sox9* expression in mice. Another long-range Enhancer (*Enh13*), was subsequently identified whose deletion completely reversed sex in mice.

Curiously, in humans *SOX9* seems not to be regulated by a homologue of TES. Several studies on human XX male sex-reversed patients with deletions, duplications and translocations upstream of *SOX9* defined three upstream regions associated with male development. When these fragments were hooked up to a transcription test system, all three showed enhancer activity, and interacted synergistically. Two upregulated *SOX9* in the absence of *SRY*. A third caused strong enhancement and contained *SRY* and *SF1* binding sites, so probably represents the core enhancer by which these genes upregulate *SOX9*. Enhancers of *SOX9* are also activated by the SOX9 protein itself, explaining how *SOX9* can autoregulate its own expression.

Turning on *SOX9* is only the beginning. It takes several protein factors to keep it active. One important factor in maintaining *SOX9* transcription is the key signalling molecule, Fibroblast Growth Factor 9 (*FGF9*). This gene plays a role in many developmental processes including cell proliferation, survival, migration and differentiation.

Duplication of *FGF9* in XX patients causes male development and mutation of its receptor in XY patients causes female development. In mice, *Fgf9* is expressed widely throughout the embryo, but it shows sex-specific expression in the gonad, finally becoming restricted to the male gonad. In mouse strains with weak sex determination, knocking out *Fgf9* produces male-to-female sex reversal: *Sox9* expression is lost, Sertoli cells fail to differentiate, testis development is aborted and the somatic cells begin to express ovarian genes.

SOX9 also needs to be shifted into the nucleus to stimulate Sertoli cell differentiation. This takes the efforts of several genes, knockouts of which reduce *Sox9* expression and translocation, and slow (but do not block) testis differentiation.

Thus *SOX9* upregulation by *SRY* via multiple enhancers is essential for testis differentiation and male development.

9.10 Functions of SOX9

SOX9 is central for turning the bipotential gonad into a testis, activating and maintaining the male gonad differentiation pathway by controlling differentiation and proliferation of Sertoli cells, and the tubular organization of the testis. How does *SOX9* achieve this?

Surprisingly, SOX9 protein has few direct targets in the bipotential gonad and affects testis determination largely by blocking ovary-differentiating genes.

SOX9 is expressed in the developing testis and maintained in the adult testis. But inactivating *Sox9* after 13.5 days of development in mice still allows the testis to develop, showing that it is not required for Sertoli cell differentiation and stability.

It has been difficult to define targets of SOX9. Like products of other *SOX* genes, *SOX9* has little effect on target genes by itself, but acts with other transcription factors, often synergistically with *SF1*. Mutations in *SF1* or any genes that affect its activity can therefore lead to XY female development, but the effects of mutation can be partially masked by other genes.

The first target of *SOX9* to be identified was not a testis-controlling gene, but one that inhibits the development of female reproductive ducts. Anti-Müllerian hormone (AMH: its alternative name is <u>M</u>üllerian <u>I</u>nhibiting <u>S</u>ubstance, MIS) is made by Leydig cells. It is a member of a gene family that sends signals by binding a receptor on mesenchyme surrounding the Müllerian duct. *SOX9* acts along with *SF1* and *WT1* (-KTS) to turn on *AMH* in XY embryos.

Mice with mutations in *Amh* retain their Müllerian ducts but develop normal testes, implying that *Amh* is not essential for testis development. Its major job in mammals is to cause regression of the Müllerian (female reproductive) ducts in males. It's not certain how AMH does this, but the process requires bone morphogenic protein (BMP).

Another target of SOX9 is the *SOX9* gene itself. It stimulates *SOX9* transcription so *SOX9* continues to be expressed in the gonad past birth. In mice, it was found that the Sox9 protein, along with Sf1, binds to enhancer elements and turns on more *Sox9* transcription (autoregulation).

Sox9 also upregulates *Fgf9* directly or indirectly. With *Sf1*, this establishes a feed-forward loop that helps to maintain *Sox9* expression, but is also critical for proliferation and differentiation of Sertoli cells in the developing testis. Other genes that are upregulated at the same time, including the growth factor desert hedgehog (*DHH*), may be regulated directly or indirectly by *SRY* or *SOX9* or both.

Other *SOX* genes can mimic the function of *SOX9*. The closely related *SOX8* and *SOX10* are co-expressed with *SOX9* and *SF1*, and appear to bind to the *Sox9* enhancer and to back up *SOX9* autoregulation and function. Mutation of *SOX8* causes DSD and infertility. *SOX3*, which is closely related to *SRY*, can also substitute for *SRY* in upregulating *SOX9* if it is misexpressed in the testis.

Thus *SOX9* is central to gonad differentiation. It has a rather weak transcriptional activation, and needs backup from cofactors to be effective. It activates *AMH* to cause regression of the female reproductive tracts (Müllerian ducts), and it autoregulates itself. But it does not directly activate other genes critical for testis differentiation – its major function is to inactivate genes in the ovary pathway (Section 9.12).

9.11 Testis Differentiation

So if SOX9 does not, after all, turn on testis differentiation genes, what controls cell migration and proliferation, the hallmark of early testis development? What promotes Sertoli cell and Leydig cell differentiation? And growth of male-specific blood vessels? (Figure 9.5).

Sertoli cells are the organizers of the developing testis. They are the first to differentiate, and they produce factors that bind to a range of specific receptors on other cell types. Leydig cells arise just after testis determination, and are responsible for androgen production that controls male development of the body.

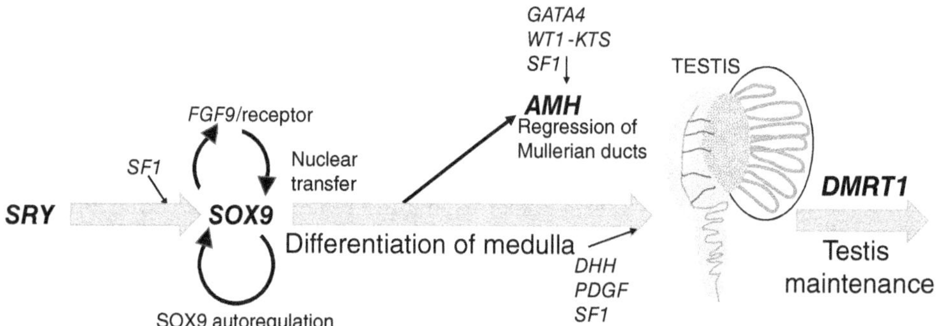

Figure 9.5   Differentiation of bipotential gonad into testis. *SRY* and cofactors activate *SOX9*, a critical and highly conserved step of testis determination. Once activated, *SOX9* self-regulates (circular arrows). SOX9 protein is involved in activating some other genes with effects on testis growth and differentiation, but its main positive action is in activating the anti-Müllerian hormone gene *AMH*, which causes the regression of the Müllerian ducts.

One of the hallmarks of early testis development is the migration and proliferation of cells from the mesonephros (embryonic kidney) to form testis cords. This requires growth factors, including *FGF9*, which itself regulated by *SOX9*.

Key to testis growth and differentiation is *DHH* (*Desert Hedge Hog*, named for the spiky projections on the cuticle of mutant fruit flies). *DHH* is a member of the *hedgehog* gene family of signalling molecules that have wide developmental roles in all animals; it is the only *hedgehog* gene expressed in the gonad. Its role in testis differentiation was revealed by *DHH* mutations in several human XY patients with gonadal dysgenesis.

DHH is secreted by Sertoli cells and regulates myoid cells and Leydig cells. *Dhh* is first expressed in somatic cells of gonads in XY (but not XX) mice at 11.5 dpc, and stays on in differentiated Sertoli cells. It operates by binding a receptor on the membrane of Leydig and myoid cells, and controls upregulation of *Sf1* in Leydig cells and the ultimate expression of steroid hormones. Knocking out *Dhh* or its receptor in XY mice produces abnormalities of testis cord structure, and reduces testosterone production, which leads to feminized development.

Two other pathways are needed for Leydig cell differentiation, governed by factors encoded by three growth factor genes and their receptors; in their absence, Leydig cells are depleted. Steroid synthesis is induced by yet other pathways.

The blood vessels that service the testis also develop at this time: a large vessel on the surface of the testis next to the mesonephros, and minor vessels between testis cords. The major vessel is formed after cells migrate from the mesonephros and proliferate under the influence of another growth factor. It is not clear whether they are controlled directly by *SRY* or *SOX9*, or via pathways that control cell migration and growth.

Another gene that is vital for testis differentiation and maintenance was revealed in XY sex-reversed patients with deletions of the tip of human chromosome 9. In the critical region was an intriguing gene, *DMRT1*, with deep evolutionary roots in sex. DMRT1 protein contains a domain (DM) that binds DNA and plays a role in sex differentiation in organisms as diverse as flies and worms. Indeed, the name of this gene comes from the *Drosophila doublesex (dsx)* gene and the *Caenorhabditis elegans mab-3* gene, which regulate sex-specific structures.

In mice, *Dmrt1* is expressed in the genital ridge of XY embryos, and stays on in the Sertoli cells of the testis throughout life. XY *Dmrt1* knockout embryos develop normal gonads, but testis differentiation fails after birth. Unexpectedly, conditional loss of *Dmrt1* from adult mouse Sertoli cells caused a total reprogramming of the testis into an ovary-like gonad that expresses several ovarian genes and even makes female-like germ cells. This remarkable transformation suggests that *DMRT1* is crucial for the maintenance of Sertoli cells throughout life.

Thus, testis growth and differentiation require a suite of genes, including many growth factors that are active in Sertoli or Leydig cells. Several key testis genes exert their effect indirectly by repressing ovary differentiation, and at least one is required to maintain the testis throughout life.

9.12 Ovary Differentiation

Until recently we knew little about the genetic pathway of ovary development. This is partly because the discovery of *SRY* focused attention on testis development, and partly due to the outmoded concept that ovary development was merely the default pathway and did not require the active role of ovarian genes.

Once again, patients with sex reversal syndromes, this time XX SRY⁻ males, provided clues, and knockouts and overexpression in mice provided detailed data on many genes involved in ovary differentiation.

The ovary develops later than the testis, and morphological differences in mouse embryos are not obvious until 18.5 dpc, when oocytes reach the first stage of meiosis. However, female-specific genes are expressed much earlier (Figure 9.6).

Sex-specific expression in mice first revealed that one member of a developmentally important gene family had a role in ovary determination. The 'wingless' (*WNT*) family of genes was first identified in *Drosophila* (hence its name, which is rather inappropriate for mice!).

In mice, *Wnt4* is expressed by 9.5 dpc in the urogenital ridge of both sexes, and it stays on in the bipotential gonad. At 11 dpc it is downregulated in the developing testis, but stays on in the female gonad and in cells surrounding the Müllerian ducts.

XX *Wnt4* knockout mice show partial female-to-male sex reversal, with masculinized gonads, degenerating oocytes and testosterone production, and male vasculature. This suggests that *Wnt4* activates ovary differentiation and controls oocyte development by repressing male development. Knockout mice retain the male Wolffian ducts and lose the female Müllerian ducts, suggesting that *Wnt4*

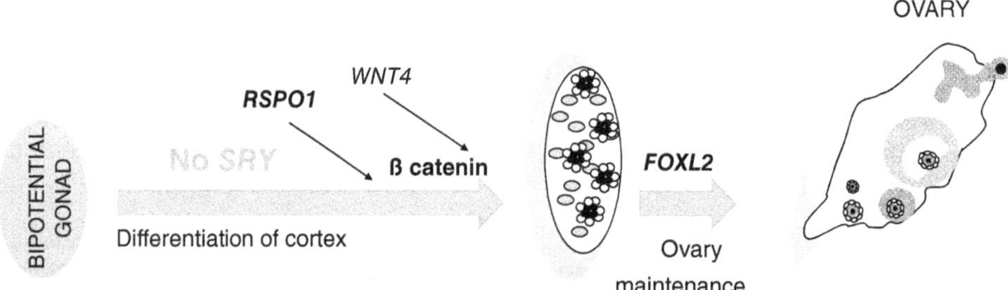

Figure 9.6   Differentiation of bipotential gonad into ovary. The female pathway is activated in the absence of *SRY*. Genes (particularly β-catenin and *WNT4*) promote the differentiation of the cortex. Maintenance of the ovary requires *FOXL2*.

also represses *Amh*. Overexpression of *Wnt4* in XY embryos does the opposite, disrupting testis-specific vasculature and inhibiting testosterone synthesis.

Some XX patients with masculinized gonads and Müllerian duct regression have mutations or deletions in *WNT4*, implicating this gene also in human ovarian differentiation.

WNT4 drives ovary differentiation by activating a protein called β-catenin (discovered by its interaction with *wnt4* in *Drosophila*), which has many roles in cell adhesion and transcription. *WNT4* binds with cofactors to a cell membrane receptor and transfers the signal to the cytoplasm. This triggers dephosphorylation and stabilization of β-catenin, which then enters the nucleus and interacts with a complex of transcription factors that activate target genes. In the absence of WNT signalling, phosphorylated β-catenin is degraded, inactivating target genes.

The incomplete sex reversal of XY *Wnt4* knockouts suggests that it is not the only gene that promotes ovary development. Another gene *Rspo1* seems to work with *Wnt4* to stabilize β-catenin.

Human R-spondin 1 (*RSPO1*) was discovered when a mutation was identified in four brothers who all proved to have XX sex chromosomes. *RSPO1*-deficient patients had low levels of *WNT4* as well as β-catenin, suggesting that *RSPO1*, with *WNT4*, interacts with membrane receptors to stabilize β-catenin via the WNT signalling pathway.

Knocking out *Rspo1* in XX mice confirms this, producing partial female-to-male sex reversal via decreased *Wnt4* expression. Sex reversal can be prevented by expressing stabilized (dephosphorylated) β-catenin in somatic cells of the developing ovary. This suggests that β-catenin acts downstream of *RSPO1* to block testis development in XX female gonads. In line with this explanation, overexpression of stabilized β-catenin in mouse gonad somatic cells causes the reverse; male-to-female sex reversal. *Rspo1* and *Wnt4* must therefore act via the same pathway to activate β-catenin.

Another gene important for ovarian development is *FOXL2*, which encodes the forkhead transcription factor 2, one of a family of transcription factors first identified in *Drosophila*. This was first indicated by sex reversal, not in human patients, but in a strain of goats in which XX males had deletions near *FOXL2*.

In XX human patients, *FOXL2* mutations occur in several syndromes with ovarian dysfunction. XX *Foxl2* knockout mice develop a testis during embryogenesis and are masculinized after birth.

Surprisingly, conditionally knocking out *Foxl2* in the ovary of adult XX mice results in trans-differentiation of the ovary into testis. Ovarian structures became more testis-like as granulosa and theca cells transdifferentiated into Sertoli and Leydig cells, respectively. Ovarian genes were down-regulated and the whole suite of testis genes were up-regulated, even producing testosterone. *Foxl2* turns out to be crucial for maintaining the adult ovary by suppressing the pro-testis action of *Sox9/Fgf9*, which it does by indirectly repressing the upstream enhancers of *Sox9*.

Another gene important for gonad differentiation was discovered in sex-reversed XY female patients with adrenal overgrowth. These patients had a duplication of the X chromosome that contained only one gene, *DAX1* (for <u>D</u>osage sensitive sex reversal, <u>A</u>drenal hyperplasia critical region, <u>X</u> chromosome, gene <u>1</u>). *DAX1* codes for a nuclear receptor. In mice *Dax1* is initially turned on by *Sf1* in the bipotential gonad of both sexes, then is downregulated in the testis and upregulated in the ovary.

The observation that overexpressing *DAX1* promotes ovarian development and suppresses testis suggested that *DAX1* is an ovary differentiation gene. However, XX *Dax1* knockout mice developed normal ovaries, and XY *Dax1* knockouts show early testis defects and spermatogenic failure, suggesting a pro-testis function. The contradictory effects of *Dax1* deficiency and duplication seem to be caused by excess *Dax1* soaking up corepressors and affecting the ability of *Sf1* binding with *Sry* to activate *Sox9* transcription, leading to XY female development.

Thus, ovary development is just as complicated as testis development. In XX embryos, *WNT4/β-catenin* signalling is crucial for precursor cells in the indifferent gonad to develop into ovarian granulosa cells, ultimately leading to a female phenotype. The transcription factor *FOXL2* is required to maintain the ovary from early foetal life through to the adult. But several of these genes, including *WNT4, FOXL2* and *DAX1*, seem to act by inhibiting testis pathways in the developing gonad.

9.13 Sex Determination – a Balancing Act

The gonad is unique in its ability to pursue two different developmental pathways (Section 8.4). It is clear that *SRY* is the switch that sends the bipotential gonad down one path or the other.

However, it is not a simple switch between fates. The choice of developmental fate is a delicate balancing act. Many gene networks are involved in gonad differentiation into testis or ovary, and several of them act antagonistically (Figure 9.7). A battle for sexual supremacy is waged between the testis and ovarian pathways from early embryogenesis, and this continues throughout adult life.

In the bipotential gonad, male and female forces are evenly matched. A principal in this tug of war is *FGF9*, induced by *SOX9*. Its main function seems to be to turn off *WNT4* and prevent it from inducing β-catenin. However, *WNT4/β*-catenin suppresses *FGF9/SOX9*, preventing the upregulation of *DHH* and suppressing *AMH* (thereby permitting Müllerian duct development). *FOXL2* negatively regulates *SF1-SOX9* and downstream factors that promote androgen synthesis. Also on the side of femaleness is *RSPO1*, which represses factors that reinforce *SOX9* activity.

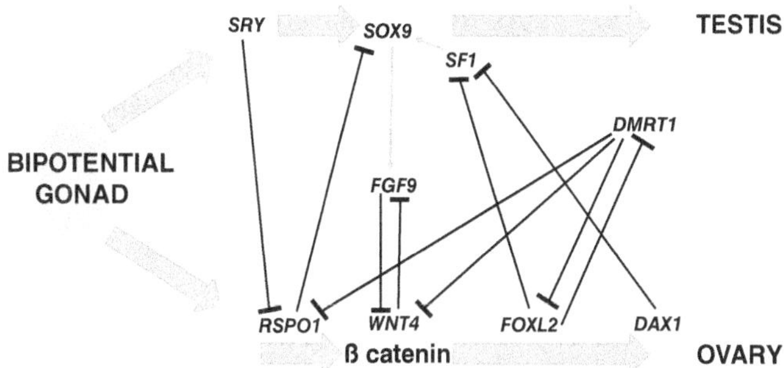

Figure 9.7 Antagonism between testis and ovary pathways. Some of the most important steps in gonad differentiation are controlled by inhibition between the male and female pathways (shown by lines and blocks between genes in the male and female pathways (e.g., *DMRT1* inhibition of *FOXL1* helps maintain testis differentiation).

The battle of the sexes continues downstream, with continuing antagonism of *FOXL2* against *SOX9/ SF1*. Also *DAX1* (on the female team) negatively regulates *GATA4/SF1* and *SOX9*, promoting the female pathway. In XY embryos *SOX9* is upregulated, and male development triumphs.

The battle has even been taken to the postnatal stages with the remarkable standoff between *DMRT1* and *FOXL2*. Ablating *DMRT1* leads to upregulation of *FOXL2*, and ablating *FOXL2* to the upregulation of *DMRT1*. *DMRT1* activates male-specific genes and represses female genes (*WNT4*/β-catenin and *RSPO1*), so its expression is essential for maintaining testis function even in the adult XY male. In XX females, *FOXL2* expression is essential for maintaining ovary function.

Nor are genes simply on or off as you might expect from a switch mechanism. Many genes have a dosage sensitive effect, so half is not enough. For instance, *SOX9* and *DMRT1* must both be present in two copies for maleness to triumph; haploinsufficiency for either causes male-to-female sex reversal, or *SOX9* duplication allows a testis to form even in the absence of *SRY*. *DAX1* duplication has a feminizing effect.

These dosage relationships are not the same in humans and mice – or even between different mouse strains. For instance, in the lab mouse, haploinsufficiency for *Sox9* and *Dmrt1* is not sex reversing, and nor is duplication of *Dax1*. At a biochemical level this suggests that many individual steps of the pathway are dosage sensitive and that dosage relationships are subtly shifted in response to evolutionary pressures.

The other revelation is that many genes in the mammal sex determination pathway act on different targets at different stages. For instance, *SF1* and *GATA4* both have targets in the bipotential gonad as well as the developing testis, and *FOXL2* and *DMRT1* are critical for gonad differentiation in the embryo as well as maintenance in the adult.

Many genes also have targets in systems other than sex determination; for instance*WT1* and *WNT4* are expressed in the kidney and *DAX1* in the adrenal glands. *SOX9* is expressed in chondrocytes and is essential for bone formation. This multiple effect (pleiomorphism) is not particularly surprising or unusual – many of the genes involved in *Drosophila* sex determination have other functions, many in the neural system.

Why so complex? As for so many biological questions, the answer lies not in the function, but in the evolution of sex-determining genes, which I will discuss in Chapter 10.

9.14 Genetic Pathways for Making Eggs and Sperm

The whole point of differentiated testis and ovary is to make the two different kinds of haploid gametes, sperm and eggs. Like gonads, sperm and eggs arise from a common pool of cells, and are shepherded to their different fates by a network of antagonistic genes (Figure 9.8).

How do primordial germ cells (PGC) decide their fate? The presence or absence of *SRY* seems immaterial since in a mosaic embryo spermatogonia were found to be half XX and half XY (Section 8.7). Also, XX germ cells in a testis undergo

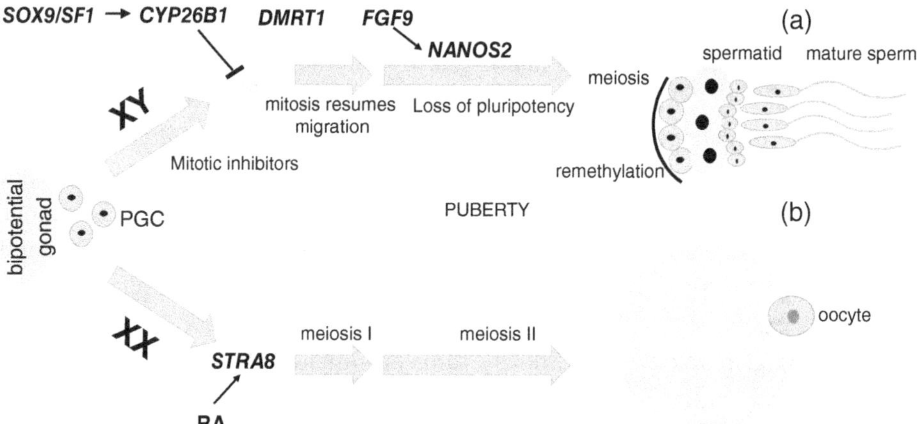

Figure 9.8   Germ cell differentiation. Pathways involved in differentiation of primordial germ cells (PGC) to sperm (a) or ova (b). Arrows represent enhancement and bars represent inhibition. In the presence of *SRY*, and the inhibition of retinoic acid, PGC divide and migrate to the genital ridge. They interact with pluripotency factors, then begin a lifelong commitment to meiosis and differentiation into mature sperm. In the absence of *SRY*, *STRA8* is activated by retinoic acid to trigger meiosis in PGC; oocytes arrest at the first stage of meiosis, which gets to the next stage when the follicle matures after puberty, and is completed when the egg is fertilized.

the male pattern of differentiation, although they do not survive male meiosis. XY PGCs in an ovary can go even further, completing meiosis and even forming Y-bearing oocytes.

So PGC differentiation must be the result of signals from the testis or the ovary. These signals are reciprocal in the ovary, which degenerates in the absence of germ cells, and begins to express male markers *SOX9* and *AMH*. Germ cells are not required for testis differentiation, which still occurs if PGCs are removed genetically or chemically.

The different timing of entry into meiosis is the first outward sign of sex-specific germ cell differentiation; meiosis is blocked in spermatogonia, but is initiated in oogonia. The key question is therefore, what regulates the timing of meiosis? The answer seems to be the presence of retinoic acid (RA), a metabolite of vitamin A. This explains early observations that spermatogenesis fails in rats fed a diet deficient in vitamin A and that vitamin A deficiency in mice caused loss of spermatocytes, which could be reversed by supplementing with retinoids.

RA is provided by the blood and enters the developing ovary via the embryonic kidney from the head end and induces meiosis in a wave from the top down. It does this by activating (or releasing inhibition of) a female-specific gene <u>St</u>imulated by <u>R</u>etinoic <u>A</u>cid (*STRA8*), the 'meiotic gatekeeper' that controls the switch between mitosis and meiosis. *STRA8* expression is somehow augmented by *DMRT1*.

Recent evidence that knocking out some RA receptors has no effect on the PGCs suggests that RA works in even more complex ways, or that it is not the only regulator of meiosis entry.

Maintaining female germ cells requires other regulatory loops. *WNT4* interacts with the products of two other genes to maintain female germ cells. If *Wnt, Rspo1, β-catenin* or downstream genes are knocked out in mice, male vasculature develops and germ cells degenerate. It is not clear whether this results from direct interaction with *STRA8*.

How do spermatogonia avoid meiosis? In the developing gonad of XY mouse embryos, germ cells are protected from RA by the Sertoli cells which surround them within the testis cords. Activated by *Sox9/Sf1*, Sertoli cells produce an enzyme called CYP26B1 which breaks down RA and prevents germ cells entering meiosis. The gene *Cyp26b* is initially expressed in gonads in mouse embryos of both sexes, but it becomes male-specific by 12.5 dpc. Knocking it out allows germ cells to enter meiosis in the foetal testis. This makes sense of early observations that migrating PGCs that stray into the adrenal or kidneys in XY embryos pursue an oocyte trajectory.

It is possible also that the male-specific *FGF9* diminishes the cell response to RA, and RNA-binding proteins help to ward off meiosis by rounding up and degrading meiotic activators. Intriguingly, these proteins include *DAZL* (the auto-somal ancestor of the Y gene DAZ (<u>D</u>eleted in <u>Az</u>oospermia) that is required for spermatogenesis in humans).

Regulating RA levels also plays a key role in bringing about mitotic arrest in spermatocytes, but this also requires *DMRT1* and several growth factors and hor-mones. Commitment to the male pathway is finally sealed by the expression of *NANOS2*.

There are many reports of miRNAs that are expressed sex-specifically by germ cells in gonads of one sex or the other (Section 8.13). Micro RNAs are small (22 bp) RNA molecules derived from hairpin loops in RNA transcripts. They silence genes by pairing with sequences in mRNAs, so it is possible that they may have a role in germ cell differentiation. In addition, DNA methylation is clearly involved; global demethylation occurs in both testis and ovary, but remethylation occurs to different extents and is involved in the downregulation of pluripotency markers in sperm.

If germ cell development is arrested at an early stage when pluripotency genes are still expressed, germ cell tumours can arise. This is a problem in many cases of DSD. The presence of a Y chromosome in a dysgenic testis poses the risk of gonadoblastoma, and the dysgenic gonad is often removed to prevent this. A 'gonadoblastoma' locus *GBY* on the Y chromosome is suspected to be the Y-borne *TSPY*. Tumour progression depends on testosterone, so gonadoblastoma, mani-festing at puberty, is a risk for sex-reversed XY females with high androgen.

Thus germ cell differentiation, though it is linked to gonad differentiation, employs a different regulatory network. At the heart of the decision of whether to develop as sperm or eggs is regulation by retinoic acid and the genes that produce it and mediate its effect on entry into meiosis.

9.15 Conclusions – Sex Is Not So Simple

In this chapter, I summarized the evidence that a single gene – *SRY* on the Y chromosome – has profound effects on the phenotype. It does this by activating a network of other genes, most of them on other chromosomes (only the master switch needs to be on the Y) to make a testis in the embryo. The testis makes male hormones and the hormones make the baby a boy. In the absence of *SRY*, the gonad becomes an ovary and the baby develops as a girl.

The developmental pathway to either a testis or an ovary is no simple switch, but depends on the balance of opposing forces, even in the adult. Neither pathway is straightforward. In fact, there are different parallel pathways, for instance, to dif-ferentiate the gonad, germ cells and the reproductive ducts. Inhibition is just about as important as activation, as opposing male- and female-promoting networks fight it out. And neither are they linear, being full of feedback loops and multi-level effects.

It is, indeed, a complex network of interactions. *SRY* turns on *SOX9*, which activates a network of male genes and suppresses a network of female genes, sending the bipotential gonad down the path to form a testis. In the absence of *SRY*, *SOX9* is not activated, and the network of female-promoting genes is allowed to do its work to make an ovary.

This network contains at least 68 genes, and we are still counting. More mutations are being unearthed by whole genome sequencing of patients with DSD, and each defines another transcription factor or cofactor that promotes or inhibits, directly or indirectly, one of the major players in sex determination.

So how can one gene, or even 68, influence the expression of a third of our genome (Section 8.1), and affect the phenotype so profoundly? Firstly, many of the genes in the sex-determining network have effects also on other pathways and tissues. More importantly, many pathways are moderated directly or indirectly by the steroid hormones produced in males and females. At least some of this sex bias is introduced by differences in alternative transcription and splicing. For instance, intron retention (which often leads to non-functional protein) is much more common in XX embryos.

The complexity of sex determination also has bearing on the variation we see in male and female phenotypes, in humans as well as other animals. Variation in any of the 68 genes in the sex-determining network, or in any of the genes that regulate their expression, or in genes involved in steroid hormone synthesis, can have effects on sexual development ranging from trivial to lethal. It is not much wonder that there is such variation in sexual development, morphology and behaviour amongst humans (Chapter 11).

Why is sex determination so complicated? Is making a testis special, or is all organ development governed by such complex networks? The answer to this question, as ever, may be found in evolution rather than in function. How does our fishy ancestry – indeed, our eukaryote ancestry – shape how we make and use gonads and germ cells? Where did all these genes come from and what shaped their interactions?

It is therefore instructive to compare sex determination across mammals and with sex determination in other vertebrates, and even invertebrates. In Chapter 10, I will widen the discussion to sex determination across divergent animals, and recount how different genes have taken the master switch role and how this has driven sex chromosome evolution. And I will finish with a discussion of vertebrates that lack sex chromosomes, which determine sex epigenetically, using environmental cues such as temperature and some of the same genes as we have met in Part II.

FURTHER READING

Classic Papers

Capel B, 2000. The battle of the sexes. *Mechanisms of Development* 92: 89–103

Foster JW, Dominguez-Steglich MA, Guioli S, et al., 1994. Campomelic dysplasia and autosomal sex reversal caused by mutations in an *SRY*-related gene. *Nature* 372: 525–530.

Harley V, Goodfellow P, 1994. The biochemical role of SRY in sex determination. *Molecular Reproduction and Development* 39: 184–193

Wagner T, Wirth J, Meyer J, et al., 1994. Autosomal sex reversal and campomelic dysplasia are caused by mutations in and around the *SRY*-related gene *SOX9*. *Cell* 79: 1111–1120

Reviews and Research Articles

Bowles J, Knight D, Smith C, et al., 2006. Retinoid signaling determines germ cell fate in mice. *Science* 312: 596–600

Eggers S, et al., 2016. Disorders of sex development: Insights from targeted gene sequencing of a large international patient cohort. *Genome Biology* 17: 243

Gershoni M, Pietrokovska S, 2017. The landscape of sex differential transcriptome and its consequent selection in human adults. *BMC Biology* 15:7 https://doi.org/10.1186/s12915-017-0352-z

Liu CF, Liu C, Yao HH, 2010. Building pathways for ovary organogenesis in the mouse embryo. *Current Topics in Developmental Biology* 90: 263–290.

Rosenfeld CS, 2017. Brain Sexual differentiation and requirement of SRY: Why or why not? *Frontiers in Neuroscience* 11:632

Sekido R, Lovell-Badge R, 2013. Genetic control of testis development. *Sexual Development* 7: 21–32

She Z-Y, Yang W-X, 2014. Molecular mechanisms involved in mammalian primary sex determination. *Journal of Molecular Endocrinology* 53: R1–R37

Spiller C, Koopman P, Bowles J, 2017. Sex determination in the mammalian germline. *Annual Review of Genetics* 51: 265–285

Wilhelm D, Koopman P, 2006. The makings of maleness: Towards an integrated view of male sexual development. *Nature Review Genetics* 7: 620–631.

10

Evolution of Sex Determination

In the preceding chapters, I discussed how sex is determined in humans and other mammals by the action of the *SRY* gene on the Y chromosome. But we know that the mammal Y chromosome is quite young (Section 4.14), and is not shared with non-mammal vertebrates, or even with monotreme mammals.

So what genes determine sex in other vertebrates? Do they act to trigger a sex-determining pathway that is conserved throughout vertebrates? How do changes occur at the genetic and chromosome level?

Here, I will examine the gathering evidence that gene networks that are basically conserved, but are triggered by many different sex-determining genes in different vertebrate lineages. And, unlike mammals, some groups seem to have amazingly malleable sex determination systems, involving novel genes and hijacking other pathways.

How can we discover unknown sex-determining genes? Formally, we can distinguish two basic chromosomal sex-determining systems; XX female: XY male systems (male heterogamety) and ZZ male: ZW female systems (female heterogamety). Each of these can be defined either by a dominant gene on the sex-specific chromosome, or by dosage differences of a gene on the chromosome that is present in two copies in the homogametic sex and one copy in the heterogametic sex. This gives the possibility of four different basic mechanisms (Figure 10.1).

However, there are many possibilities for molecular action within these four categories. A dominant male determining gene might act by supplying a unique male-determining substance, or by increasing the dose of a male-determining substance – or by inhibiting a female-determining substance. This accords with the view that the sex-determining pathways are full of balancing and opposing male- and female-determining steps (Section 8.13).

It will be no surprise to find that evolution has made use of all of these strategies in one lineage or another.

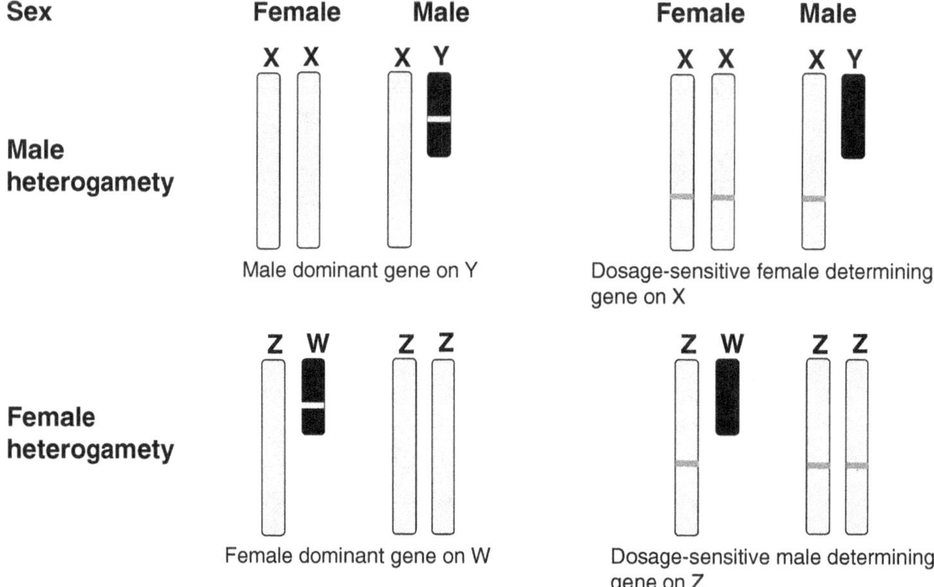

Figure 10.1   Different sex-determining systems Top row: XX female:XY male systems of male heterogamety (in which the male makes two sorts of gametes) can work by either a male dominant gene on Y (like mammals), or a dosage sensitive gene on the X (e.g., *Drosophila*). Bottom row: ZW female: ZZ male systems of female heterogamety (in which the female makes two sorts of gametes) can work by either a female-specific gene on W (e.g., the toad *Xenopus laevis*) or a dosage sensitive gene on Z (e.g., birds).

10.1 Sex-Determining Genes in Eutherian Mammals

Mammals subscribe to a system of male heterogamety and *SRY* has been shown to act as a domiant male determiner (Figure 10.1) in at least humans and mice (Section 8.13). Is *SRY* sex determining across all eutheria?

A gene with a function as critical as sex determination might be expected to be highly conserved between different mammals. Indeed, Southern blotting with a human *SRY* probe revealed a male-specific *SRY* gene across placental mammals, including primates, rodents, sheep and cattle, cats and lions and even elephants.

The structure of *SRY* in all mammals – except for rodents – is rather similar, with a fairly well-conserved HMG box of about 70 amino acids containing sequences that bind and bend DNA, as well as two nuclear localization signals that allow *SRY* to get into the nucleus. The HMG box is flanked by an N-terminal domain of 30–60 amino acids and a C-terminal domain of 70–100 amino acids.

Functional studies, knockouts and transgenesis are not practicable in most mammal species, but the similar amino acid sequence within the HMG box suggests that the binding and bending function of this gene is conserved in placental mammals. Sex reversal in horses (surprisingly common because female show ponies with male characteristics were favoured for breeding) can be the result of *SRY* deletion. Goat and rabbit *SRY* are male determining in transgenic mice, pointing to a conserved function in sex determination. And although the sequence outside the HMG box is quite poorly conserved between mammals, *SRY* flanking sequences from pig and human *SRY* both work to direct reporter transgene expression to the genital ridge of transgenic mice.

The expression profile of *SRY* in a range of mammals is also similar. In dog, pig and sheep, as well as human, *SRY* is expressed in genital ridge throughout foetal development as well as in many non-target tissues.

Rodent *Sry* is startlingly different. Located in a large inverted repeat, mouse *Sry* codes for a protein that lacks the N-terminal domain of other *SRY* proteins. It also has a long glutamine-rich ('poly Q') tail. *Sry* in other mouse species have poly Q tails of varying length, and the length of the tail varies even within species.

This variation and instability might suggest that the peculiar poly-Q tail lacks function. However, its requirement for sex determination is shown by transgenesis of mouse *Sry* from which various lengths were deleted (Figure 9.2). Poly-Q domains of different lengths were shown to contribute to TESCO binding and *Sox9* activation. One possibility is that the tail replaces some of the protein cofactors that mouse lacks.

Sry copy number also varies in different rodent species. There is one copy in mice, but seven copies on the rat Y chromosome, and even more in some old world rodents. Multiple copies also exist in rabbits, in large palindromes that swap sequences by gene conversion. *SRY* seems to have gone rather crazy also in *Microtus* (vole) species, in which both males and females have many inactive *SRY* pseudogenes.

This variation is unexpected, given the crucial function of *SRY*. But it is far from unusual among Y-borne genes, which are frequently amplified. Multiple copies are often partly or fully inactivated, suggesting (to me at least) that amplification evolved to compensate for mutations that lower *SRY* activity (Section 4.8).

The downstream pathway seems to be highly conserved among mammals, even rodents. Expression of candidate genes ('the usual suspects' *SOX9, DMRT1, AMH, FOXL2* and *WNT4, RSPO1*), in rabbits and sheep, and functional studies in the rabbit generally mirror findings in mouse.

This suggests that the fundamental biochemistry of sex determination is conserved, even though the gene that switches it all on might be structurally and even functionally different, especially in rodents.

10.2 Y-less Rodents and Sex Gene Turnover

There are two rodent lineages in which the Y chromosome has entirely disappeared (Section 4.9), although other genes of the sex-determining pathway are still there. Has *Sry* jumped to another location? Or has it been replaced?

Three of four species of the mole vole (genus Ellobius) of eastern Europe and Russia have lost their Y chromosome. In *Ellobius lutescens* both sexes are XO, and in two other species (*Ellobius tancreii* and *Ellobius talpinus*) both sexes are XX. Another species, *Ellobius fuscocapillus* retains a Y chromosome. Genome sequencing shows that *E. talpinus* is much more closely related to *E. fuscocapillus* than either is to *E. lutescens*, which implies that the loss of the Y occurred twice, independently.

Sry is present in the *E. fuscocapillus* males but there is no trace of it in males of the other species. The discovery that *SRY*'s target (*SOX9*) has a promoter mutation suggests that a weakened *SRY* system was abandoned in a common ancestor, allowing novel systems to arise and the Y to decay and disappear independently. Several attempts have been made to discover what gene has taken over from *Sry*. Candidates including *Sox9, Sf1, Sox3, Foxl2, Ar* and *Dmrt1* were tested but none is sex linked in *E. lutescens*. A full genome screen for sex-linked markers revealed no credible sex locus.

Other Y genes that are required for male fertility were, however, found in the Y-less species. *Zfy, Eif2s3y* and *Uspy* were present on the X and were expressed in the testis. Relocation of these genes to the X may have been the precondition that permitted – twice – the evolution of a new sex gene and the ultimate loss of the Y.

The other group of Y-less mammals are the spiny rats of genus Tokudaia that eke out an uncertain existence on small islands in Japan. Again, of three species, two have entirely lost the Y. Both sexes are XO. There is no sign of *Sry* in these Y-less species, but other Y genes, or copies of them, are found in a bloc on the X, as well as on two autosomes.

A decades-long search for the gene that has taken over sex determination in spiny rats has at last been successful. A comparison of the full genome DNA sequence of males and females revealed a tiny region (170 kb) on an autosome that is duplicated in all males, but in no females (Figure 10.2). This male-specific duplication lies in a very interesting region that is homologous to part of mouse chromosome 11, which contains the crucial male-determining gene *Sox9* and sequences that regulate it. One of these enhancer sequences is included in the 170 kb duplication, and transgenesis of XX mice with this sequence was shown to upregulate mouse sex genes. It seems, then, that the male-specific duplication of an upstream enhancer that can turn up *Sox9* without the benefit of *Sry* became the new male determining gene, and defined a new Y chromosome.

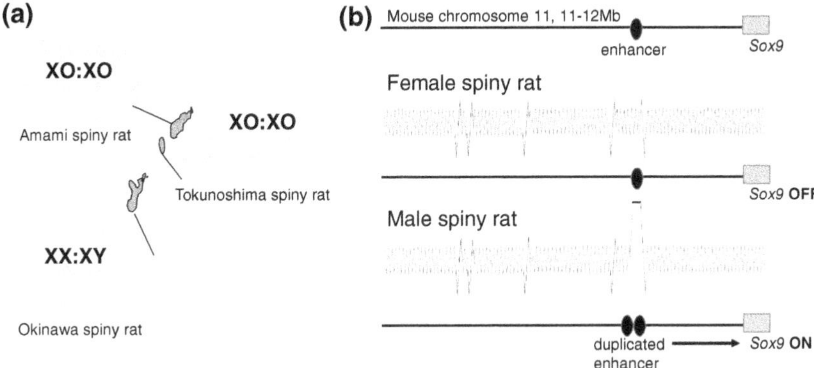

Figure 10.2 Identification of new sex gene in Y-less spiny rat. (a) Three species of spiny rat occur on three islands in southern Japan. The Okinawa spiny rat has XX females and XY males (with *Sry*), whereas in the other two species, both males and females have a single X chromosome (XO), and no *Sry*. From Terao et al, 2022. (b) A genomic difference between male and female spiny rats was detected on chromosome 11, one Mb upstream of *Sox9* by a male-specific doubling of read depth. The difference consists of a small duplication (170 kb). This region aligns with 11–12 Mb of mouse chromosome 11 that contains *Sox9*, and the duplication coincides with an enhancer element on mouse chromosome 11. This suggests that in the spiny rat, the male-specific duplication contains an enhancer element, which, in double dose, activates the *Sox9* gene in the absence of *Sry*.

Sry is present in the spiny rat species that retains a Y, but may be inactive because its amino acid sequence contains a substitution that is sex reversing in humans. *Sry* has amplified to more than 100 copies on this Y chromosome. Perhaps a partially inactivated *Sry* mutation in a common ancestor was compensated for in one population by making more copies of it, and in another by replacing it. Again, amplification looks to me like the last gasp of a dying Y chromosome, and the evolution of male-specific upregulation of *Sox9* heralds the birth of a new Y.

Curiously, there are other mammal lineages in which *SRY* is present but ineffective. Several polar mammals have fertile XY females as well as XY males and XX females. For instance, classic studies of the collared lemming in Sweden show that an X* chromosome somehow suppresses testis determination by a Y chromosome that bears a normal *SRY* gene. A preponderance of females might be a selective advantage in a polar mammal with a short breeding season.

In a group of akodont rodents in Patagonia, three species have XY as well as XX females. A variant Y* is passed down from XY* females to their XY* daughters, implying that it is no longer male determining. Sequencing revealed that the Y* chromosome bears a normal *Sry*. Since other related species lack XY females, it was proposed that Y* females had evolved independently in these three species in response to selection for a female bias in these rapidly breeding species. However,

I suspect that a structural or epigenetic modification (inversion next to heterochromatin?) occurred in a common ancestor, but is modified epigenetically in different species.

Such peculiar mammals may seem to be mere sexual curiosities. However, studying them will provide answers to some very deep questions, such as the means by which sex genes (and the sex chromosomes they define) turn over in evolution.

10.3 Sex Determination in Marsupials and Monotremes

The function of *SRY* in sex determination seems to be very stable, at least in eutherian evolution. Is it shared by more distantly related mammals, marsupials and monotremes?

In marsupials, we know that the Y chromosome determines testis because XXY animals have testes, and XO animals lack them. And a gene homologous to *SRY* lies on the Y chromosome in marsupials (Section 8.14). However, the amino acid sequence and three-dimensional structure of the wallaby and opossum SRY protein suggest that its binding to DNA is weak, and bending of the target region might be insufficient to trigger *SOX9* upregulation. There is no direct evidence that it is male determining, given that we have no sex-reversed mutants to study, and we cannot yet conduct transgenesis or knockouts in marsupials.

SRY also has a potential rival on the marsupial Y. *ATRY*, the homologue of the X-borne *ATRX* (mutations within which are sex reversing in humans) lies on the Y in all marsupials, but was lost in eutherians. Could *ATRY* determine testis in marsupials? Probably not – it is expressed in the developing testis, but only in the germ cells, not the pre-Sertoli cells, which we expect to be the site of action. So *SRY* is our best candidate for the marsupial sex-determining gene.

Thus, *SRY* appears to be present on the Y in therian mammals, and probably initiates sex determination through a conserved pathway of gonad differentiation.

But what about monotremes, which represent the base of the mammalian radiation? The bizarre multiple XY chromosomes of platypus and echidna (Section 4.4) were initially believed to have homology to the XY chromosomes of therian mammals, so an *SRY* gene was expected to lie on one of the Y chromosomes. Many efforts to find a male-specific *SRY* homologue by Southern blotting and PCR yielded autosomal *SOX* genes, but none were male-specific.

One of these genes was the orthologue of *SOX3*, the proposed ancestor of *SRY*. In echidna this gene, along with other human X-borne genes, lay, not on the sex chromosomes, but on chromosome 6 that was later shown to be homologous to the conserved region of the therian X. This confirms that the therian XY pair was an autosome in the ancestor of all mammals, as predicted by the theory of sex chromosome differentiation (Section 4.5).

If there is no *SRY*, what determines sex in monotremes? The multiple XY chromosomes of platypus have no homology with those of therian mammals – but instead share genes with sex chromosomes of birds. This includes *DMRT1*, which appears to be vital in two doses for male determination in therians, as well as in birds (see below). However, in platypus and echidna, *DMRT1* lies on the large X_5 chromosome, which is present in two copies in females and a single copy in males, and is not inactivated. Thus, the dosage is 2 females: 1 male, the wrong way around for dosage-sensitive male determination. No male-specific *DMRT*-related genes were detected in platypus.

The hunt was on for other candidate genes on the platypus and echidna sex chromosomes, made more difficult because there are 10 of them. Were we looking for a Y-borne male-dominant gene, or a dosage-dependent female determining gene on an X? The most promising location is in the tiny Y_5 at the end of the chain, which was probably the original sex chromosome that became serially translocated to other autosomes.

Searching Y_5 is not easy because, like other degenerate Y chromosomes, it is full of repetitive sequences. However, it contains some intriguing genes, including the anti-Müllerian hormone gene *AMH*, which is presently the best candidate for platypus sex determination.

Thus marsupials share *SRY* and a *SOX9* with eutherians, suggesting that *SRY* turns on *SOX9* through a similar enhancer in all therian mammals. But monotremes have entirely different sex chromosomes, more closely related to those of birds than therian mammals, and defined by a different sex-determining gene (*AMH?*). This is splendid evidence for turnover of the sex gene and the chromosomes it defines, and raises the question of how *SRY* evolved its sex-determining function.

10.4 The Evolution of *SRY*

SRY lies on the Y chromosome of therian mammals, but does not exist in monotremes – or in other vertebrates as we will see. This dates its emergence to between 166 Mya (when monotremes and therians diverged) and 148 Mya (when marsupials and placental mammals diverged).

SRY emerged on an autosomal pair that is represented by platypus and echidna chromosome 6, syntenic with the ancient region of the therian XY. *SOX3* maps to this region in a conserved synteny group, and its location and similarity to *SRY* (Section 8.12) suggested, in the 1990s, that it was the ancestor of *SRY*.

How could *SOX3* gain its sex-determining function? *SOX3* is not testis determining; knocking it out has no effect on sex determination in mice. It is expressed in the brain and neural tube, so it would seem to be a brain determining, rather than a testis-determining gene. It is expressed in the developing testis, but is specific to germ cells and is absent from Sertoli cells, which are the drivers of testis determination.

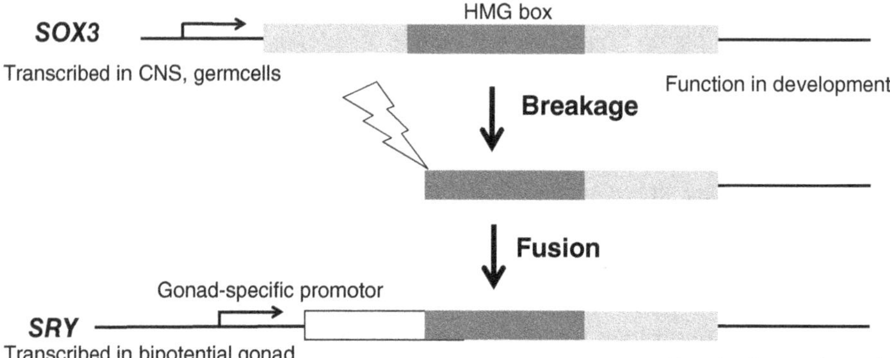

Figure 10.3   The evolution of *Sry* from *SOX3*. *SOX3* is an ancient HMG box-containing gene transcribed in the central nervous system and germ cells, and has an important role in early development in all vertebrates. It lay on an autosome pair in an ancient mammal. The *SOX3* allele on one of the pair suffered breakage near the border of the HMG box, then fusion to a gonad-specific promoter, which drove its expression into the bipotential gonad, where it took on the role of activating the testis developmental pathway. This newly minted *SRY* gene defined a new Y chromosome.

The origin of *SRY* from *SOX3* remained speculative for a decade. Then two XX baby boys were described who lacked *SRY*, but expressed *SOX3* in the somatic cells of the gonad. Several patients with XX testicular DSD were later shown to have genomic rearrangements both upstream and downstream of *SOX3*. At the same time, a line of transgenic mice with XX males was accidentally created in which *Sox3* had been relocated next to a gonad-specific promoter and was expressed in the genital ridge. So evidently *Sox3* was able to substitute for *Sry*, up-regulating *Sox9* and inducing testis development (Figure 10.3).

This strongly supports the theory that *SRY* evolved when one of the pair of *SOX3* alleles was rearranged close to a gonad-specific promoter that drove its expression into the supporting cell lineage of the developing gonad. There is some evidence of parts of another X-linked gene embedded in upstream non-coding sequences upstream of *SRY*.

Thus, a change in expression profile was all that was needed to create a novel sex-determining gene, and define a new Y chromosome. How deliciously simple it is to create a novel sex-determining gene!

10.5 Sex Determination in Birds

Birds display some of the most amazing sexual dimorphisms among vertebrates. In some species males are extravagantly coloured and go through complex mating displays, while females are dull and circumspect.

As for mammals, bird sex determination is genetic, but sexual differentiation is largely up to the hormones. Testosterone is produced by the developing testis, and oestrogen by the ovary. As for other egg-laying vertebrates, oestrogen – and the enzyme (aromatase) that makes it from testosterone – is paramount (Section 10.14).

In bird embryos, as for mammals, gonads develop from a thickening on the surface of the embryonic kidneys. The bipotential gonad comprises an outer layer of epithelial cells (cortex) and an inner layer (medulla) of cords of somatic cells. Primordial germ cells (PGCs) arise from outside the embryo and migrate into the bipotential gonad via the blood stream. Reproductive tracts (the female Müllerian duct and the male Wolffian duct) are both present in embryos of both sexes.

In male embryos, Sertoli cells are the first to differentiate. They surround germ cells in testis cords and suppress meiosis. *AMH* produced by Sertoli cells causes Müllerian duct regression, and Wolffian ducts develop under the influence of androgens.

At about the same time in female embryos, the cortex thickens and PGCs are surrounded by follicle cells. PGCs are stimulated by retinoic acid (RA) to enter meiosis, which then arrests. Curiously only the left gonad develops into an ovary; the right gonad produces no RA, so no meiosis ensues and the gonad regresses. The inner layer (medulla) of both gonads make oestrogen, which is essential for ovarian development, but only the left Müllerian duct develops into an oviduct and Wolffian ducts regress in the absence of androgen.

Removing the left ovary from chicken embryos triggers development of the right gonad – into a testis-like structure. This is often interpreted to mean that in birds, unlike mammals, male is the default sex, but is probably better understood as a disturbance of antagonistic pathways.

Developmental pathways for gonad differentiation in birds employ orthologues of many genes in the mammalian pathway. *GATA4* operates upstream in male and female embryos to promote the bipotential gonad; however, another gene *hemogen* (*HEMGN*) seems to have been recruited to help; in mammals this gene is involved in blood cell differentiation but not sex. *SOX9* (aided by *PGD2*) directs Sertoli cell differentiation in male embryos.

Female development in chickens, as in mammals, involves both the *FOXL2* and *RSPO1/WNT4/β-catenin* pathways. *CYP19*, which encodes aromatase (which makes oestrogen), is upregulated in females by *FOXL2* and *SF1*. Oestrogen is essential for ovary development; blocking its action with drugs causes male development and even spermatogenesis.

Thus, except for the more dominant role of oestrogen and the participation of hemogen, the pathways for testis and ovary differentiation in birds has many similarities to those in mammals. However, it has been known since the early 1990s that birds lack *SRY*. Some other gene must kick-start bird sex determination.

10.6 *DMRT1* and Dose-Dependent Sex

Birds have a system of chromosomal sex that is completely different from that of mammals (Section 4.11), subscribing to a system of female heterogamety (Figure 10.1). Male birds have two copies of a large, conserved, gene-rich Z chromosome. Females have a single Z and a W chromosome that is a degraded version of the Z. There is no homology between the bird Z chromosome and the mammal X. Not surprisingly, *SOX3* is autosomal and there is no *SRY*.

So what gene triggers sex determination in birds? Is it a W gene with a female-dominant effect? Or a dose-dependent gene on the Z? (Figure 10.1) This should be easy to sort out from the sex of chickens with ZZW or ZO sex chromosomes. However, decades of searching found no diploid birds with aberrant sex chromosomes, and it is likely that these chromosome constitutions would be lethal because they upset gene dosage balance. However, triploid birds with ZZW chromosomes were found to be viable: they were initially female, but after a few months developed male plumage and started crowing at dawn, suggesting antagonism between Z dosage and a feminizing effect of the W.

Answers were sought from gene mapping and sequencing. A frustrating search of the highly repetitive W chromosome netted several W genes. These were initially greeted with rapture, but they all turned out to have exact homologues on the Z, or were not on the W in other birds, or were expressed in the wrong place. A promising candidate was jettisoned when its introduction into ZZ embryos failed to induce ovary development.

Could it then be a sex gene with a copy on the Z chromosome but not the W? This would set up a 2:1 dosage difference between males and females. Two copies in a ZZ bird would make a threshold amount of Z gene product required to determine testis and a male phenotype, and an insufficient supply in ZW birds would promote female development.(Figure 10.4).

Searching the large chicken Z chromosome without a deletion map or sex-reversed mutant chickens was hard, so the most effective strategy was to examine candidate genes.

Chief among these was *DMRT1*, which encodes a transcription factor containing a zinc-finger motif (DM domain) that binds to DNA. *DMRT1* mapped at the tip of human chromosome 9, homologous to the chicken Z. This region was deleted in XY female patients, suggesting that a single copy of *DMRT1* was insufficient to support testes development in mammals (Section 9.11).

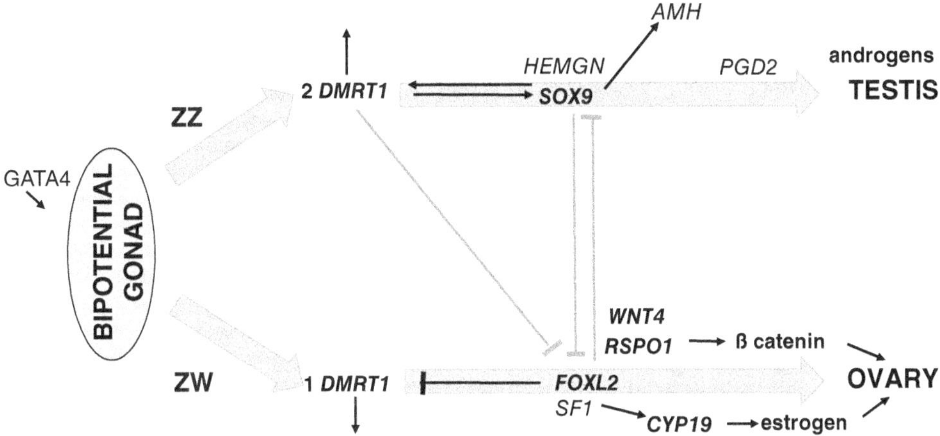

Figure 10.4   Genes involved in bird sex determination. The bipotential gonad develops under the influence of several genes that have homologues in mammal sex determination, but their roles and times of action may differ. In ZZ eggs, two copies of the Z-borne *DMRT1* induce *SOX9*. The bird-specific hemagen feeds back to increase *DMRT1* output. *DMRT1* and *SOX9* both repress *FOXL2* and suppress ovarian development. *AMH* is activated and Müllerian glands degenerate: androgens stabilize Wolffian ducts. In ZW eggs, a single copy of *DMRT1* is further downregulated by *FOXL2*, which represses *SOX9* and allows the *WNT4-RSPO1-β catenin* pathway to develop into ovary. Aromatase (*CYP19*) triggered by *FOXL2* and *SF1*, converts testosterone to oestrogen.

DMRT1 was found to lie on the Z but not the W in chickens, and even emus, in which only a small region of the W is differentiated. In ZZ embryos *DMRT1* is expressed early (two days before *SOX9*). It is more active in male bipotential gonads than female, and remains stronger throughout embryonic development. Its original 2-fold dosage difference is amplified by upregulation in the male gonad by *SOX9/HEMGN* and downregulation in the female gonad by *FOXL2*.

Knocking down *DMRT1* using interfering RNA in early chicken embryos was found to produce partial sex reversal. *SOX9* expression decreased and aromatase was activated, driving ovary differentiation. In ZW embryos, over-expressing *DMRT1*, or ectopic expression in female gonads, antagonized the female pathway. This was confirmed by specific *DMRT1* mutation by CRISPR/Cas9; ZZ embryos carrying only one active copy of *DMRT1* developed ovaries rather than testes, following the same timetable as normal ZW embryos.

It is not yet clear what are the direct targets of *DMRT1* in chicken, but two doses upregulate *SOX9* and *AMH* (with help from *SF1*) and testis cord-like structures develop. This suggests that *DMRT1* acts upstream of *SOX9* and *AMH* in birds, rather than downstream as in mammals. As in mammals, *AMH* causes degeneration of Müllerian glands, but knockout experiments do not support a role for *AMH* in testis differentiation. In addition, two doses of *DMRT1* antagonize *FOXL2*, thereby suppressing oestrogen expression and ovary formation.

In the presence of only one active *DMRT1* allele, the ovary pathway is activated; *FOXL2* is upregulated and suppresses the testis pathway (by inhibiting *DMRT1* and *SOX9. WNT4* and *RSPO1* stabilize β catenin. However, as well this, bird embryos need oestrogen (made by upregulated *CYP19*) to develop ovaries. Inhibition of oestrogen caused testis development in normal ZW and ZZ heterozygous mutant embryos.

So it is clear that *DMRT1* plays a central role in bird sex determination, although the reporting of gynandromorphs means that another Z-borne gene also plays a role (Section 10.14). Does it do anything else? ZW embryos with a mutant *DMRT1* are quite viable and develop an ovary, implying this gene has no critical function in somatic tissues. However, no oocytes develop, suggesting that *DMRT1* is required for oocyte maturation, probably because a step of meiosis is missing.

Thus, sex determination arose independently in birds and mammals and employs different master genes to trigger a conserved pathway; *SRY* in mammals and *DMRT1* in birds.

It is intriguing that monotreme sex chromosomes have homology to the bird ZW. Does this mean that *DMRT1* was the ancestral amniote sex-determining gene or is *DMRT1* just a good choice as sex determiner? To explore the origin and turn-over of sex genes, I will describe sex determination in other vertebrates.

10.7 Searching for Reptile Sex Genes

Sex in reptiles is triggered by an amazing variety of genetic systems (GSD), including sex chromosomes of all descriptions (XY, ZW with sex chromosomes at every stage of differentiation) as well as environmental cues such as temperature (TSD), which I will discuss in Section 10.15. We might therefore expect to find an amazing variety of sex-determining genes.

It has been difficult to identify sex-determining genes in reptiles because, without variants and complete gene maps, positional cloning is impracticable. Investigation of candidate genes has therefore been at the forefront.

The most extensive studies of a candidate gene have focused on *dmrt1*, since this has a controlling role in bird sex determination. However, *dmrt1* is autosomal in snakes, as well as in the few lizards that have been examined. Exceptions are one gekko and one turtle species, whose Z chromosome resembles that of birds; however, studies of related species show that these Z chromosomes evolved only recently. *Amh* lies on the sex chromosomes shared by some groups of lizards, but there are no data on its differential expression in males and females.

Genome sequencing could identify reptile sex-determining genes. Comparing sequences between males and females can reveal sex-specific sequences (on the

Y or the W) that might include male- or female-dominant sex determiners, and a halved read depth can identify unpaired regions of a Z or X that might harbour dosage-sensitive sex genes.

This approach was used in snakes, which supposedly all had a system of female heterogamety (ZZ male: ZW female), in which the W is a more or less degraded relic of the Z (Section 4.12). A triploid ZZW snake was male with testes, suggesting that sex determination relied on a dose-dependent gene on the Z rather than a female-determining W gene. Sequencing identified large differential regions of the Z chromosome in 'advanced snakes' (colubrids and viperids), but they contained too many genes to examine. And boa and python, whose indistinguishable sex chromosomes were assumed to constitute the same pair which differed in a tiny region, turned out to have unrelated XY systems (Section 4.12), which are likely to be defined by quite different sex-determining genes. Snake sex-determining gene(s) remain elusive.

In turtles, many genes from the conserved vertebrate sex-determining pathway (*dmrt1, amh, sox9, foxl2*, and *cyp19a1*) were identified, and sex-specific expression of genes including *amh* – and many non-coding RNAs – was described. *Dmrt1* was expressed early in the ZZ:ZW soft shelled turtle, and knockdown in ZZ embryos caused male-to-female sex reversal; overexpression in ZW embryos caused ectopic expression of *sox9* and *amh*. *Dmrt1* also seems to be sex determining in the red-eared slider turtle. *Amh* was expressed just after *dmrt1* in gonads of ZZ embryos and high expression continued in Sertoli cells of the testis cords. Knocking it down cause male to female sex reversal, with lowered expression of *sox9* and upregulation of *cyp19a1*. Upregulation of *amh* caused masculinization of ZW embryos, and *amh* was strongly upregulated by female-to-male sex reversal caused by inhibiting aromatase. This suggests a pathway *dmrt1–>amh–>sox9* in turtles. However, *dmrt1* does not lie on the Z or W, so there must be a preceding step governed by a master sex gene.

A few lizard species have now been sequenced. The green anole was chosen as a model lizard species, and several genes known from sex-determining pathways in birds and mammals were identified. However, anoles have complex sex chromosomes which have recently rearranged with different autosomes in several species, so no sex-determining gene jumped out.

Sex determination has recently been resolved in the crocodile lizard, of the family Anguimorpha, charismatic lizards that mostly have a very stable ZW system that involves Linkage group 7. However, this species has evidently jettisoned this chromosome for Linkage Group 3. Whole genome resequencing and use of DNA markers whittled the sex-determining region down to 1 Mb, homologous to a part of chicken chromosome 9 that contains only 10 genes. One of these is the female-promoting *foxl2*, which has been duplicated in the crocodile lizard, and would seem to be a good candidate sex-determining gene.

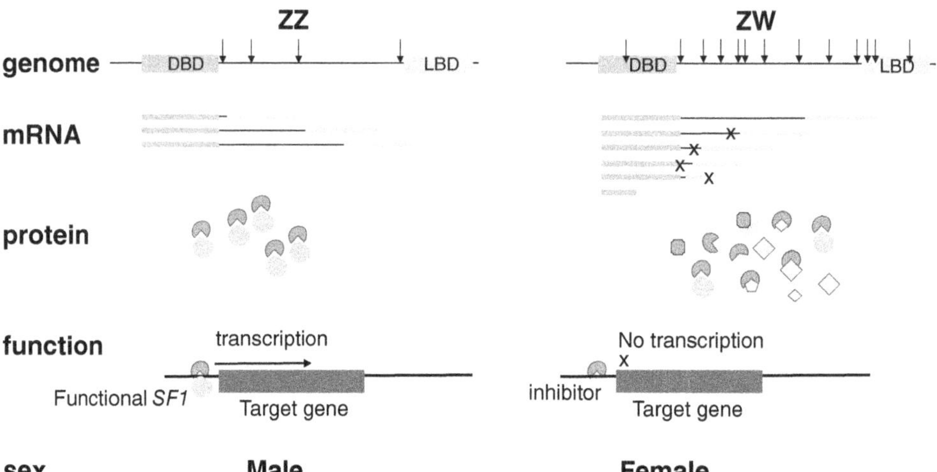

Figure 10.5   Sex determination by epigenetically controlled splicing of *sf1* in the dragon lizard. The *sf1* gene encodes a protein with a DNA-binding domain (DBD) and a ligand-binding domain (LBD), separated by a hinge. Identical alleles lie on the Z and W chromosomes. In ZZ embryos (left), the gene is transcribed into three alternative transcripts that all encode functional SF1 protein with a variable hinge region. However, in ZW embryos (right), the gene is transcribed into 18 different transcripts, most of which contain stop codons or are truncated. The protein made by ZW embryos contains truncated proteins that lack a ligand-binding domain, which may act to suppress transcription in ZW embryos.

A tractable lizard species is the Australian central bearded dragon, which has Z and W micro chromosomes. Mapping of Z and W fragments produced a strong candidate gene *sf1*, which is intimately involved in sex in mammals and birds. A problem was that *sf1* had alleles on both the Z and W chromosomes, and their genomic sequence was identical. However, sequencing transcripts showed that in ZZ lizards *sf1* is transcribed into full-length transcripts, but in ZW animals there are 18 alternative transcripts most of which encode a DNA binding, but not a ligand-binding domain, which may repress *sf1* binding to target genes (Figure 10.5). Epigenetic regulation of *sf1* splicing may be determined by the conformation of the W chromosome, which is full of repetitive sequences. Thus, epigenetic control seems to be involved in the GSD system of dragon lizards, adding to the known mechanism of sex gene action (Figure 10.5). Full genomic sequencing has since identified an extra copy of *amh* and its receptor *amhr2* that have been fused to the Z, but not the W, and these genes appear to act upstream of *sf1*.

There must be many different sex genes acting in many different ways among reptiles, and many sex-determining genes may be added to the list (Table 10.1). Now that the genomic tools are at hand, we can capitalize on the great variability of reptile sex-determining systems.

10.8 Sex-Determining Genes in Amphibians

Amphibians – at least frogs and toads – subscribe to genetic sex determination. Most frogs lack differentiated sex chromosomes, but genetic studies in many species revealed sex linkage, and both male and female heterogamety.

Many genes involved in mammal sex determination are conserved in frogs. A highly conserved *sox9* is upregulated in tadpole testis but not ovary, and remains on in testis cords. *Dmrt1* is expressed in gonads of both sexes, then becomes male-specific and confined to Sertoli cells in testis cords. Oestrogen is expressed early and may control early steps.

Identification of sex linkage in 33 frog species pointed to a remarkable diversity; at least six chromosomes of the conserved frog karyotype acted as sex chromosomes in one or other species (Section 4.13). This was initially puzzling, but the answer became clear when it was shown that the most popular frog sex chromosome bore two genes that are sex determining in many birds, reptiles, frogs and fish (*dmrt1* and *amh*). Other genes from the sex-determining pathway (*cyp10a1, sox3*) lay on other frog sex chromosomes (Figure 10.6). Thus, these genes independently acquired a sex-determining role and defined novel sex chromosomes in multiple frog species.

It has been difficult to identify genes at the head of sex-determining pathways. Candidate genes have been mapped, but the usual suspects are autosomal. However, there are two particularly interesting situations in which an amphibian sex-determining gene has been identified.

The African clawed toad *Xenopus laevis* lacks differentiated sex chromosomes, but female heterogamety (ZZ male: ZW female) was demonstrated genetically. A hunt for *dmrt1* homologues produced a surprise; although the orthologous *dmrt1*

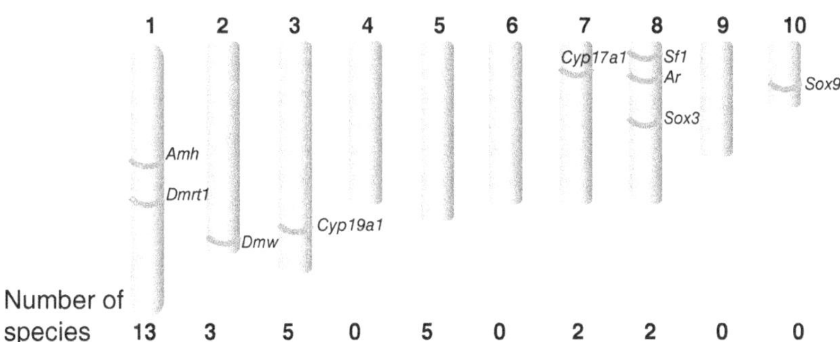

Figure 10.6 Frog sex chromosomes defined by different sex-determining genes. A group of 30 frog species have identical chromosomes (10 pairs), but use one of five different pairs as sex chromosomes in different species. These pairs harbour genes known to be sex determining in other vertebrates (grey rings), and by far the most popular is Chromosome 1, which bears *dmrt1* and *amh*.

is autosomal, a female-specific copy lay on the W chromosome. This *dmw* was expressed strongly in the ovary during sex determination, and became concentrated in the somatic cells that surround primordial germ cells. Transgenesis of ZZ tadpoles with *dmw* upregulated the female-promoting *foxl2* and the aromatase gene *cyp19*, producing primary ovarian structures. Conversely, knockdown of *dmw* in ZW embryos produced testis structures.

It was proposed that *dmw* competes with *dmrt1* for common targets, diminishing the concentration of autosomal *dmrt1* below the threshold needed for male determination. Indeed, its structure and expression are consistent with a role in competitive inhibition (10–12). Sequence comparisons showed that *dmw* originated from a partial duplication of *dmrt1*. This must have occurred recently, since closely related *Xenopus* species lack *dmw*.

Thus, a small genetic change altered the dosage of a sex-determining gene to produce a female-dominant (dominant negative) anti-testis gene that took over the job of triggering the sex-determining pathway, and defined a new W chromosome (Figure 9.1). Again, it is astonishing how easy it is to make a new sex gene!

Another candidate sex-determining gene comes from the remarkable Japanese wrinkled frog. *Glandirana rugosa* has an XY system on one island and a ZW system on another, with a hybrid zone in the middle (Section 4.14). These XY and ZW chromosomes have identical g-band patterns and gene content, and are both homologous to the ancestral frog chromosome 7. Remarkably, this ZW/XY chromosome has homology to the mammal XY, and contains homologues of mammalian *SOX3*, the ancestor of *SRY*, as well as *AR*, which encodes the androgen receptor that is required for androgen function.

In ZW tadpoles the W-borne *sox3* allele is overexpressed, and drives aromatase expression. The *cyp* promoter contains a *sox3* binding site, removal of which ablates aromatase activation. *Sox3* therefore seems to have a role in directing indifferent gonads to develop into ovaries in *G. rugosa*.

A rival candidate gene in *G. rugosa* is the androgen receptor, which also lies on the Z and W chromosomes. The *ar* allele on the W is expressed only weakly, and the Z allele is not dosage compensated, so *ar* expression is greater in ZZ than ZW tadpoles. ZW frogs transgenic for Z-*ar* showed upregulation of male promoting *dmrt1*, which produced some testis development.

Another amphibian sex gene was isolated from the axolotl *Ambystoma mexicanum*, which has a giant genome (10× the size of human). This species has undifferentiated ZW chromosomes, but 300 kb of W-specific sequences were generated by comparing reads from males and females. The only active gene recovered was a duplicated copy of *atrx*, mutations of which cause sex reversal in humans, and which has a Y-borne homologue in marsupials. This *atrw* becomes a candidate for a dominant female sex determination gene.

Thus amphibians showcase recently evolved systems that re-use genes, or copies of genes that are in the conserved vertebrate sex differentiation pathway (Table 10.1). We need to know more about frogs!

10.9 Fish Sex Determination – Usual Suspects and Unique Strategies

Fish display almost every type of sex determination system (Section 4.14). Most show genetic sex determination, with XY or ZW chromosomes defined by many different sex-determining genes (Table 10.1). However, the downstream gene network appears to be quite conserved, containing several of the same genes as in other vertebrates.

Expression of candidate genes has been studied in many fish species. For instance, quantitative PCR was used to follow gonadal expression of 17 genes in the Nile tilapia before and after gonad differentiation. *Sox9* was shown to be involved in testis determination, and *wnt4* in ovary determination. *Cyp19a1* was expressed very early in XX embryos, along with the *foxl2* gene it upregulates. *Dmrt1* was expressed early in XY gonads, but *sox9* and *amh* were expressed later in testis development. This implies that aromatase (encoded by *cyp19a1a*) is essential to trigger the female pathway and *dmrt1* to trigger the male pathway in tilapia.

These genes seem to be associated with sex determination in all fish studied, but their roles are not always the same. For instance, *amh* is initially expressed in

Table 10.1 *Sex-determining genes in vertebrates.*

Species	Sex chromosomes	SD gene	Ancestral gene	Modification
MAMMALS				
Theria (typical)	XY	*SRY*	*SOX3*	Fusion with gonad-specific promoter
Spiny rat	XY	*SOX9Y*	***SOX9***	Duplication of upstream enhancer
Monotremes	X5Y5	*AMHY*	***AMH***	Chromosome fusion, allelic mutation
BIRDS				
All species	ZW	*DMRT1*	***DMRT1***	Null mutation on W sets up dosage inequality
REPTILES				
Dragon lizard	ZW	*sf1?* *amh,* *amhr2*	***sf1***	W conformation controls *sf1* splicing Duplication, fusion with Z

Table 10.1   (*cont.*)

Species	Sex chromosomes	SD gene	Ancestral gene	Modification
Crocodile lizard	ZW	*foxl2*	**foxl**	Candidate female-determining gene
Black marsh turtle	ZW	*wt1?*	**wt1**	
AMPHIBIANS				
African clawed toad	ZW	*dmw*	**dmrt1**	Transposition of truncated copy to become *dmrt1* inhibitor
Japanese wrinkled frog	XY, ZW	*sox3*	SOX3	
axolotl	ZW	*atrw*	**atrx**	Duplication and transposition
FISH				
tongue sole	ZW	*dmrt1*	**dmrt1**	Allele mutation, epigenetic control
medaka	XY	*dmrtbY*	**dmrt1**	Duplication, insertion into autosome
Luzon rice fish	XY	*gsdfY*	*gsdf*	Allele mutation
Indian rice fish	XY	*sox3*	*sox3*	Allele mutation
killifish	XY	*gsdfY*	*gsdf*	Allele mutation
sablefish	XY	*gsdfY*	*gsdf*	Allele mutation
fugu	XY	*amhr2Y*	**amhr2**	Allele mutation
pejerry	XY	*amhY*	**amh**	duplication
stickleback	XY	*amhY*	**amh**	duplication
herring	XY	*bmpr1bbY*	*bmpr1bb*	Duplication, truncation
catfish	XY?	*bcar1Y*	**bcar1**	Male-specific binding to oestrogen receptor
trout	XY	*sdY*	*irf9*	Duplication, neofunctionalization

Ancestral genes that are involved in the sex-determining pathway in other vertebrates are in bold.

females in some species, as is *sf1* in tilapia. In cichlid fish *cyp19a1* is not ovary specific. This is somewhat confused by the fact that some of these genes (e.g., *sox9* and *cyp19*) exist in duplicate because bony fish underwent a genome doubling, so a spare copy can acquire a different function.

A candidate gene approach combined with genetic mapping identified some of these sex pathway genes as the master regulator. For instance, in a flatfish (the half-smooth tongue sole, *Cynoglossus semilaevis*) with cytologically distinct ZW sex chromosomes, sequencing male and female genomes identified a region that was present in double dose in males and a single dose in females (Z), and a paralogous region that was female-specific (W). Remarkably, this region of the sole Z had homology with the bird Z chromosome, and included a homologue of the bird sex gene *DMRT1*. Its partner on the W chromosome was an inactive pseudogene.

Dmrt1 was transcribed only in the somatic cells of the undifferentiated male gonad at the sex-determining stage, as well as the adult testis. This suggested that, as in birds, two copies of *dmrt1* in a ZZ fish specify male development and a single copy in a ZW fish leads to female development (but see Section 10.10 for an epigenetic twist to this story).

A stunning discovery came from a search for *dmrt1* in the medaka (*Oryzias latipes*). This revealed that, as well as an orthologous *dmrt1* gene on linkage group LG9 (an autosome), there was an extra copy on LG1 which is the sex chromosome pair in this species. This *dmy* gene lay on the Y but not the X chromosome, in a sex-determining region with low recombination.

This male-specific *dmy* was expressed only in somatic cells surrounding the germ cells in XY medaka embryos, starting just before gonad differentiation. Mutation or knockdown of *dmy* produced XY female sex reversal, whereas *dmy* transgenesis of XX eggs produced male development. Thus *dmy* acts as a male-dominant sex-determining gene. The amino acid sequence encoded by *dmy* was almost identical to that of autosomal *dmrt1*, making it likely that *dmy* simply increases the concentration of DMRT1 protein, to exceed a threshold that determines maleness.

Comparison with closely related species showed that *dmy* arose as a duplication of a small region on LG9 that contains *dmrt1*. The copy was transposed to LG1, and outfitted with a *sox9* binding site to become the master sex gene.

Another familiar gene was found to be sex determining in one fish species. In mammals, *AMH* lies downstream of *SOX9* in the testis determining pathway, but in reptiles it is expressed before *SOX9*. Mammalian AMH inhibits the (female) Müllerian ducts; fish lack Müllerian ducts, and *amh* has been implicated in germ cell proliferation, probably reflecting its ancestral function.

The Patagonian pejerry (a South American freshwater fish) has male heterogamety but no heteromorphic XY pair. Genetic crosses identified a sex-linked region, and molecular cloning and FISH mapping revealed that it contained a male-specific copy of *amh*. This *amhy* was expressed in pre-Sertoli cells of XY embryos before visible sexual differentiation. *Amhy* knockdown in XY embryos upregulated *foxl1* and *cyp19a1*, resulting in ovarian development and confirming *amhy* as the sex-determining gene.

Consistent with this, the receptor for AMH appears to be sex determining in the distantly related puffer fish *Fugu rubripes*. Genetic studies identified a male heterogametic system, and linkage mapping defined a small sex-determining region, containing only two genes. One of these genes encoded AMH receptor II. This *amhr2* is embedded in a region conserved in other fish, implying that it is a true orthologue, rather than a copy. *Amhr2* became the sex-determining trigger only recently since it exists in only a few species that shared a common ancestor 5 Mya.

DNA sequencing revealed that *amhr2* acquired its sex-determining role via the tiniest possible molecular change; a single base substitution (C to G) that was heterozygous in all males and absent from all females. The base change resulted in an amino acid substitution in the kinase domain that interferes with its binding function. Similar mutations in medaka disrupt germ cell proliferation and meiosis and produce XY females: in humans they produce partial hermaphrodites. Thus, the male-specific *amhr2* acts as a dominant male determining gene in *Fugu*.

The discovery that sex can be triggered by a single base change in undifferentiated 'young' sex chromosomes illustrates the utility of genome sequencing, but particularly underscores the power of genetics to pick out the one-in-a-billion significant base pair. And again, it emphasizes how easy it is to make new sex genes.

Thus, most of the same genes affect sex determination in all vertebrates, and some of these have independently taken over a role at the head of the pathway in several fish species.

10.10 Sex-Determining Genes in Fish – Variety and Change

A wider search for sex-determining genes in fish has uncovered remarkable diversity, and reveals pathways that found a new function in sex determination.

Fish related to the medaka first yielded extraordinary information about the variety and rapid turnover of sex-determining genes. This group of small Asian freshwater fish share a conserved karyotype, but sex linkage points to different chromosomes functioning as sex chromosomes.

Comparing sex determination of medaka and related species showed that a sister species *Oryzias luzonensis* that diverged only 5 Mya has only an inactive *dmy* pseudogene on LG 1. Sex markers mapped to a different chromosome, LG 12 and mapping defined a sex region. Candidate genes were identified; one of these was more active in XY than XX embryos, and transgenesis into XX female eggs produced fertile males. Surprisingly this gene was not a transcription factor. Rather, *gsdf* (*gonadal soma derived growth factor* on the Y) is a member of the TGFβ superfamily of growth factors.

The X and Y alleles of *gsdf* encoded an identical protein but differed by a few bases in an upstream regulatory region. This Y borne regulatory sequence produced higher expression of a reporter gene than did the X allele.

Gsdf was also found to be sex determining in an outgroup, as well as the more distantly related sablefish, suggesting that this teleost-specific gene is intimately connected to sex determination in fish, and was probably the original sex gene in this fish group. In medaka, *dmy* seems to work by turning on *gsdf*.

In a more distant medaka relative, the Indian rice fish *O. dancena*, sex-linked markers mapped to yet another chromosome, LG 10. The sex-determining locus was positionally cloned, and turned out to be an orthologue of mammalian *SOX3*, the ancestor of *SRY*. *Sox3y* expression was sex-specific in developing gonads, and probably also works by regulating *gsdf*. Mutation of *sox3* produced ovary-like tissue in XY embryos. The X and Y alleles of *sox3* encoded the same protein, but differed in upstream sequence, suggesting that sex determination requires regulation by a distant cis-element.

Several genes with widely different functions have now been linked to sex determination in other fish species. Trout and salmon, intensively studied because they are so delicious, have male heterogamety, but sex linkage points to different chromosomes in different populations. A gene *sdy* (for sexually dimorphic on the Y) was discovered by expression studies in the rainbow trout, and seems to be shared by other salmonids (remarkably, transposed to different chromosomes). *Sdy* was highly expressed in XY males, and transgenesis produced XX males, some of which were even fertile.

Surprisingly, *Sdy* encoded a 192-amino acid protein with homology, not to other sex determining or differentiating genes, but to part of a gene *IRF9* (Interferon Regulatory Factor 9) that acts through a well-known signalling pathway (called STAT) to regulate the immune system in mammals. *IRF9* has no known association with testis differentiation. So how did *sdy* get involved with sex? Comparing *sdy* sequence with *IRF9* showed that it lost its DNA-binding domain and nuclear localization signals, but retained and modified its protein-protein interaction domain. It no longer binds STAT proteins, but binds strongly to *foxl2*, inhibiting its female-determining function by blocking *cyp19a1* and suppressing oestrogen production.

Other unexpected sex-determining genes have also emerged. In the Atlantic cod sequencing identified a 9 kb sex region that contained two RNA-binding zinc knuckle genes that are upregulated in the male bipotential gonad. Yet another outlier came from comparing sequences between XX and YY catfish (obtained by doubling up haploid genomes). Full X and Y sequences pointed to a homologue of *BCAR1*, for Breast Cancer Anti Resistance 1) that was expressed in the gonad before differentiation, and knockout was sex reversing. X and Y alleles of *bcar1* encode proteins that differ by just two amino acids, and *bcarx* binds to the oestrogen receptor and inhibits *sox9* to promote the female pathway.

This variety leads us to wonder if just about any gene can take up a career in sex determination. Why, then do we see the same gene popping up again and again?

10.11 Conservation – or Re-use – of Sex-Determining Genes

Amidst all the variety of master sex-determining genes, we see some that are sex determining in widely divergent species. Does this signify identity by descent of an ancient sex-determining gene? Or reuse of a particularly handy sequence?

The discovery that the multiple X_5Y_5 system of platypus shares homology with the bird Z chromosome excited speculation that a 310 million year old *DMRT1* gene might be an ancestral sex determiner in amniotes. The discovery of *DMRT1* on the Z chromosome in a gekko (*G. hokuoensis*) heightened these expectations. However, platypus *DMRT1* lies on an X, so has the wrong dosage relations, and the absence of *dmrt1* from sex chromosomes in related gekkos implies that the *G. hokuoensis* Z evolved recently. The revelation that *dmrt1* is also sex determining in the half-smooth tongue sole tells the same story, since it is not shared with species that diverged a mere 30 Mya. Perhaps, then, *DMRT1* is simply good at determining sex. This idea is reinforced by evidence that *dmrt1* homologues are intimately involved in sex determination in crustaceans, as well as in insects and worms.

This conclusion is borne out by the finding that two other sex-determining genes are popular. As well as being the ancestor of mammalian *SRY*, *SOX3* is the sex-determining gene in a frog (*G. rugosa*) and a fish (*O. dancena*). The anti-Müllerian Hormone *AMH* and its receptor *AMHR2* are sex determining in two quite distantly related fish, in the bearded dragon, and may also determine sex in the platypus.

Thus these three well characterized genes, *DMRT1*, *SOX3* and *AMH/AMHR2* appear as sex determiners in widely divergent vertebrates. They can't all be identical by descent. The alternative is that they each present very favourable raw material for making novel sex genes, and they are selected repeatedly for their function.

However, these popular genes are never very far away from sex. The bird sex-determining gene *DMRT1* has a dosage-dependent function maintaining testis in all vertebrates, and *AMH* has downstream function inhibiting the female pathway. The HMG box of *SOX3* is similar to that in several *SOX* genes (as well as *SOX9*) that are involved in sex.

What common function could these popular genes have? *DMRT1* and *SOX3* are transcription factors, but *GSDF* and *AMH/AMHR2* are part of the TGFβ signalling pathway, and *SDY* is related to an immune gene. So it seems that there are more than one different pathway that can be used to control the cascade of genetic interactions that lead to gonad differentiation.

Recruitment of novel genes as the master sex determiner can also be understood by their interactions with these common pathway genes; for instance signalling factors in medaka species interact with the *DMRT1/SOX9* pathway, salmon *SDY* and catfish *BCAR1* interferes with the *Cyp19a1* production of oestrogen.

How, then, are novel sex-determining genes made?

10.12 Making New Sex-Determining Genes

It is remarkably easy to make novel sex-determining genes (Table 10.2). The starting material is usually genes that already have roles in sex determination, or genes in pathways that interact with them (Section 10.11).

Genes can be simply translocated or transposed to a new site, defining a new sex chromosome. For instance, a chromosome region containing *AMH* was added to an original bird-like sex chromosome to become the new sex determiner in monotreme mammals. And the trout sex gene *sdy* appears to have transposed to other chromosomes in other salmonids.

Table 10.2   *Evolution of new vertebrate sex-determining genes.*

MODE	MECHANISM	EXAMPLE
Fusion, translocation, insertion, of sex gene	New sex gene is added to one member of autosomal pair	*AMH* in region translocated to the platypus Y5
Mutation	Null mutation of one allele sets up 2:1 dosage difference of new sex gene	Bird *DMRT1* on Z but lost from W, sole *dmrt1* on Z but pseudogene on W
Copy (active)	Extra copy on neo Y sets up 3:2 dosage difference	*dmy* on medaka neo-Y bolsters autosomal *DMRT1*
Copy (inactive)	Truncated copy inhibits action of autosomal gene	*dmw* on Xenopus neo-W inhibits autosomal *dmrt1*
Regulatory change in expression	New regulatory sequence changes amount of sex gene product	Duplication of enhancer in spiny rat increases *SOX9* expression on neoY
Expression profile change	New regulatory sequence changes tissue of expression	Translocation or insertion of gonad-specific promoter to make mammal *SRY*
Epigenetic change	DNA methylation/chromatin conformation represses sex gene on Y or W	Methylation of *dmrt1* on sole W represses transcription

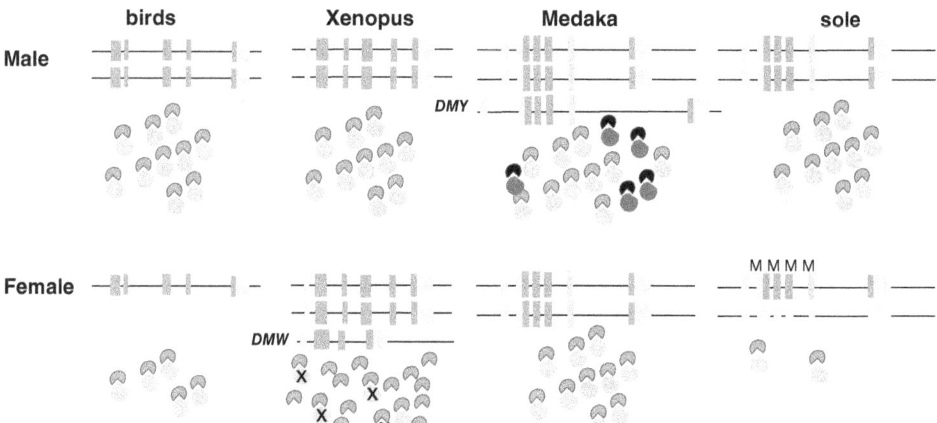

Figure 10.7　Sex determination by *DMRT1* and its copies in birds, frogs and fish. *DMRT1* has a similar sequence in all vertebrates, encoding a protein with DNA binding (dark grey) and transactivating domains (light grey). Variations in structure, number and location allow this gene to determine sex in different ways in birds, frogs and fish (medaka and sole). In birds, *DMRT1* lies on the Z but not the W chromosome. The 2-fold dose in males activates testis determination. In *Xenopus laevis*, *DMRT1* is autosomal, so both sexes have two copies. However, a truncated copy lies on the W chromosome: it encodes a truncated protein with the DNA binding domain but no transactivating domain; this acts as a competitive inhibitor, (inhibited *DMRT1* indicated by crosses) and produces female development. In medaka fish, *DMRT1* is autosomal. An active copy was transposed to a neo-Y, which encodes an extra dose of *DMRT1* (black), which boosts DMRT1 concentration and promotes male development. In the half smooth tongue sole, *DMRT1* lies on the Z chromosome, an inactive pseudogene is on the W (pale grey). Methylation represses transcription of the single active copy of *DMRT1*, so there is little DMRT1 protein and development is female.

Many new sex-determining genes arise as copies of conserved sex-determining genes, particularly those that work by dosage differences. For instance, *dmrt1* copies have been independently inserted into new chromosomes in a frog and a fish (Figure 10.7). They have completely opposite effects; in medaka, *dmy* acts as a dominant male determining gene that defined a neo Y, whereas in *Xenopus laevis* *dmw* is a dominant female-determining gene that defined a neo W.

Analysis of sequence shows that in medaka the male determining gene *dmy* encodes an active copy of *dmrt1*, so with two copies of the autosomal *dmrt1* locus in both sexes, the dosage relationships become 3:2 between XY and XX, evidently sufficient to determine sex. However, in *X. laevis* the female determiner *dmw* turned out to be a truncated version of *dmrt1*, encoding a highly homologous DNA-binding DM domain, but with a stop codon preventing translation of the transactivating domain. It acts as a competitive inhibitor of autosomal *dmrt1* by binding to the same target site, but failing to activate downstream genes

Interestingly, *dmrt1* dosage is also regulated in the half-smooth tongue sole, this time by repressing it in ZW females by DNA methylation. If methylation is disrupted by higher temperature, ZW hatchlings are 'pseudomale'. Evidently the 2:1 ZZ male to ZW female difference is not sufficiently robust, and was tweaked by epigenetically suppressing *dmrt1* in the ZW female. Evolution is ever inventive!

Duplication can also offer a gene copy ripe for neofunctionalization. For instance, *amh* was duplicated in two fish species (pejerry and pike) and one copy became male-specific. Whole genome duplication occurred in bony fish, creating an extra copy of all genes. Most of these were lost, but some (such as *sox9a* and b) have acquired novel functions.

Other new sex genes are derived from mutations of genes in situ. For instance, *amhr2* became sex determining in fugu when a missense mutation occurred on one homologue; homozygosity for the disabling mutation determined female and heterozygosity for the active wild type allele became male determining and defined a novel Y chromosome.

The classic case is the bird *DMRT1*, which must be the original gene because it sits in a conserved synteny group. All it took to become the head of the sex-determining pathway was a null mutation that set up a 2:1 dosage difference and defined a new female-specific W chromosome. In the tongue sole, an equivalent mutation left an inactive *dmrt1* pseudogene that defined a W chromosome.

Mutation or rearrangement of upstream regulatory sequences also accounts for the generation of several sex-determining genes. The classic case is the mammalian *SRY*, which evolved in situ from the ancient *SOX3* gene simply by altering its expression pattern (Section 10.4). Sequence comparisons suggest that a rearrangement occurred upstream of the gene to put it under the influence of a promoter that drove its expression into the developing gonad. This process was re-enacted in the XX male babies in whom *SOX3* is ectopically expressed in the gonad, and also transgenic mouse lines in which *Sox3* expression is driven into the developing gonad and turns on *Sox9*. The same thing happened more recently in spiny rats when a duplication of an enhancer upstream of *Sox9* produced a new sex-determining gene (and a new pair of sex chromosomes).

This happened independently 20 Mya in medaka relative *O. dancena*; a cis-regulatory element was acquired upstream of *sox3* that drives its expression into the developing gonad and turns on *gsdf*. Increase in *gsdf* expression that renders this gene sex determining is driven by a few nucleotide changes in the *gsdf* regulatory region in other medaka relative, and by an insertion of a transposable element in the *gsdf* promoter in sablefish. In fugu, a single nucleotide change in *amh2* weakens the receptor's binding to *amh*, leaving the original allele to accomplish male determination.

Thus making new sex genes is surprisingly easy. What is more difficult to understand is how these novel genes can take over the job of determining sex from an established sex-determining gene, and what selective forces are at work.

10.13 Turnover of Sex Genes

How can a new sex-determining gene take over from the old one? Surely there will be a war of the sex genes when two rivals co-exist in the same population. What will happen in a double heterozygote? Will one be dominant (epistatic) to another? Or will there be intersexual development and infertility in hybrids? Such a 'fitness valley' that a species must cross before a new sex gene becomes fixed has been long predicted.

Tricky though turnover may be, it has happened repeatedly in vertebrate evolution. Some events, such as establishment of the bird *DMRT1* and the therian mammal *SRY*, are very ancient. Others, like the loss of *SRY* in two rodent lineages, and the switches in medaka, are recent. Such recently evolved systems might provide clues to turnover.

Indeed, turnover in medaka and its relatives is easy to understand because the three sex genes *gsdf, dmy* and *sox3* lie in the same pathway. Comparisons of the sex-determining gene in the medaka group and other fish suggest that *gsdf* has a fundamental role in fish sex determination. Both *dmy* and *sox3y* upregulate *gsdf* expression, so either could take over control if a mutation conferred male-specific upregulation. Male-specific repeats inserted upstream of *sox3y* apparently provided this capacity. Neither change would require major rejigging of the downstream pathways, so could be accomplished without disrupting sex determination.

Modifications may arise that subvert one sex. For instance, several polar mammals have sex reversal systems that shift the sex ratio in favour of females, an advantage where there is a short breeding season. Lemmings and pygmy mice independently evolved a female-determining switch on an X* that disrupts *SRY*-driven male determination.

We also have several examples of species in which different sex systems coexist. For instance, the wrinkled frog *Glandirana rugosa* has XY and ZW populations, and there is a hybrid zone in which animals may have practically any combination of X, Z, Y and W chromosomes. The sex gene here is likely to be *sox3* in both populations, suggesting a hierarchy of dominance amongst alleles.

There are many fish species in which different sex chromosomes coexist. For instance, some platyfish populations have both a male determining Y and a female-determining W segregating alleles at the same (unknown) locus, just like the wrinkled frog.

Populations in which different sex-determining genes coexist are often referred to as 'polygenic'. However, these examples do not seem to represent a stable state that requires two or more independent genes. Nearly all claims of polygenic inheritance describe a transitional stage as one gene takes over from another, or an abnormal situation in a hybrid between species with different sex-determining genes.

Who wins in these competitions for supremacy? Two closely related species of toad in France have XY and ZW systems, and one is moving in on its sibling. They can hybridize, and introgression of markers is occurring on both sides of the front. Even more remarkably, XY and ZW systems also coexist in the toad *Xenopus tropicalis*. Males may be ZZ, YZ or YW, females may be ZW or WW. This would require three alleles at a single (unknown) sex-determining locus with dominance relationships Y>W>Z. In some Lake Malawi cichlid fish a female-determining W gene overpowers the male determining Y.

But war between competing sex determination systems can be averted by joining forces. Two populations of the common European frog *Rana temporaria* have different sex chromosomes bearing different but complementary sex genes. One population has as its sex pair LG2, containing *dmrt1* and *amh*, whereas the other (LG7) contains the AMH receptor *amhr*). But in an alpine population LG2 and LG7 segregate together with sex, and are probably physically joined by a translocation.

Most remarkably, zebrafish have undergone sex gene turnover over decades rather than millions of years. Laboratory populations of this model fish established 30–40 years ago yielded contradictory results for the identity of sex chromosomes. This was a puzzle that was eventually resolved when a ZW system was found in wild zebrafish, implying that one of the sex chromosomes was inadvertently lost during lab breeding. Novel sex-determining genes are being independently selected in an evolutionary eyeblink. None of these work very well; sex ratios are aberrant and the environment (particularly temperature) has a big influence. It will be exciting to dissect these brand new sex-determining systems.

What evolutionary forces drove turnover of sex-determining genes? In some species it was a last ditch response to an existential crisis. For instance, *X. laevis* underwent tetraploidization, which would have delivered chaotic sex chromosome segregation from XXYY males. The solution – making a new gene from an inactive *dmrt1* copy to define a neo-W. Many fish species (e.g., some sticklebacks) got their start as species hybrids, and incompatibility between their sex chromosomes may also have required a fresh start.

An alternative for such species is abandoning sexual reproduction altogether. Several all-female reptile and fish species, in which females make eggs from combinations of their own genes and chromosomes, are hybrids or polyploids.

What makes some sex determination systems extremely stable (e.g., birds and mammals) and others crazily labile (as for many reptiles and fish)? The degree of sex chromosome differentiation is critical. There is a tipping point, beyond which the sex-specific chromosome (Y or W) is too degraded to reinstate its function as an autosome. However, at the end of the degradation cycle, a dysfunctional Y chromosome, or mutated sex-determining gene could drive selection for an all-new system; for instance the Okinawa spiny rat still retains a Y bearing many copies of a mutant *Sry* that doesn't work very well, while its sister species have dispensed with it and started again.

At the other extreme are species with sex chromosomes that are differentiated only over a small, non-critical region, even a single gene, even (as in *fugu*) at a single base pair. Takeover by a new sex master gene can occur with little loss of fitness, as it seems to have in frogs, with their 'ever young' sex chromosomes.

Thus turnover of sex genes can be very frequent and their effects trivial where sex chromosomes remain little differentiated. However, as soon as a Y or a W is significantly degenerated, replacing it occurs only on as a result of a compelling evolutionary drive. Terminal degradation of the sex-determining gene is one such reason; major chromosome change or hybridization is another.

10.14 Evolution of Sex Hormone Action

There is evolution, too, of downstream components of the sex-determining pathway. One of the most important is sex hormones.

Sex hormones, particularly oestrogen, have a powerful effect on sex in most vertebrates Egg-laying vertebrates (birds, reptiles, frogs and fish) are extremely responsive to oestrogen, to the extent that (almost) complete sex reversal can be attained by treating chromosomally male embryos with oestrogen. Upregulation of oestradiol and the enzyme that makes it, aromatase, appears to be an early and critical factor in female development in fish, amphibians and reptiles.

Sex in frogs is easily reversed by steroid hormones. Androgens (and oestrogen inhibitors) induce female to male sex reversal, and oestrogens (and androgen inhibitors) induce male to female sex reversal. Androgens are upregulated in genetically male tadpoles and oestrogens in females. Aromatase is highly expressed in female gonads and is promoted by *foxl* and *sox3* expression. Intriguingly, steroidogenic enzymes are expressed before the indifferent gonad shows any sign of sex differentiation, so it is possible that one or more hormone genes act as the sex-determining trigger.

Reptile sex determination is very sensitive to exogenous hormones. Oestradiol and aromatase induce ovary development in reptile embryos, and testosterone induces testis development. Inhibitors of oestradiol and androgens have the opposite effect, disrupting expression of pathway genes such *as sox9, cyp19, rspo1* and *amh*.

In birds, the influence of oestrogen is profound – but not absolute. Blocking oestrogen in female embryos leads to male gonad development and even spermatogenesis. Sex reversal of genetically male birds can be accomplished by injecting oestrogen into the egg. But ZZ sex-reversed females retain the male song, attesting to some unknown Z or W gene that controls the influence of oestrogen on cells of the brain.

This might also explain two remarkable chickens with male characteristics (larger muscle mass, feather pattern and plumage colour) on one side and female on the other, and gonads to match. The male side was largely composed of ZZ cells and the female side by ZW cells, so these 'gynandromorphs' probably arose from mistakes of meiotic segregation or fusions of male and female embryos in a single egg. The spectacular difference in sex traits on two sides of the same bird implies that the external phenotype is not directly hormone driven; a cell autonomous factor, encoded by a gene(s) on the Z or W, must mediate hormone action in somatic cells.

Thus sex hormones are not just downstream products of sexual differentiation, but are intimately involved in sex determination in egg-laying vertebrates. This poses problems because environmental contaminants (e.g., birth control pills) that mimic or inhibit steroid hormones can reverse sex of fish, alligators and turtles in the wild.

Mammal sexual development, on the other hand, is very resistant to oestrogen. This makes sense because it would be difficult for an XY embryo to develop testes within the ocean of oestrogen afforded by his mother's circulation. Oestrogen has therefore been forced to take a back seat in mammals.

Instead, androgens have a major influence on sex determination and differentiation in mammals. In eutherian mammals, XX embryos with a sex reversing mutation (e.g., in *RSPO1* or *SOX3*) may develop testes, and produce androgens. Cells of both males and females express the X-borne androgen receptor, so the tissue response is a fully male phenotype.

Marsupials don't entirely fit this picture. Oestrogens do have an influence on sexual development if they are delivered when the gonads are still differentiating. Gonad development occurs after birth of the underdeveloped young; if birth is brought forward one day, oestrogen injection produces XY female development.

Marsupials also differ from eutherians in that not all male traits are specified by the Y chromosome, *SRY* and androgens. Although XXY animals have testes, they are internal and animals lack a scrotum and gubernaculum for testis descent. In their place is a pouch with mammary glands. Conversely, XO animals lack testes, but have a scrotum in place of a pouch. (The scrotum is not homologous to the scrotum of eutherians, but develops from the same patch of abdominal epithelium as the pouch). It seems that scrotum/pouch switch is specified, not by androgens unleashed by *SRY*, but by a factor on the X chromosome. Either the number of X chromosomes (two for a pouch, one for a scrotum), or the specification of pouch by a paternally derived X, may control these sex dimorphisms in marsupials.

Even in therian mammals, there is some remarkable variation in hormone levels, apparently driven by the necessity of females of some species to defend themselves and their young. For instance, female moles lead a tough life underground, and this has evidently selected for a high androgen level that promotes aggression. This is accomplished by development of a region of the gonad into testis-like tissue composed of steroid-manufacturing Leydig cells. The genetic specializations that allow this make use of downstream genes, including a rearrangement that affects the timing of expression of the testicular growth factor *FGF9* and a triplication of *CYP17A1*, coding for aromatase that controls androgen synthesis, modifications that have the effect in female transgenic mice. Female hyenas, too, are aggressive and highly androgenized to the extent that the genitalia is male-like; the long phallus complicates mating and birth, such that firstborn offspring usually don't survive. Thus sex hormone balance is behind many behavioural differences among mammals species.

10.15 Environmental Sex Determination and Epigenetic Control

Many reptiles lack sex chromosomes and rely on environmental cues to determine sex. For instance in crocodylians and sea turtles the sex of an embryo is determined by the temperature at which eggs are incubated (temperature sex determination, TSD); if it's hot you get all male alligators, or all female turtles. Some species are female at both temperature extremes, suggesting that there is an optimal temperature range for male development, and which is the high temperature sex just depends on which side of this optimum the viability limits lie (Figure 10.8).

TSD seems to have been adopted particularly by long-lived species. There are many hypotheses regarding the relative advantages of TSD, perhaps as a means to optimize sex ratio. One such is that female alligators prefer cool nest sites, which are female producing, so that when the population is low most hatchlings will be female and the population will grow rapidly. At higher population densities, many females have to settle for high-temperature sites so more males are produced, and must compete for females.

Temperature-dependent sex has been observed in 60 fish species from diverse lineages, and some species use other environmental cues such as pH and oxygen tension. Other species are serially male then female, or the reverse, pursuing different strategies. Fish species which adopt a dominant male protecting a harem tend to go for large aggressive males and generally develop first as females, whereas more solitary fish do better with larger females that can produce more eggs, so tend to develop first as males.

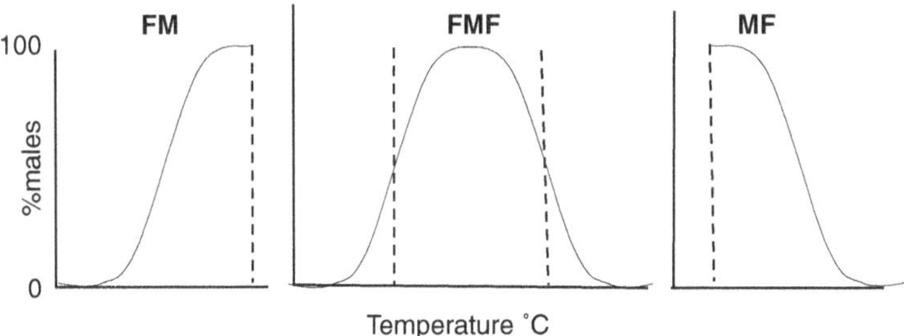

Figure 10.8   Patterns of temperature sex determination (TSD). The proportion of males in the population varies with temperature to an optimum of 100%, on either side of which male development falls off and females predominate (FMF pattern, observed in some species of lizards). However, at very low temperatures the eggs do not develop, and at high extremes they die; these viability limits (dotted lines) mean that in some species like alligators only the cooler side of this curve is observed as a FM (female-male) pattern, whereas in other species (e.g., marine turtles) only the hotter side is observed as a MF pattern. The temperature at which half the eggs develop as males (the pivotal temperature) varies between species.

Most extraordinary are fish in which social cues bring about sex change in adults. Most famous is the blue head wrasse, a coral reef species, in which a dominant (blue) male protects a harem of (gold-striped) females. Removal of the male induces the largest female to become male. She changes her behaviour in minutes, her colour in hours, morphology in a few days, and develops testes and produces sperm by 10 days.

Attempts to understand how environmental cues trigger male or female development go back 50 years. Temperature shift experiments revealed a window during development in which temperature acts. Intersex development is rare, suggesting that some gene product builds up to a threshold and trips a male-determining switch.

The molecular biology of TSD has been investigated by studying orthologues of genes known to be involved in genetic sex determination (GSD), finding that the usual suspects are active in alligators and turtles. RNAseq has been used to follow the activity of all genes during the thermosensitive period in the TSD painted turtle, confirming the upregulation of genes, including *amh, ar, gata4* and *fog2* at male temperature, and *foxl2, β-catenin* and *aromatase* at the female-producing temperature, as well as differential expression of many noncoding RNAs. Thermosensitivity produced a long list of candidates for genes that could mediate the temperature signal (e.g., heat-shock and RNA binding proteins), but, like all such comparisons, could not disentangle the effects of sex and temperature.

Particularly useful are reptiles with GSD, but a temperature override. A GSD dragon lizard, the central bearded dragon *Pogona vitticeps*, has ZW sex chromosomes, and likely sex genes *amh/sf1* (Section 10.7). At moderate temperatures, the sex ratio is 50:50 as predicted from sex chromosome segregation. However, when dragon eggs were incubated at higher temperatures, surprise – all the offspring were female. Half of these were ZW as expected, and the other half were ZZ sex reversed females.

Comparisons of transcriptomes of adult ZW normal females with ZZ sex reversed females revealed a splicing change in the gene *jarid2* (which encodes a component of the master regulator PRC2, polycomb repressive complex) and another member of the 'Jumonji' family (*kdm6b, aka jmjd3*) that is associated with epigenetic remodelling by demethylating histone 3 lysine 27 (H3K27), and removing repressive marks. In embryos, intron 11 of these two related genes was spliced out in sex reversed ZZ females but retained in normal ZW females and ZZ males. The retained intron contained chain terminators in every phase, so the mRNA could not be translated into active protein in the sex reversed females. Sex-specific retention of Intron 11 in these two genes was also found in alligators and turtles, suggesting this peculiar method of turning off epigenetic modifier genes is used throughout TSD reptiles.

The link with epigenetic change was strengthened by RNAseq of red eared slider turtle embryos. This identified temperature-sensitive and sexually dimorphic expression of *kdm6b*, which demethylates histone H3K27 and activates *dmrt1*. Knocking down *kdm6b* produced male-to-female sex reversal, which was rescued by *dmrt1* overexpression. This suggested that higher (female-producing) temperatures act to repress *dmrt1* by way of epigenetic modification (Figure 10.9).

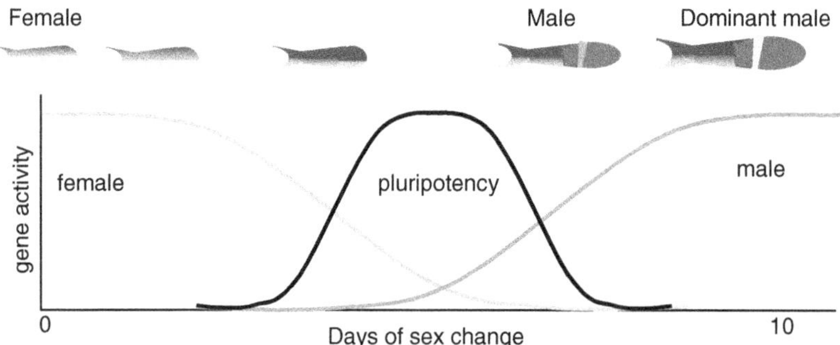

Figure 10.9   Sex change in the blue wrasse. A dominant blue-headed male guards his harem of many smaller, golden females. But if the male is removed, the largest female becomes male – in just 10 days (top row). She changes behaviour and colour and by 10 days her ovaries are replaced by functional testes. Transcriptomics at daily intervals shows that the activity of female typical genes decreases rapidly. Many genes with functions in the pluripotent embryo are turned on as the ovary is repatterned into a testis. Male-typical genes are activated as the testis starts to make sperm, and the sex-reversed male becomes dominant.

Epigenetic factors, as well as pluripotency genes are also involved in the adult sex change in the blue wrasse. RNAseq during the transition period documented rapid decline of female genes, and increased activity of male genes. But particularly interesting was the activation of a group of pluripotency genes known from mammal stem cells that evidently deprogrammed the ovary. Also genes encoding PRC2 components and enzymes that affect histone methylation and acetylation were expressed dynamically during transition, implying that chromatin modification accompanies sex change.

Sex reversal may be more common in nature than had been thought. For instance, breeding a common frog in the laboratory in the absence of environmental stress produced 5% of males with female sex chromosomes, and in a wild population in Hungary, this proportion was 20%, perhaps in response to environmental stress.

10.16 Flip-Flops between GSD and TSD

It seems now that many reptiles and fish have both genetic and environmental sex determination, and even species with differentiated sex chromosomes may show some environmental influence. Even the hard-wired mammalian sex determination relies on many epigenetic signals (Section 9.15).

This turns out to be significant for the turnover of sex genes, because TSD can greatly accelerate turnover and widen the possibilities. This was first apparent from breeding sex reversed ZZ female dragon lizards hatched at high temperature, which turned out to be viable and fertile. Matings with normal ZZ males could produce only ZZ offspring, and their sex completely depended on incubation temperature. Amazement greeted the demonstration that eliminating the W chromosome could change a GSD into a TSD system in a single generation. And concern greeted the demonstration that this is happening – fast – in the wild as ambient temperatures rise in their desert habitat (Figure 10.10).

In groups of reptiles and fish with TSD and GSD, there are many examples of evolutionary flip-flops, sometimes within closely related groups. For instance, the group of Australian agamid lizards changed from TSD to GSD to TSD to a different chromosomal GSD within 17 million years. Some reptiles and fish (for instance the Atlantic silverside), display both systems, which are selectively advantageous at different latitudes. In other fish like the sea bass, an environmental component seems to mitigate the negative effects of takeover by a new sex-determining gene.

The ability to lose a W (or a Y) chromosome and become TSD opens the way for an entirely new sex-determining gene to evolve, and this seems to have happened repeatedly. Modelling also suggests ways in which ZW systems can transition to XY in response to environmental pressure that inactivates a Z-encoded dosage dependent product.

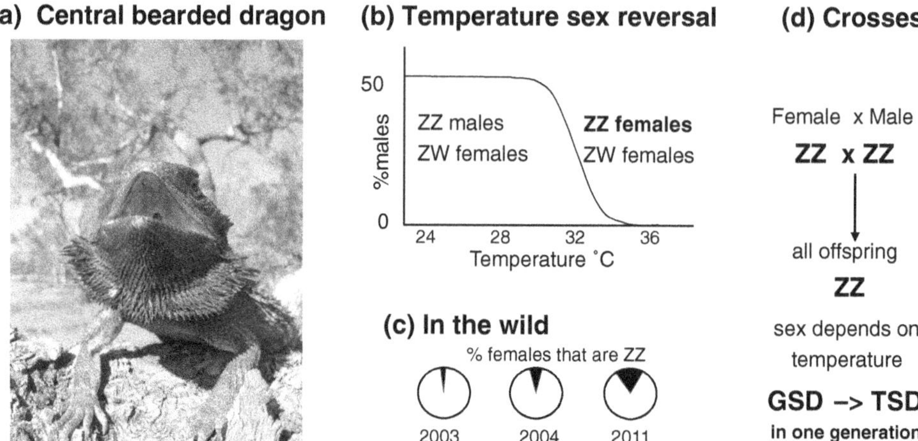

Figure 10.10   Change from GSD to TSD in a dragon lizard. (a) The central bearded dragon, *Pogona vitticeps*. Photo supplied by Arthur Georges, the University of Canberra, Australia. (b) At physiological temperatures (24–32°C) sex is determined by sex chromosomes (ZZ male, ZW female). However, at higher temperatures ZZ as well as ZW eggs develop as females, until at 36°C all offspring are female (half ZW, half sex reversed ZZ). (c) In the wild, the proportion of phenotypic females that are ZZ (sex-reversed females) has risen from 6% to 26% over eight years of sampling. (d) ZZ sex-reversed females are fertile (in fact they lay more eggs and have a better hatch rate than their normal ZW sisters). If ZZ females are mated to ZZ males in the laboratory, all eggs are ZZ. Their sex depends completely on temperature (with a lower pivotal temperature than the original population). Thus the whole sex-determining system has been shifted – in just one generation – from GSD to TSD.

There has been much debate about whether GSD or TSD is ancestral to amniotes. I think that the extreme ease by which these systems can interconvert, and the frequency with which they do so among closely related species, makes this question meaningless.

Thus environmental and genetic sex determination are not, as first supposed, two entirely different pathways, but a continuum. Epigenetic regulators translate the environmental cue into gene products that interact with the conserved sex-determining network. Environmental sex determination, as in alligators, is at one end of the spectrum; strictly genetic sex determination as in mammals and birds (who are, in any case, homeothermic and therefore cannot make use of temperature cues) are at the other.

10.17 The Big Picture

Overall, sex determination in vertebrates can be seen as a network – hardly a pathway – of reactions, some promoting, others antagonizing, male or female development. This network is conserved in essence, but not in detail, across vertebrates.

The most blatant differences between species are the identity of the gene that kick-starts the entire process.

It has been popular to consider the sex-determining network in terms of master-slave relationships, following the old adage that 'masters change, slaves remain'. I prefer to think of it more as a ringmaster-performer relationship in view of their coordinating roles and interdependence.

Indeed, ringmasters do change, but some (*SOX3*, *DMRT1* and the *TGFβ* growth factor family that includes *AMH*, *AMHR*, and *GSDF*) seem to pop up repeatedly. This cannot arise from common descent, but means that some genes are particularly well suited for the job. In addition, most ringmasters have some conserved performer role in the network, although others can be recruited from other pathways, as is the case with the salmon *Sdy* gene.

In addition, not all the performers remain. The conserved genes of the downstream pathway can change their position and function, and even trade places with their ringmasters. New performers can be co-opted, such as the bird-specific hemogen.

Can we explain the evolution of sex-determining pathways by a 'bottoms up' evolution, in which pathway is built up from a conserved downstream gene by recruiting regulators, then regulators of regulators? Indeed, there are several situations where this applies, including the displacement of *gsdf* in medaka fish by two genes that regulate it, and the independent evolution of X-borne regulators of the mammalian *SRY* in fast-breeding mammals.

Many themes – and even genes – in the sex-determining pathway have their parallels – maybe their roots – outside vertebrates. For instance, the intensively researched sex-determination systems of model invertebrates like *Drosophila* reveals sex initiated by dosage differences and effected by alternate splicing.

In *Drosophila*, a male heterogamety XX/XY system works by dosage of genes on the X (Figure 10.1) The initial trigger is the ratio of X chromosomes to autosomes (the Y is a dummy). A ratio of 1:1 is read by an X-borne gene *sex lethal* (*sxl*). *Sxl* induces female-specific splicing of RNA transcribed from another gene *transformer* (*tra*), and the product of this gene favours female-specific splicing of another gene *doublesex* (*dsx*), which makes a female type protein that upregulates downstream genes necessary for female development. In males the X:autosome dosage ratio of 1:2 prevents *sxl* action, leading to male-specific splicing of *tra* which is not translated into functional protein; in the absence of Tra product, the male form of *dsx* is produced which activates downstream male pathways.

Many of these elements are conserved in many insects, in which sex determination involves a cascade of alternate splicing, as well as sex-specific splicing of *dsx* by *tra*. The silk moth does the same thing differently: a female dominant W-borne gene makes a short piRNA that cleaves a long noncoding (masculinizing) RNA that favours male-specific splicing of *dsx*.

A homologue of *dsx* is also essential for male determination in some crustaceans.

Have we underestimated the role of alternate splicing in mammal sex determination? RNA seq in developing mouse embryos documents a plethora of alternative transcripts that are sex-specific. Particularly stunning is the heavy preponderance of female-specific intron-retention. Perhaps this is an ancient and time-honoured way to inactivate male-promoting genes, not just a peculiarity of TSD lizards.

Gene dosage differences lie at the core of many vertebrate sex-determining systems, especially those, like birds, that subscribe to a ZZ:ZW system. Even where the sex-determining trigger is female- or male-dominant (as in mammals), many downstream performers like *SOX9* and *DMRT1* are dosage sensitive. Sex determination is intimately tied up with dosage compensation in *Drosophila* and many other invertebrates, as it is in some vertebrates.

The participation of epigenetic modifiers is also a common thread; very overtly in the case of environmental sex determination, but still there even in male dominant systems such as mammal *SRY*, which requires histone modification and DNA demethylation for its activation.

Not only are these themes in common, but so are some of the genes. In particular DM domain genes – *doublesex (dsx)* in *Drosphila* and *DMRT1* in vertebrates are quite conserved. So is *SOX3*, the ancestor of SRY, which is also involved in *Drosophila* (and frog) oogenesis.

Other insects boast a huge variety of ringmaster genes, some recruited from the conserved network, others recruited from outside, some bizarrely (such as the pillbug sex determiner kidnapped from an invertebrate parasite *Wollbachia*). But again, several genes are deeply conserved and recur in widely distinct lineages. For instance, the *DMRT1* homologue *mab3* that is critical for sex determination in *Drosophila* and *C. elegans*, is also the primary gene in sex determination in flatworms and waterfleas.

As for vertebrates, ringmasters in one pathway can be performers in another; for instance *tra* is downstream in *Drosophila*, but is the sex-determining trigger in the housefly and a wasp. Also in the honeybee, with a haplodiploid mechanism, sex is determined by homo or heterozygosity of a gene that is a homologue of *tra*. In the mosquito too (with an XY system), the male determining geneis a *tra* homologue, as are other *Drosophila* downstream genes including *dsx*.

10.18 Conclusions

When we examine sex determination right across vertebrates, we see that the choice between male and female development is governed by a fairly conserved, but very complex network of reactions that cause differentiation of a bipotential gonad to either a testis or an ovary. This network is composed of a web of genes (68 discovered so far, and still counting) that include interlocking pathways promoting testis, or ovary, as well as pathways opposing testis, or ovary. These are finely balanced, so that if one component is missing, or present in the wrong dose, sex reversal may occur.

At the head of this pathway is a master (I call it ringmaster) gene that directs the pathway down one way or another by activating and repressing slave genes (I call them performers). The human (mammalian) *SRY* is but one of many genes to take on this role. Its male-dominant action, binding to and activating the conserved male-promoting gene *SOX9*, defines a Y chromosome that is specific to therian mammals.

Birds use another ringmaster gene and a different strategy: *DMRT1* on the Z chromosome has a dosage-dependent effect on *SOX9* and downstream performers. Other reptiles, frogs and fish use both these strategies, and many more, including growth factors, to activate the sex-determining network. Many reptiles and some fish use environmental cues to interact with the same network through epigenetic modifications that affect *DMRT1*.

It is striking that these ringmaster genes, or copies of them, appear all over the vertebrate phylogeny and their ringmaster status cannot be identical by descent; they are simply good at the job and have been selected for independently. Strikingly, most of them are drawn from the ranks of the performers, although some are ring-ins from other pathways that have been co-opted. It is striking, too, that making new sex-determining genes seems to be quite straightforward. Performer genes may be relocated, mutated to set up dosage differences, copied into active or interfering ringmasters, their expression altered by mutating or rearranging upstream enhancers or even epigenetic repression.

Thus evolution has worked within broad themes, using elements from a store of genes with highly conserved functions, and building networks of genes that promote or oppose male or female determination. New genes are being constantly added from without, or being repurposed from within.

This echoes my suggestion of a molecular toolbox available for dosage compensation (Section 7.15). In fact, these two toolboxes share tools, such as the epigenetic regulators that control *SRY* in humans and *DMRT1* in TSD reptiles. Indeed, sex determination and dosage compensation are intimately linked. Dosage differences are exploited to determine sex in birds. In *Drosophila*, the same genes that read the X:autosome ratio to determine sex also trigger an efficient dosage compensation pathway.

Like many other biological systems, sex determination makes no functional sense, but can be readily understood in terms of evolution. Why do lemmings have an X* chromosome that represses *SRY*? X* was an accident that was selected for because it is good for these polar mammals to have more females to shore up the population during the short arctic breeding season. Why do platypuses have 10 sex chromosomes? It works, and translocation happened, and isn't likely to un-happen. Why do cichlid fish have bizarrely convoluted sex genes and chromosomes? Maybe they subdivide species that can all find subtly different ecological niches in Lake Malawi. Why do alligators go for temperature-sensitive sex when this puts them at risk from global warming? It works, maybe as a population regulating device; evolution can't think ahead. Why are some lizard species female only? The were created by a chromosomal accident (hybridization or polyploidization) and parthenogenesis made the best of a bad job.

Indeed, sex determination offers a wonderful view of the ingenuity of evolutionary fixes. That's why I love it.

FURTHER READING

Books

Beukeboom LW, 2014. *The Evolution of Sex Determination*. Oxford Academic
Bull JJ, 1983. *Evolution of Sex Determining Mechanisms*. Benjamin/Cummings

Classic Papers

Foster JW, Graves JAM, 1994. An *SRY*-related sequence on the marsupial X chromosome: implications for the evolution of the mammalian testis-determining gene. *Proceedings of the National Academy of Sciences of the United States of America* 91: 1927–1931
Holleley CE, O'Meally D, Sarre SD, et al., 2014. Sex reversal triggers the rapid evolution of temperature dependent sex. *Nature* 523: 75–82
Matsuda M, Nagahama Y, Shinomiya A, et al., 2002. *DMY* is a Y-specific DM-domain gene required for male development in the medaka fish. *Nature* 417: 559–563
Smith CA, Roeszler KN, Ohnesorg T, et al., 2009. The avian Z-linked gene *DMRT1* is required for male sex determination in the chicken. *Nature* 461: 267–271

Reviews and Research Articles

Bakloushinskaya I, Matveevsky S, 2018. Unusual ways to lose a Y chromosome and survive with changed autosomes: A story of mole voles Ellobius (Mammalia: Rodentia). *OBM Genetics* 2 https://doi.org/10.21926/obm.genet.1803023
Capel B, 2017. Vertebrate sex determination: Evolutionary plasticity of a fundamental switch. *Nature Reviews Genetics* 18: 675–689
Curzon AY, Shira A, Ron M, Serouss E, 2023. Master-key regulators of sex determination in fish and other vertebrates – a review. *International Journal of Molecular Science* 24: 2468; https://doi.org/10.3390/ijms24032468
Gemmell NJ, Todd EV, Goikoetxea A, et al., 2019. Natural sex change in fish. *Current Topics in Developmental Biology* 134: 71–117

Herpin A, Schartl M, 2015. Plasticity of gene-regulatory networks controlling sex determination: Of masters, slaves, usual suspects, newcomers, and usurpators. *EMBO Reports* 16: 1260–1274

Major AT, Smith CA, 2016. Sex reversal in birds. *Sexual Development* 10: 288–300.

Mawaribuchi S, Yoshimoto S, Ohasi S, et al., 2012. Molecular evolution of vertebrate sex determining genes. *Chromosome Research* 20: 139–151

Parma P, Veyrunes F, Pailhoux E, 2016. Sex reversal in non-human placental mammals. *Sexual Development* 10: 326–344

Pennell MW, Man JE, Peichel CL, 2018. Transitions in sex determination and sex chromosomes across vertebrate species. *Molecular Ecology* 27: 3950–3963

Sarre S, Georges A, Quinn A, 2004. The ends of a continuum: Genetic and temperature-dependent sex determination in reptiles. *BioEssays* 26: 639–645

Terao M, Ogawa Y, Takado S, et al., 2022. Turnover of mammal sex chromosomes in the *Sry*-deficient Amami spiny rat is due to male-specific upregulation of *Sox9*. *Proceedings of the National Academy of Sciences USA* 119: e2211574119

Thépot D, 2021. Sex chromosomes and master sex-determining genes in turtles and other reptiles. *Genes* 12: 1822 https://doi.org/10.3390/genes12111822

And Something Really Crazy …

Leclerque S, Thézé J, Amine Chebbi M, et al., 2016. Birth of a W sex chromosome by horizontal transfer of Wollbachial symbiont genome. *Proceedings of the National Academy of Sciences USA* 113: 15036–15041

11

Sex, Genes and the Human Condition

Sex is the most dramatic normal human polymorphism. It is remarkable that so many physical differences – in gonads and gametes, in anatomy and behaviour – are triggered by a single gene that activates the whole network of genes that induce a ridge of cells to become one of two very different organs; a testis or an ovary. And that these two organs make chemical messengers that activate other whole networks of genes in far-flung tissues and organs.

What do these genetic, anatomical and physiological differences mean for humans? Here, I explore the role of sex differences in health, longevity, gender, reproduction and evolution, and the ethical dilemmas thrown up by variation.

We can't escape the fact that men and women are genetically very different creatures. In most mammals, birds, reptiles, males and females have evolved to fill different roles. In human societies through the ages, men and women have been treated very differently (usually to the great advantage of males). What does sex mean for roles in society? In careers?

In humans, as in other mammals, development usually goes down one of two alternative channels toward male and female development. The system is robust to environmental interference, and even to genetic variability; we call this channelling 'canalization'.

However, as for every kind of development, there is variation, genetic and environmental. Many babies are born with variations of anatomy and physiology because one or other step in sex determination is less or more active than usual. There are dilemmas of how to treat, and whether to treat, babies with atypical sexual development. And how to create a 'level playing field' for sport.

Some variants may be behind dilemmas of sexual identity, and the age-old question of mate choice. Here, I will argue that there are strong genetic bases for both transgender and homosexuality, which may both be seen as simply the edges of normal curves of male and female sexual identity, and of mate choice.

And what of the future? Are we heading for a more enlightened world in which variation in sexual development and sexual identity is accepted, even celebrated? A world in which reproduction has been decoupled from sex?

And what of the long-term future? Assuming that the human race can resist its self-destructive tendencies and survive thousands, if not millions of years, do we seriously have to worry about the extinction of the Y chromosome and the emergence of new hominid species?

11.1 Genetic Differences between Men and Women

There are obvious genetic differences between men and women. For a start, men have a Y chromosome and women don't (Chapter 2). But there are very few genes (only 27 protein-coding genes) on the male-specific portion of the Y, and several of these are expressed only in the testis.

Another source of genetic difference between men and women is the number of X chromosomes; women have two copies and men only one. This has the obvious effect that males, but rarely females, suffer from many diseases caused by mutations of genes on the X (Section 2.5). Women have the backup of a normal allele on the other X, but men don't. Although one X is silenced in females, it is a different X in different patches of cells, so heterozygous women are mosaics and usually have enough normal tissue to get by. This is one reason that males die at a higher rate than females at every age.

Although one X is silenced in women (Chapter 4), some 200 of its 1000-odd genes escape inactivation partly or fully, producing dosage differences between men and women. There is now evidence that the number of X chromosomes, independent of sex, has large effects on many metabolic pathways. 'Four Core' strains of mice have been engineered to be sex reversed; XY males can be compared with XX males containing a male-determining gene, and XX females with XY females lacking the male determining gene, as well as XO females. Large differences in metabolic pathways – for instance, those involved in obesity and diabetes – have been found between animals of the same sex but with different numbers of X chromosomes.

In addition to this, the possession of an inactive X chromosome itself is now blamed for a direct hit to women's health. New studies show that the ncRNA made by the *XIST* gene is attacked by misguided antibodies, and contributes to women's higher frequency of autoimmune diseases. What else might it be doing?

In addition to dosage differences, the two X chromosomes of women may carry different alleles, so that heterozygous women may be mosaic for different forms of a protein. This may be a great advantage, especially for the 60-odd immune response genes which confer a wider range of immunity, perhaps part of the reason that women are less susceptible to infectious diseases and live, on average, six years longer than men.

These genetic differences amplify to produce sex differences in many pathways in many tissues. Scanning the RNA profiles of tissues from many men and women detected major sex differences in nearly one-third of our 20,000-odd genes in one tissue or another. These were predominantly in gonadal and sex-specific tissues, but many were in organs such as the heart, liver and even brain. Other studies reveal major sex differences in the expression of genes in muscle, and sex bias in 18% of genes in prenatal brain and 30% in adult brain. A vast analysis of genetic pathways in hundreds of men and women also concluded that there were major sex differences in gene action in all somatic tissues and organs.

Thus, there are extensive genetic differences between men and women that may explain differences in growth, morphology, health and lifespan, as well as differences in gonad differentiation and reproductive role. Most of these are coordinated by a single gene on the Y, though the X chromosome contributes variations in gene dose, and there are complications engendered by an epigenetic silencing system.

11.2 Sex and Health: Women Are Not Just 'Small Men'

We have known for decades – probably millennia – that men and women are differentially vulnerable to many diseases. For instance, men are much more prone to Parkinson's disease and women to Alzheimer's; men are more prone to viral infections and women to autoimmune diseases like multiple sclerosis.

Sex differences in health and disease are not at all surprising, given the genetic differences between men and women, and their amplification of differential action to a third of our genome. The data on gene activity suggest that every tissue and organ in the body is directly or indirectly sexualized by differential gene action, hormone responsiveness and different networks that achieve the same endpoints. Even hearts and kidneys show differences in gene action that may predispose to some conditions differentially.

What is surprising is that sex differences in disease frequency and treatment efficacy were ignored by the medical establishment until quite recently. A landmark report by the US National Institutes of Health in 2001 first brought the data together on a host of diseases and treatments, making the point loud and clear that 'women are not just small men', and trials of new treatments should include women.

Some of the largest discrepancies are in cardiovascular problems. There are major differences in the types of strokes presented by men and women. For instance, most (68%) atherosclerotic strokes occur in men, whereas cardioembolic strokes are more frequent in women. Recovery from stroke takes longer in women.

Treatments may also be more or less effective in men and women. For instance, aspirin is cardioprotective in men but not women, yet it does lower stroke risk in women. It will be important to discover what sex differences underlie these differences in order to target prevention and treatment.

Other discrepancies are less obvious. For instance, there is good evidence that men and women sense pain differently (women are more sensitive and more discriminatory), yet nearly 80% of pain research in humans before 2005 studied only males.

Drug development has followed research, which has mainly involved men, and this means that drug tests tend to be sex biased. The effects of sex in studies of model mammals have also been poorly studied, and several drugs have passed phase II tests, but have later been shown to be ineffective in women.

Thus, there are major differences between men and women in susceptibility to many diseases, and to the efficacy of treatments, and it makes no sense to ignore them. There is now an Office for Research on Women's Health (ORWH) at the NIH that advocates for research and testing to include women, as well as ethnic minorities. As well as ensuring appropriate diagnosis and care for both men and women, studies that compare the response to disease and treatment in men and women are sure to unearth new fundamental data on sex differences.

11.3 Sex and Lifespan: Are Men the Weaker Sex?

Worldwide, the sex ratio at birth favours boys (about 105:100), but women outlive men by six years on average. Men die disproportionately at every stage of life. This has been attributed to environmental factors (dangerous occupations and risk-taking behaviour fuelled by androgens), but also seems to have a biological basis.

The most obvious sex difference that might account for the fragility of men is their lack of a second X chromosome (Figure 11.1). This leaves them vulnerable to the effects of mutations of X-borne genes, such that essentially only males suffer from diseases such as haemophilia and muscular dystrophy. The high concentration of X genes involved in brain development and function also means that they suffer disproportionately from X-linked intellectual disability conditions like Fragile X.

But doesn't X chromosome inactivation level the playing field between men and women? No. Firstly, for rare diseases, there will be few homozygous women. Heterozygous women are mosaics of normal and mutant bearing X chromosomes, and for a recessive condition like haemophilia or muscular dystrophy, patches of normal tissue make enough blood clotting factor or dystrophin to get by.

Also, about 270 of the 1000-plus genes on the human X escape inactivation, and it is likely that the dosage difference plays out to the advantage of females. We know from studies of four-core mice (with XX and XY males, and XX and XY females) that the number of X chromosomes has many effects on basic metabolism independent of sex.

Other major sex differences in health and longevity may also be connected to the differences in X chromosome dosage. For instance, 60 immune response genes lie on the X, and it seems that having two different versions of these genes give XX women a broader spectrum of defences. This may partly explain why women have a stronger immune system than men.

Figure 11.1   My XX-boosting button celebrates the beneficial heterozygosity of XX women.

The possession of a Y chromosome may also act against the best interests of men. For instance, there is growing evidence that the *SRY* gene is expressed in the brain and may have a direct role in Parkinson's disease.

Sex hormones are also highly involved in many health states – and diseases. Androgens in both men and women promote development of bone and muscle. However, they may be indirectly responsible for man's susceptibility to some viral diseases; for instance, androgen binding to a cell surface receptor (AR) is required for initiation of the enzyme that cleaves the spike protein of corona virus and allows it to enter lung cells.

Are sex differences in disease susceptibility and mortality simply the by-catch of genetic and hormone differences between men and women? It is likely that, like many other traits, health and longevity were selected differentially in males and females because of differences in life strategy (Section 11.11).

11.4 Sex on the Brain

Are there sex differences in brain development and function in cognition? This has been a fraught question for decades, and there is a huge literature arguing every which-way.

As I have related, every organ and tissue in the body, including brain, shows sex difference in gene action and gene networks. Recent comparisons of transcriptomes reveal hundreds of genes that are differentially expressed in the brain, starting from the earliest developmental stages. Similar results in monkeys mean they have been entrenched for 70 million years.

Decades of work on brain anatomy, neural networks and behaviour have claimed myriad subtle sex differences at every level. But the whole field of sex differences in brain anatomy and function has become controversial. To be sure, there are problems in assessing differences. Human brains show great variability in almost every aspect, and sample selection and small sample sizes bedevil comparisons between men and women.

Another problem is that the environment – nutrition, trauma, obesity, for example – affects brain anatomy and performance. Genetic, hormonal and societal influences all interact, and may dwarf sex differences.

The third and perhaps the most egregious reason is that it has become politically impossible to measure sex differences and have a sensible debate about what they might mean. At one extreme are deeply ingrained beliefs that sex is paramount in determining your aptitudes and preferences, from toys and colours to careers and seniority. At the other extreme there are many 'myth-debunking' books claiming that there are no sex differences in the brain, 'no more than the kidney or the heart'.

There is little agreement even on the question of structural differences. It is obvious that on average, men have bigger brains, but it is not clear what this means (after all, elephant and whale brains are even bigger). Brain anatomists have gone to great lengths to demonstrate sex differences in distribution of grey matter and more cortical folding in female brains and more axon bundles in the corpus callosum. These anatomical differences have been disputed, although sophisticated modern imaging techniques have been brought to bear on enormous samples and there is some consistency across studies and across cultures. A recent paper claims that 88% of imaged traits show sex differences.

But it is not clear how any of these differences relate to function, although some sex differences do occur in areas of the brain that control motor functions (men the winners) and memory (women the winners), and differences occur in areas known to be damaged in neuropsychiatric conditions.

Sex hormone differences in brain development and function have also been observed when hormone balance is disrupted. Androgen may have an organizing role in development, and seems to promote differential brain growth (as it does differential digit growth such that the ratio of index to ring finger length is lower in men). Oestrogen receptors abound in regions of the brain (hypothalamus, pituitary gland and hippocampus) and oestradiol seems to enhance learning and memory. Again, many of these differences have been disputed.

Behavioural differences between males and females have been charted at every stage. They start early, with baby boys and girls reacting differently to physical or social images. At every stage, boys appear to perform better at spatial manipulation, girls at word and memory tasks. Neural imaging of men and women shows that neural activity patterns may be different even when performing the same task, and that men and women had different responses to empathy and revenge. Boys and men appear to be more aggressive. But again, there is everlasting controversy about the size, and even the existence of sex differences, given the large effects of environment on behaviour.

General intelligence has been compared many times in males and females of every age cohort, with inconsistent results. Distributions largely overlap, although the male distribution is wider with longer tails, accredited to their having a single X, bearing (good or bad) variants of many intelligence genes. There appears to be no difference in average intelligence, but differences in spatial ability and verbal ability are reported fairly consistently.

Personality traits, likewise, have been strongly differentiated. There is some agreement that women display more agreeableness, warmth and openness, males more assertiveness, as our high school cohort observed (Figure 8.1). Women report more fear, PTSD, more intense joy and sadness, guilt and shame than men. Or maybe they are just more likely to admit to it. Again, a strong role of education and environment makes it difficult to nail biological differences, although twin studies have confirmed a large genetic component of personality traits.

There is no denying a large sex difference in the frequency of neurological and psychiatric conditions. For example, autism, ADHD, Tourette syndrome, language impairment and dyslexia are more common in males, whereas depression, anxiety disorder, anorexia are more common in females. But again, the origin in biological differences is disputed because social factors have a large effect.

Although it is hard to draw conclusions from the mass of contradictory data on male and female brain anatomy and function, the transcriptome data are quite clear. There are major differences in gene activity that are established even before gonads have formed. It would be surprising if the genetic differences between men and women were not manifest in major differences in cognition and behaviour. Genetic differences are unlikely to stop at the neck!

11.5 Sex and Sexism: Careers and Roles in Society

Evolutionary psychologists have proposed that biological differences in brain and behaviour have been selected differentially in mammalian males and females. Superior spatial ability, risk-taking and assertiveness of males have been ascribed to their need to disperse and find new niches, as well as compete with other males, kill large game and engage in warfare, whereas stay-at-home females have been selected for superior spatial memory required for foraging, and qualities that aid long-term care for young.

Could differences in brain and behaviour between men and women, selected over aeons, mandate different roles in society today? Specifically, could they – should they – restrict men and women to certain roles and careers?

It is obvious from global data that women are underrepresented in certain occupations; engineering and science are the most often cited.

And women are severely underrepresented in the managerial class. Much has been written about 'the leaky pipeline', starting with equivalent numbers of male and female graduates, and suffering losses of females at every stage of advancement, most obviously at peak child-bearing age. Many of us have spent our lives trying to stem the leaks and encourage more women to enter science and engineering, to stay the course and climb to the top of the tree.

Why should we be limited by our evolutionary past that selected males and females for different traits? After all, humans were selected for many qualities that are no longer relevant; for instance, our predilection for sweet things was selected so that we might choose the ripest and most nutritious fruit. But we temper our sweet tooth in the interests of health and do not generally gorge on candy if we know what's good for us. Evolution has got us where we are, but does not mandate that we trudge in the same traces.

We think nothing of correcting mistakes of nature; children born with extra fingers are routinely offered surgery (rather than being shunned and neglected), and infertile men and women are helped to conceive by any number of medical interventions (Section 11.10). Genetic determinism is being tempered every day by our decisions.

Although there have been great improvements in women's access to male-dominated careers, participation is still far from 50:50. Does this mean that we have far to go to achieve gender equity? Or does it mean that there is an innate preference among women for humanities and biology over engineering and computer science? Curiously, the most sexually accepting countries (Sweden, Norway and Denmark) show an accentuation of sex differences in career; a lower sex ratio in engineering and a lower percentage of female board members.

Thus, sex differences in career choices would not be at all surprising, and I personally will be satisfied when the opportunity to pursue chosen careers is truly equitable and there are plenty of role models, like female engineers and male nurses, to demonstrate that the gates are open. But there is no case for coercing women and men into particular channels just to balance the numbers.

What would be much more useful would be an understanding and embracing of biological differences behind children's choices of educational paths to follow, and a de-sexing of these choices so that engineering is as welcoming to women as men, and nursing and primary education as welcoming to men as women. This requires role models of both sexes, and some dismantling of sexist professional structures, as well as a thousand practical measures to share family responsibilities more equitably.

It seems to me ridiculous to deny that there are differences in male and female brain structure or function, and that these do not affect career choices. We should acknowledge the differences, but, given the huge variation within sexes and the huge overlap in skills and talents between the sexes, we should avoid channelling of men and women into different pathways.

11.6 Variation in Sex Determination

Many babies are born with a genetic variant that leads to atypical sexual development (Section 9.6). This is deeply troubling for well-meaning doctors, and anxious parents who want the best life for their child.

Just how to achieve this best life has caused anxiety in parents, angst in patient groups and antagonism toward the medical profession. Is surgery and/or hormone treatment the best option? Or is it child abuse? What are the consequences of early treatment? Inappropriate treatment? What are the consequences of doing nothing?

If we consider the whole pathway from genes and chromosomes to the expression of sex and sexuality, we can see enormous scope for variation (Figure 11.2). There are at least 68 genes (and still counting) in the network that interprets

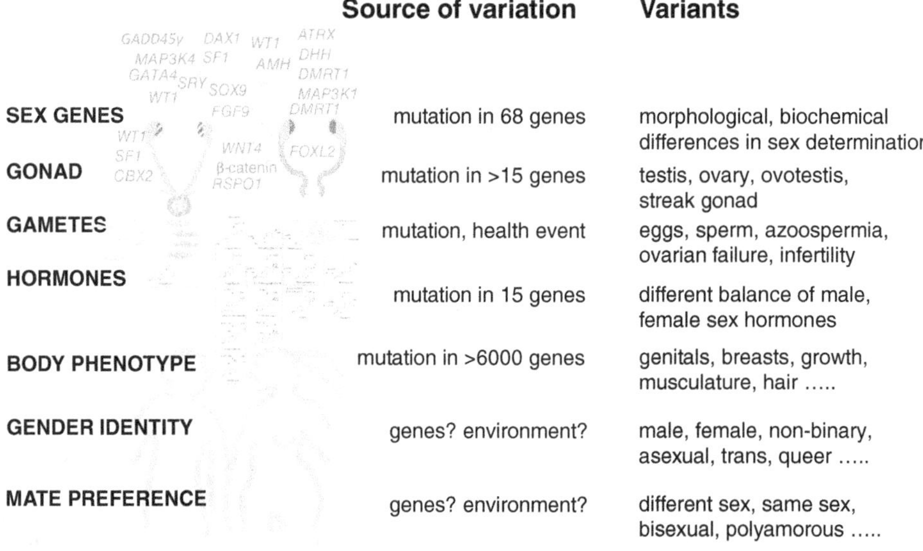

Figure 11.2   Sources of variation in sex and gender. Variation can be at the level of variation in any one of the many genes in the sex determination network, in any one of the interactions of sex hormones, or thousands of target genes, and (with environmental variation) can control variations in anatomy, fertility, behaviour, gender identity and mate choice.

the signal (*SRY* or no *SRY*) that directs development of an ovary or testis in the embryo, and many more genes that make sex hormones and interpret their signals. Hundreds of genes control secondary sexual characteristics like fashioning genitalia and breasts and controlling voice pitch and hair distribution. A mutation in any of these sex determining or sexual differentiation genes may affect sexual development of the embryo.

There is a great range of effects. At one end are quite common (~1/400) and minor anatomical differences like undescended testicles and a urethral opening that is displaced from its usual position at the tip of the penis.

But other, much less common (<1/2,000) mutations lead to atypical gonad differentiation, which can have a wide range of effects. Some mutations may be undetectable at birth and be picked up only when problems at puberty or infertility are investigated years later. For instance, relocation of the *SRY* gene onto the X chromosome permits testis determination and sexual differentiation of the embryo, but the absence of Y-borne fertility genes leads to azoospermia in these 'XX males'. A mutation in the androgen receptor gene ('androgen insensitivity') permits female development of XY babies, but AIS girls don't undergo typical puberty and don't menstruate.

Several mutations lead to 'ambiguous genitalia' in healthy children. Mutations in the sex steroid pathways may provide inadequate hormone signals; for instance, deficient 5α-reductase produces insufficient dihydrotestosterone and leads to female-like genitalia. Babies are born apparently female, or with genitals somewhat masculinized. Many of these children become more masculine when hormones rev up at puberty. This condition is quite common in some populations and has even been accorded status as a third sex.

Mutations in genes that promote or maintain testis development (e.g., *DMRT1*) impede virilization of XY babies. Conversely, a mutation in one of the genes (e.g., *RSPO1*) that promote female development by keeping testis genes at bay may lead to male-like development in XX babies.

Some disorders are less benign. For instance, children with congenital adrenal hypoplasia (CAH) may make too much androgen but too little cortisol and aldosterone because of a block in the biosynthetic pathway. This can cause a fatal loss of salt. At the tragic end of the spectrum are gene mutations that disrupt more than sexual development. For instance, *SOX9* is essential for bone formation as well as testis development, and deletion of one copy produces, as well as sex reversal, bone deformities so severe that babies cannot breathe and usually die within a few hours. This spectrum makes it impossible to lump affected children into a single category.

There has been much discussion in advocacy groups, as well as the medical profession, even on what to call people with these conditions. 'Intersex' describes some patients but not others (for instance, there is nothing 'inter-' about the sexual phenotypes of XY women with AIS and XX men with *RSPO1* mutations). 'Disorders of Sex Development (DSD)' is a term adopted medically but, not surprisingly, affected people quail at labels like 'disorders', 'syndrome' or descriptions of 'mutants' – even 'patients' – that denote a disease and invite discrimination. 'Differences in Sex Development' (also DSD) has been suggested, but there is little general agreement. I shall use 'Atypical'.

Indeed, there is considerable normal genetic variation in sexual development, as there is in all development (Figure 11.2). Perhaps variation is even wider than, for instance, in liver development, because variations in gonad function generally have less impact on health. You can live perfectly well without a functional testis, but not without a functional liver.

The border between normal variation (think short and tall people) and abnormal variation (think dwarfism and gigantism) is often fuzzy. Do some variations in gonads or genitalia lie beyond the boundary of normal variation? Does it matter as long as the child is healthy?

It does matter from the point of diagnosis. Knowledge of what step in the pathway is blocked is essential for predicting the developmental trajectory. Some conditions, such as salt-losing CAH are life-threatening if not recognized and treated early. Other conditions will require treatment, but there is no great hurry – for instance XY girls with androgen insensitivity may develop a slow-growing cancer (gonadoblastoma) of the internal testes. There are now many molecular methods for obtaining rapid and accurate diagnoses of more than 60 conditions, allowing doctors to provide sensible advice to parents.

For many conditions the children are perfectly healthy but show varying degrees of atypical genital development. Some can be surgically modified; for instance, the opening of a penis can be moved to its typical position and undescended testes lowered in boys, and a vagina can be fashioned for XY girls with androgen insensitivity.

More controversial has been surgery to assign 'babies of uncertain sex' to 'boy' or 'girl' according to their chromosomal or genital sex, and to reinforce this choice with hormone treatment. For some decades, surgeons advocated operating early for a better medical outcome, to give the child a consistent gender identity. This was often deemed successful, and many XY sex reversed patients reported living satisfying lives as women.

However, early surgery raises insoluble problems of informed consent, because the parents must make a life-changing decision on the child's behalf. Which sex? And what if you get it wrong so that the child grows up hating who they are? For some people with atypical sex development, sex assignment surgery as babies was a disaster. They tell heart-rending accounts of never feeling comfortable or accepted in their gender role (usually female), particularly if they didn't know they had been born with a Y chromosome.

Some patient advocate groups aver that it is more appropriate to do nothing. Why not wait till the child can participate in decisions about whether or not to opt for surgery, and which sex to become?

But this is where there are really no universal answers. In a perfect world, it shouldn't matter if a child is a boy or girl or a happy little intersex. But parents fear that their child will be teased and bullied because of their atypical anatomy. And sadly, their fears are not misplaced; children can be very cruel to those who are even trivially different. Adults are sometimes no more accepting; in some countries, families with children with atypical sexual development are ostracized and the children may be neglected or even killed.

Why should sex be so important anyway? Can't we de-sex classrooms and avoid gender stereotypes in the media? Can't we accept, even celebrate, diversity? Many of us despair that the media – and society generally – seems to be getting ever more sexualized.

11.7 Sex and Sport

Sex differences in body size and shape, lean muscle mass and physiology are undisputed, although all these have wide ranges of variation and there is considerable overlap for most traits. There are, for instance, plenty of short men and tall women. Venus or Serina Williams would defeat most men on earth at tennis.

However, elite performance must favour men for many or most sports, largely as the result of androgen effect on bone growth, heart size, muscle development and function. It has therefore been almost universal to separate men and women in athletic competition.

There have been men who have taken advantage of this and posed as women for competitions, historically and right up to the modern Olympics. For this reason, sex tests were introduced; initially physical examination, then genetic tests. The first genetic tests consisted of a smear of cheek epithelial cells, which was examined for sex chromatin bodies (Figure 5.3) under the assumption that all XX women would have an inactive X chromosome. Alternatively, a chromosome test could be performed to eliminate the presence of a Y chromosome.

There were problems with this testing regime, with false positives and false negatives. However, a more fundamental problem is that the sex chromatin test misclassifies as men women with only a single X chromosome; XO women with Turners syndrome, XY women with androgen insensitivity or disorders of androgen metabolism. Also, such a test would not eliminate XX*SRY*$^+$ men with an *SRY* gene transferred to the X chromosome.

Testing directly for *SRY* using polymerase chain reaction was suggested, and has recently been adopted by World Athletics. This would identify XX*SRY*$^+$ men, but still fails to correctly classify athletes with atypical sex development, including AIS women and XY women with an inactive *SRY* gene.

All of these tests were deemed invasive and embarrassing to athletes. Failures of successful athletes were very public and there was at least one suicide of a female athlete who was banned. Physical or genetic tests invited medical and surgical interventions, including genital and gonadal surgery that was unwarranted from a health perspective. The UN regarded this as an infringement of human rights.

Androgen treatment has spectacular effects on the performance of female athletes, and anti-doping rules are in place. If androgen levels are the most closely correlated with athletic performance, why not just measure androgen levels? This does not solve anything, since androgen levels vary widely among both male and female populations and there is considerable overlap (13% of women overlap with 16% of men). Androgen testing was introduced in 2009 in response to complaints about a female (by any measure) runner with extremely high androgen levels (hyperandrogenism). A limit of 5nm/L was imposed in 2018. However, setting limits is arbitrary, and encourages use of potentially harmful anti-androgen drugs.

Complicating sex testing further, there may be non-androgen mediated male advantage. Curiously, the frequency of complete androgen insensitivity syndrome AIS is much higher (1/429) among elite women athletes than it is in the general population (1/20,000–50,000), suggesting that although AIS girls have no androgen advantage, activity of *SRY* or other Y genes may confer direct benefit to tissues or organs.

Reassignment of sex (Section 11.8) poses new problems with transgender athletes. Transwomen whose gender reassignment was after puberty would certainly have had growth and muscle advantages; larger and stronger bones, larger heart and lung capacity. However, competitive sport has been a lifesaver for many transitioning athletes, and until recently, inclusivity has won over strict fairness. The ruling was that transwomen could compete as women as long as their androgen levels stayed less than 10nm/L for the last year (compared with the typical 1.7 nM for cis-women). There are now many transgender females competing at all levels in many sports, although there have been several challenges.

How far should we go in pursuit of a level playing field? Given the tremendous variability in physique and hormone levels among XX women, further complicated by atypical sexual development of XX women with masculinizing mutations and XY women with testes and elevated hormone levels, there seems to be no really fair way of constructing a level playing field in women's sport.

So what do we do? Removing coercion and the incentive for harmful interventions would seem to be the most important factor. Perhaps the stakes are too high and we would do better to put less emphasis on international competition and performers who are, whatever their androgen levels, at the extreme edge of distributions of all sorts of traits: height or muscle mass or hormone levels – they are certainly not average.

Perhaps we should handicap athletes as we do racehorses? Or should we go for categories of competition, as is done for children's sport and the ParaOlympics? For instance, we could hold separate competitions for women with shorter or longer legs, or more or less androgen. I await my chance to represent Australia in the female over-80 medium-leg low-androgen Olympic netball team.

11.8 Sex and Gender

You have maybe noticed that the word 'gender' has hardly got a mention so far in this book. That is because whereas sex is a biological description, gender is a social construct, and I have stuck to biology. But perhaps gender, as well as sex, has a biological basis?

The distinction between biological sex and socially assigned gender doesn't matter much to a fish or a bird, but for humans it has been of great interest for millennia, and is now a mainstream concern. History is replete with legends of women posing as men (opera is full of women singing the part of men disguised as women). And there are many stories, ancient and modern, of men masquerading as women and discovering their feminine side.

But it is very evident that such gender-swapping has a much more profound aspect, in that significant numbers (1/200) of biologically XY men are convinced they should have been born woman, and significant numbers of biologically XX women (1/400) believe themselves to be men. This apparent mismatch of biological sex and gender identity can lead to severe gender dysphoria. Coupled with school bullying and family rejection, it can make lives a torment for young people, and the rate of suicide is frighteningly high.

Feelings of being born in the wrong body often manifest early in childhood. As they move into adulthood, nearly half of these children continue to feel strongly that they were born in the wrong body. Many seek treatment – hormones and surgery – to transition into the sex with which they identify. Genital surgery is reasonably straightforward for fashioning female genitalia and even a vagina in transwomen; breast removal and fashioning male genitalia for transmen is more challenging. Hormone treatment; androgens for female-to-male and oestrogens for male-to-female transitioning is essential to maintain the gender of choice. As yet, there is no easy route to conferring fertility to transgender females or males, although there is much promising research on preserving and remodelling germ cells (Section 11.10).

Transitioning is more effective if done before puberty. That poses problems, because about half the children who had expressed strong feelings of being the wrong gender change their mind by this time. And with the topicality of transgender and some strong role models among entertainers, curious teens see 'what sex would I *like* to be' as a legitimate question. There are concerns that young people might make life changing decisions that they later regret. Although male to female and female to male transitions are now much more available and accepted, the road to transition is still fraught with uncertainty and is hard to reverse.

So, is there a biological basis for transgender? Mismatch between biological sex and gender identity, culminating in its severest form as gender dysphoria, was originally ascribed to childhood trauma, family dysfunction or mental disease. A large component of family and social expectations can be seen, for example in Samoa, where a youngest son could be expected to play the role of a female carer, defining a third sex fa'afafine (literally 'in the manner of a woman').

Many attempts have been made to sheet such gender mismatch home to prenatal hormone influence, particularly a low level of androgen during pregnancy that might affect brain development.

However, another possibility is that there might be a genetic predisposition toward transgender. There is some evidence that identical twins are somewhat more likely to be concordant for transgender than fraternal twins. This idea was explored early in Melbourne by a psychiatrist who was revered in the transgender community for his willingness, in the 1990s, to authorize sex change surgery. Dr. Herbert Bower tried to persuade me to look for variants in the sex gene (*ZFY* at the time) in his transsexual patients. This did not seem to be a fruitful area for research, since his patients had normal testes, but the idea has recently been pursued by looking for variants of other candidate genes.

Several studies have sought an association between transgender and particular gene variants. Researchers have concentrated on likely candidate genes, focusing on the sex hormone pathways. They sampled hundreds of transwomen who had, or planned sex change operations, finding that transwomen had a high frequency of particular DNA variants of four genes that would alter sex hormone signalling during development in utero. It will be interesting to look for variants of other genes in unbiased whole genome association studies, which may reveal sightings for many – perhaps hundreds – of genes that affect gender identity.

Genes that affect sexual identity need not be on the Y chromosome, so they will not necessarily be in synch with having a Y chromosome or an *SRY* gene. Thus, gender identity may not accord with biological sex. This would mean that among both sexes there would be a spread of more or less masculine, or more or less feminine identities. Among men you would expect to see a range of identities from strongly masculine to more feminine – and at the extreme to transwomen. And among women you would expect to see very feminine women at one end of the range and transmen at the other.

Thus, there is no doubt that gender identity is a powerful trait, and transgender is relatively common and can be a serious factor in health and human happiness. There is increasing evidence for a biological basis and a strong genetic component.

11.9 Gay Genes and Mate Choice

Does sexual behaviour have a genetic component? Genetic techniques such as twin studies were used over decades to search for a genetic component of male homosexuality. This was hard to do because homosexuality was so stigmatized that many gay men preferred not to be open about their preference. It was even harder because many studies showed that shared genes could be only part of the story; hormones, birth order and environment played a role too.

In 1993, Dean Hamer found families that contained several openly gay males on the mother's side, suggesting a gene on the X chromosome (Figure 11.3). He studied families in which two brothers were openly gay and might therefore be expected to share a gay genetic variant on the X. Brothers inherited different combinations of genes on their mother's two X chromosomes. If they share a gay gene variant, they would probably also share bits of the X on either side of the gay gene.

Hamer screened their DNA for genetic markers on the X and found a piece of the X chromosome that was identical between pairs of gay brothers, and was also present in the genome of some of their gay uncles. He reasoned that this small piece near the bottom of the X chromosome must contain a gene that influences sexuality, and proposed that a variant of this gene predisposes a man to homosexuality.

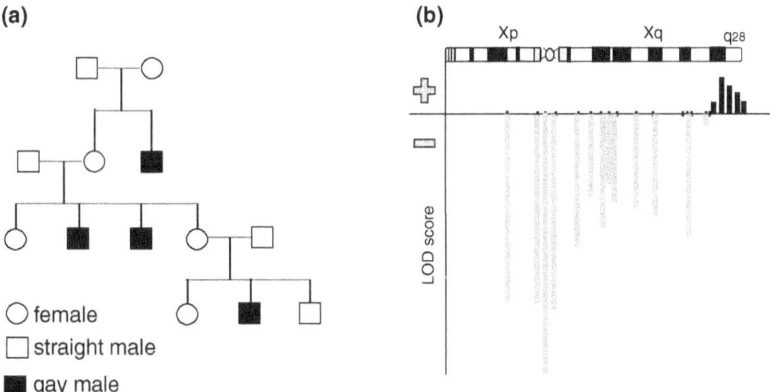

Figure 11.3   Identification of a 'gay gene'. (a) Families were chosen that contained two openly gay brothers. For instance, in this family two gay brothers had a gay uncle and a gay nephew, all on the maternal side, suggesting X-linked inheritance. (b) The degree to which related gay men shared markers (black bars) covering the X chromosome (banded X shown) can be measured as a LOD score. This was negative except for markers at the terminus of the long arm at Xq28, suggesting a gene here influences mate choice (after Hamer 1993, *Science* 261, 321).

The publication of this work caused a furore. Hamer was challenged at every turn by people unwilling to accept that homosexuality was at least partly genetic, rather than a 'lifestyle choice'. Religious conservatives argued that it couldn't be genetic because 'homosexuality is a sin, and God could not create a sinner'.

Gay men were divided on whether it was or was not helpful to know about gay genes. It was helpful in the sense that it vindicated the oft-repeated claims that 'I was born this way.' But they worried that it might present the means to identify gay gene variants even in embryos, and discriminate against their bearers.

Other studies with a similar design gave contradictory results, but later searches of the whole genome found associations with genes on three different chromosomes. A much larger study of gay brothers, using the many genetic markers available through the Human Genome Project, confirmed the original finding, and also identified another genetic variant on chromosome 8 that predisposed to male homosexuality. Further, increasingly enormous studies suggest that there may be hundreds of genes, scattered all around the genome, that affect sexual choice in humans.

Why was there such a furore when we well know that there are mate choice gene variants in other species, from flies to mammals? For instance, in the fruit fly, there is a single mutation that makes males court other males instead of females. Homosexuality is quite common in the animal kingdom, especially among primates and marine mammals, and there are variants that influence mating preference in mice. Indeed, mating preference is one of the most heavily selected traits in all animals, responsible for one of the most potent evolutionary drivers, sexual selection.

Thus, the evidence for a biological basis for homosexuality is strong, and there is increasing evidence that many different genes affect mate choice.

11.10 Why Are Transgender and Homosexuality So Common?

That transgender and homosexuality are determined, at least in part, by genetic variants is not surprising. What is surprising is that, although these conditions confer low fertility (thus far, transwomen and transmen are infertile, as are same sex unions) they are very common in every society in the world. You might expect that transgender and gay genes would decrease in frequency and eventually fizzle out, but they haven't.

I argue here that gene variants that predispose to transgender identity, or homosexuality, have been selected for in the opposite sex. Thus, I see both as classic examples of sexual antagonism.

I submit that 'gay genes' should more correctly be thought of as 'male-loving genes' (or rather, allelic variants). Likewise, there are sure to be 'female-loving genes' in lesbian women. Boys and girls inherit a mixture of these variants, giving rise to a continuous distribution of mating preference in both sexes. Their distributions overlap, just as for human height (there are short men and tall women). Gay men and women are simply the two ends of this distribution of normal variants.

It is easy to see that a 'male loving' allele in a female could increase her genetic fitness. A woman who bears alleles that confer a strong attraction to a male sexual partner is likely to mate earlier and have more children. If this were so, we might expect that female relatives of gay men, who would share this allele, have more children. If their sisters and mother and aunts have more kids, it would make up for the fewer children of gay males.

They do. A study from an Italian group showed that the female relatives of gay men have 1.3 times as many children as the female relatives of straight men. This is a huge selective advantage that 'male loving' alleles confer on women, and offsets the selective disadvantage that it confers on men. This suggests that homosexuality, like many other traits, is subject to sexually antagonistic selection. I am surprised that this work is not better known, and its explanatory power is neglected in the whole debate about whether homosexual behaviour is a normal genetic variant.

We have no idea whether they are alleles of the same gene, or of hundreds of different genes. It is interesting that Hamer found the original 'gay gene' on the X, because this chromosome has more than its fair share of genes that affect reproduction and behaviour ('brains-and-balls genes'). However, I would expect that many genes all over the genome contribute to mate choice in humans, as suggested by recent whole genome studies.

I would also predict that we will find 'female loving' alleles of many genes in lesbian women that increase genetic fitness in males, who will mate earlier and produce more children. I am waiting for someone to look at the numbers of children fathered by male relatives of lesbian women.

If there are male-loving and female-loving alleles of tens or hundreds of genes battling it out in the population, people will inherit a mixture of different variants. Combined with environmental influences, it will be hard to detect individual genes. This should offer some reassurance to people concerned about genetic discrimination.

A similar argument may be made for alleles of one or many genes that influence gender identity. I suggest genes that influence gender identity are positively selected in the other sex. Feminine women and masculine men may partner earlier and have more kids, to whom they pass on their gender identity gene variants. Looking at whether the female relatives of transwomen, and the male relatives of transmen, have more children than average, would test this hypothesis.

These variants of gender identity and behaviour may therefore be considered examples of what we call 'sexual antagonism', in which a gene variant has different selective values in men and women. It makes for the amazing variety of human sexual behaviours that society is beginning, if reluctantly, to recognize as normal variants.

11.11 Sex and Human Reproduction

The whole point of human sex genes and sex chromosomes is to ensure the continuation of our species. Genetic input from two parents and a mechanism to recombine bits of the two genomes provide humans with enormous genetic variability to survive in many environments and cope with change.

The necessity for a male and a female parent is reinforced by genome imprinting in therian mammals, including humans. Just providing the fertilized egg with two sets of chromosomes is not enough. There are at least 30 genes that are active in the embryo only if they come via an egg, and a similar number that are active only if they come via a sperm. Genomic imprinting rules out parthenogenesis, which occurs in some female-only species of lizards and some sharks, as well as occasionally in captive boa constrictors and Komodo dragons that have no other options. But 'virgin birth' is excluded in mammals (including humans).

To what extent can modern reproductive medicine subvert this sexual system, and does it matter?

Humans are inefficient reproducers compared to other mammals. One in four or five couples all over the world have trouble conceiving, and one in ten require medical assistance to reproduce. Fully 10% of women have trouble conceiving or carrying a pregnancy; their failure to make functional eggs may be the result of age-related decline in oocyte production, accident or disease. About 7% of men have semen with no or few or abnormal sperm as the result of blockage, disease or accident, and 2% have Y chromosome deletions (Figure 11.2).

The development of in vitro fertilization (IVF) in 1978 was a breakthrough for infertile couples. Eggs can be flushed from the ovary after hormone-induced ovulation, and fertilized with the partner's sperm in a Petri dish. Donor eggs can be used, or donor sperm. One high-quality embryo is inserted into the uterus, and spare embryos may be frozen for later use or donation. The extension to single women and same-sex couples has swelled the ranks of those seeking IVF. Once considered exotic, IVF is now responsible for about 2% of all births in the USA, 4% in Australia, and more than eight million births worldwide.

Variations of IVF include isolating single sperm (which can be selected for normal morphology) from an oligospermic man and injecting it directly into the egg cytoplasm (intracytoplasmic sperm injection, ICSI).

Another variation is to remove a single-cell from developing embryos and screen for genetic diseases, then implant only genetically healthy embryos. This technique is used mainly where one parent carries a damaging gene variant or rearranged chromosome that would cause devastating illness in the child.

Do these artificial reproductive techniques affect the genetic make-up of human populations? IVF by itself may have little effect on the genetic variants that make it into the next generation (except for the over-representation of the genomes of some unethical doctors), and embryo selection might lower the incidence of severe genetic abnormalities. However, sons conceived by ICSI will perpetuate Y chromosome mutations that cause male infertility.

The most obvious genetic skewing aided by artificial reproductive technology is sex selection, which is practiced (legally or illegally) in many countries. Sex selection, either in the testing of embryos and implanting only the desired sex, or in sexing the developing foetus and aborting the unwanted sex, seriously distorts the sex ratio, usually in favour of boys. For instance, the sex ratio at birth increased to 120 boys:100 girls in China, resulting in a million excess young men and threatening severe social problems. However, sex ratio in many parts of Asia and north Africa was skewed for decades, if not centuries, by infanticide and neglect of baby girls; it is estimated that there are 130 million 'missing women' worldwide.

Artificial reproductive techniques are widely accepted in many countries. But there is concern about what new understanding and new techniques may bring. For instance, CRISPR technology has made gene manipulation a real possibility, and the prospect of gene therapy on eggs or gametes is on the horizon. This technology could potentially be employed for correcting deleterious mutations, or replacing deleted genes in a child or an embryo carrying a mutation. However, there is concern that it might be used to engineer trivial traits that could enhance appearance or boost performance or intelligence, or even to introduce genes from another species. Though fanciful, these are now possible and must be discussed before someone does something outrageous.

Cloning is another technology that is in hand for our model mammals (remember Dolly the cloned sheep?) and could potentially be used to replicate (or 'enhance') the genome of a donor. Transferring the nucleus of a somatic cell into a human egg has a low success rate, but inducing pluripotency offers a new path to cloning. To date, induced pluripotent stem cells derived from a human donor have proved to be genetically and epigenetically unstable, but the field is developing at a great pace and there is enormous motivation to produce stable human iPSC for regenerative medicine.

Manipulation of gametes is also developing apace. Reproductive tissue can be transplanted from a donor to restore fertility (though not genetic continuity) to a recipient.

Developing gametes from human induced pluripotent stem cells is the goal of several studies. As well as providing a new window into germ cell development this could enable infertile patients to reproduce. Perhaps it will also be possible to reprogram XX primordial germ cells into sperm or XY into eggs to allow same-sex or trans couples to bear children with the genes of both partners. Perhaps it will soon be possible to subvert the imprinting system to allow an XX embryo with two female pronuclei to develop. Disrupting one key gene in mice has been shown to derail the imprinting process, leading to viable embryos with two mothers.

Thus reproductive technology continues to offer more ways to surmount reproductive problems, as well as means to avoid or repair genetic mutation. But inevitably, new knowledge brings some scary new prospects for genetic mischief and abuse.

11.12 Sex Chromosomes and Human Evolution

Sex-specific selection has had profound effects on our species by driving the inexorable degradation of male-specific Y chromosome, and the unique gene content and activity of the X chromosome that was shaped by its different representation in men and women.

Ohno's 'perils of hemizygosity are manifest as a higher risk of genetic disease in males because a mutation of an X-borne gene is immediately exposed. In the same way, beneficial mutations of X genes will be immediately exposed to selection in males but not females. This may explain the disproportionate involvement (at least 5-fold) of X genes involved in male reproduction.

Does it also explain the disproportionate involvement (something like 7-fold) of X genes in brain development and function? It is not obvious why intelligence might be a male-advantage trait, and an intriguing alternative hypothesis is that human intelligence resulted from sexual selection of X genes that control brain development and function.

Recessive alleles of genes on the X that build a bigger and more complex brain would first show up in men because they have only one X. They would confer a direct benefit in hunting success, but could also be selected by females who admire a mate who not only brings home the mastodon but also has time and wit to paint on the cave walls. Since these advantageous alleles would be inherited also by females, this set up a runaway selection for a larger and more complex brain. The speed of sexual selection could explain the extraordinarily rapid increase in brain size in early hominids.

At an even deeper genetic level, it can be argued that the course of sex chromosome evolution played a part in driving the major divergences of mammal evolution (Section 4.15, Figure 4.9). Sex chromosome change has been uncommon in mammals, and it seems significant that the major changes in mammal sex chromosomes all coincide with major divergence.

The initial event was the creation of the *SRY* gene by repurposing a conserved *SOX* gene (Section 10.4). In all therian mammals (eutherians and marsupials) *SRY* defines a homologous Y chromosome. However, monotremes lack *SRY* and have multiple sex chromosomes with homology to the bird ZW system. This dates the evolution of the therian mammal XY system to between their divergence from monotremes 166 Mya, and the divergence of placentals from marsupials 148 Mya.

The emergence of new sex chromosomes around 150 Mya must have caused great upheaval in the world of early mammals. The first males with an *SRY*-based sex-determining system would have had to mate with females who retained an ancestral system. We do not know what this system was (oh for fossilized chromosomes!). Hybrids might then have waged an internal war between male- and female-dominant genes, or between male-dominant and dosage systems.

The upshot of these battles was likely to have been intersexuality and infertility. This would have posed reproductive barriers between populations with the original system and the new *SRY* system which were sufficiently severe to drive divergence between monotremes and therians.

Thus, about 150MY ago, therians adopted the new *SRY* system whereas monotremes embarked on four rounds of sex chromosome-autosome fusions that produced the bizarre multiple XY chromosomes. The two extant monotreme families differ in the last of these fusion rounds, suggesting that the last rearrangement might have also driven this later divergence.

In therian mammals, a second round of major sex chromosome change came with the fusion of an autosome region to the XY pair. Since this region is fused to the XY in all placental mammals but remains autosomal in marsupials, fusion must have occurred between the marsupial-placental mammal divergence 148 Mya and the placental radiation 105 Mya.

Could this event have driven divergence between marsupials and placental mammals? Rearrangement of sex chromosomes with autosomes causes severe infertility because half the gametes of a hybrid will have too many, or too few copies of the fused regions. Many observations and experiments in humans and mice attest to the infertility caused by sex chromosome-autosome fusions and translocations, so rearrangement could well have driven divergence of marsupials and placental mammals.

The idea that chromosome change could drive speciation was popular fifty years ago. But it was dismissed by evolutionary geneticists in favour of the idea that speciation must occur in populations already separated by a physical barrier (like a river or mountains) or behaviour (such as mating time). Small mutations would accumulate slowly and the two populations would be selected for different traits by different environments. Eventually they would become so different that they could no longer mate with each other and would form two species. This 'allopatric speciation' relied on external factors.

The alternative view, that 'sympatric' speciation can happen within a population because of intrinsic genome changes is now making a comeback. We now have several examples of closely related fish species with different sex genes, or differing by a sex chromosome-autosome rearrangement. Mammals may provide a similar example.

Thus, differential selection for traits advantageous in one or other sex may explain many of the sex differences we see in the human population. But the bizarre rules of sex chromosome evolution may be responsible for enhanced intelligence of the 'wise ape', and may have even driven the divergence of monotreme, marsupial and placental mammals, which gave apes their start.

11.13 Sex Chromosomes and the Future of Men

And what of the distant future of human sex chromosomes, and of the human race?

We know that the mammalian Y chromosome has degraded over its 166MY lifetime. We can surmise that it started with about 1000 genes that remain in its erstwhile partner, the X chromosome, and degradation has left it with only 51, including the genes in the pseudoautosomal region. Therefore, about 950 genes must have been lost over 166 million years, a rate of about six per million years. If we assume a linear rate of loss, we can predict that the last 51 will be all gone in about eight million years.

Of course, the loss of genes from the Y is unlikely to be linear. The evidence from flies is that loss occurs very rapidly at first, then slows down. The 27 male-specific genes on the human Y are functional – some dosage-sensitive genes balance non-inactivated genes on the X, and others have critical male-specific functions in spermatogenesis. So perhaps their loss would be strongly selected against, protecting the last few genes. Indeed, comparison of the Y between humans and other primates shows little mutation and loss from the human Y over the last five million years, compared with higher loss in chimpanzees.

But this protection cannot last forever. The last stages of Y evolution can leave the Y very unstable, loaded with repetitive sequences that can undergo deletion and rearrangement. The human Y is very prone to deletions (the most common reason for genetic infertility in men). The rodent Y seems to be even more unstable, and there are many variant sex-determining systems that appear to be attempts to replace a poorly functioning Y.

Indeed, in two rodent lineages, the Y has entirely disappeared (Section 10.2), showing that, unlikely or not, loss of the Y can and does occur. *SRY* is absent, and other useful Y genes have been relocated to autosomes. The spiny rat Y survives in one species, but is terminally disabled, amplifying a mutated *SRY* gene hundreds of times in a last ditch attempt to retain sex-determining function.

What does this mean for our own lineage? The primate Y seems to be unusually stable. One explanation that has been advanced is that the Y chromosome, although it does not recombine with the X over most of its length, can undergo gene conversion within the immense palindromic loops that contain amplified copies of spermatogenesis genes. A gene copy that is mutated may thereby be replaced by an active copy. However, it is hard to see why conversion between a mutant and wild type gene would favour mutants reverting; the reciprocal should be equally as frequent, giving rise to a chromosomal double or quits. Thus conversion would be expected to change the kinetics of loss, but not its rate or eventual outcome.

So, what of the future of the human Y chromosome? Observations on a range of animals lead to the conclusion that loss of the Y chromosome is inevitable, but do not say whether this will occur in thousands or millions of years. Perhaps it has already occurred in some small, isolated human populations. How would we know without examining their chromosomes or screening their genomes?

And what will happen when the human Y is lost? One scenario beloved of some groups is that humans could become an all-female species, like some parthenogenetic lizards. However, genome imprinting blocks such a process (Section 11.10). Humans and other mammals cannot make viable embryos without a male pronucleus because at least 30 imprinted genes are active only if they come through sperm.

So the choice, eight million years hence, is between extinction (unless we have already extincted ourselves by other means long since), or evolution of a new sex gene defining new sex chromosomes. Humans may go the way of the spiny rat, evolving a novel sex-determining gene that defines new sex chromosomes.

We have seen that evolution of novel sex chromosomes is surprisingly easy (Section 10.12); a simple null mutation is all that is required to set up a dosage system in birds, transposition of a gene copy enough to define a new Y in Medaka fish, and a simple change in tissue of expression enough to set the mammals on their *SRY*-directed course.

However, the process of transitioning to a new sex gene is likely to be very fraught. Introduction of a novel sex gene will set up incompatible matings, half of whose offspring is infertile.

So even if humans escape Y-degradation-driven extinction, we might have to worry about hominid speciation. As discussed above, sex chromosome turnover can pose reproductive barriers that drive speciation. Just this has happened to spiny rats, which diverged into three species. In the mole vole, too, three species are separated by different sex chromosome constitutions; XY males/XX females, XO males/XO females and XX males/XX females.

Thus anyone who returns to earth in seven million years is likely to find either no humans – or several hominid species, which might challenge our attitude to sex chromosome diversity (Figure 11.4). I'm not sure which is the scarier prospect.

Figure 11.4   Human sex chromosome and anatomical diversity predicted by toilet signs spotted at the Tokyo Medical and Dental School.

11.14 Conclusions: The Future of Sex

We must accept that there are profound genetic differences between men and women. Sex differences in genes on the X as well as the Y chromosome are amplified as different suites of genes are turned on and off in downstream pathways or as the result of hormone action, so that almost a third of our 20,000-odd genes are differentially expressed in men and women.

What these differences mean in terms of morphological, functional and behavioural differences is less clear. There are huge overlaps in every measure, even where averages are distinctly different. There are tall women and short men.

Genetic differences are expressed not only in the developing gonad but also in all tissues and organs. These differences have immediate effects on the different health outcomes of men and women, and it is gratifying to see that different vulnerabilities to diseases and efficacy of treatment are now taken into account in research on diseases and testing of treatments.

There are also genetic differences in brain anatomy and function. It would be remarkable if these did not lead to differences in behaviour and perhaps even to career preferences.

These differences have been selected in mammals for aeons, and we need to acknowledge them. However, as for many other qualities, we need not be limited by our evolution. Many technical advances remove physical barriers for women, for instance in operating heavy equipment, and many medical advances, for instance, in offering a life beyond bearing and raising multiple offspring, provide new opportunities for different career choices.

Indeed, the overlap between men and women in all measurable differences in morphology, physiology, behaviour and preferences means that it makes no sense to channel men and women into different career trajectories. Although providing opportunity – and role models – for entry into traditionally male and female careers has still far to go, the endpoint is not necessarily a 50-50 sex ratio in every field.

There is enormous variation in sexually dimorphic characteristics (Figure 11.2). Atypical sexual development is relatively common. And although atypical gonad development, unlike liver or kidney abnormalities, is medically benign, the effects of variations in anatomy can be distressing way beyond their medical severity.

So, too, are variations in sexual identity and behaviour, which I argue here have a strong genetic component, and may have been selected by enhanced fertility in the opposite sex. It makes for the amazing variety of human sexual behaviours that we are beginning to recognize as normal variants. The problems faced by transgender and homosexual men and women are largely social, and so must be the remedy.

Why is sex deemed to be such a crucial part of an individual anyway? Most of the problems raised by sex differences and variants are social. Sex is the first thing that people recognize when meeting somebody new, so perhaps recognition of sex is something hardwired into us all. Is this why attempts to desex schools and toys have been less than successful? Is this why the media and entertainment industry is more and more obsessed with sex?

What role will sex play in our evolutionary future? Is it a problem that a large fraction of the human population requires medical intervention to reproduce? Is the make-up of the human population likely to be altered by artificial reproduction and genetic manipulation? We have seen how altering the sex ratios is technically simple and has been taken up with enthusiasm in many countries, which now face significant social problems as the result of a surfeit of boys; could other manipulations, now seen as exotic and unlikely, become mainstream, and is this a problem?

Or are we ultimately at the mercy of the crazy rule-defying sex chromosomes, which could potentially cause human extinction or speciation?

FURTHER READING

Books

Davy Z, 2021. *Sex, Gender and Self-Determination. Policy Developments in Law, Health and Pedagogical Contexts*. QBD Books

Neigh G, Mitzelfelt M (eds), 2016. *Sex Differences in Physiology*. Elsevier, Academic Press.

O'Keefe T, 2021. Suicide in intersex, trans and other sex and/or gender diverse groups. Australian Health and Education centre (O'Keefe & Fox Industries)

Prum RO, 2023. *Performance all the Way Down: Genes, Development, and Sexual Difference*. University of Chicago Press.

Wizeman T, Pardue M-L (eds), 2001. *Exploring the Biological Contributions to Human Health – Does Sex Matter? National Academies Press (US)*, Washington (DC).

Classic Works

Aitken RJ, Graves JAM, 2002. Human spermatozoa: The future of sex. *Nature* 415: 963

Hamer DH, Hu S, Magnuson VL, Pattatucci AM, 1993. A linkage between DNA markers on the X chromosome and male sexual orientation. *Science* 261: 321–327

Reviews and Research Articles

Berman S, 2017. Androgens and athletic performance of elite female athletes. *Current Opinion in Endocrinology, Diabetes and Obesity* 24: 246–251

Benoit-Pilven C, Asterlijoko JV, Leinonen JT, Karjalainen J, Daly MJ, Tukianen T, 2025. Early establishment and life course stability of sex biases in the human brain transcriptome, *Cell Genomics* https://doi.org/10.1016/j.xgen.2025.100890

Camperio-Ciani A, Corna F, Capiluppi C, 2004. Evidence for maternally inherited factors favouring male homosexuality and promoting female fecundity. *Proceedings of the Royal Society B: Biological Sciences* 271: 2217–2221

Clayton JA, Collins FS, 2014. Policy: NIH to balance sex in cell and animal studies. *Nature* 509: 282–283

Del Giudice M, Puts DA, Geary DC, Schmitt DP, 2019. Sex differences in brain and behavior: Eight counterpoints. *Psychology Today* www.psychologytoday.com>blog> sexual-personalities

Dotto G-P, 2019. Gender and sex – time to bridge the gap. *EMBO Molecular Medicine*: e10668

Graves JAM, 2022. Making evolutionary sense of sex and gender. *Iconoclast: Ideas that have Shaped the Culture Wars*. M. Halloran (ed), Academic Press, Washington, pp. 179–198

Luders E, Gaser C, Narr KL, Toga AW, 2009. Why sex matters: Brain size independent differences in gray matter distributions between men and women. *Journal of Neuroscience* 29: 14265–14270

Morrow EH, 2015. The evolution of sex differences in disease. *Biology of Sex Differences* 6: 5 http://doi.org/10.1186/s13293-015-0023-0

Poani, 2010. *Animal Homosexuality: A Biosocial Perspective*. Cambridge University Press.

Reeser JC, 2005. Gender identity and sport: Is the playing field level? *British Journal of Sports Medicine* 39: 695–699 http://doi.org/10.1136/bjsm.2005.018119

Xircostas ZA, Everingham SE, Moles AT, 2020. The sex with the reduced sex chromosome dies earlier: A comparison across the tree of life. *Biology Letters* 16: 20190867 http://doi.org/10.1098/rsbl.2019.0867

Glossary

I have avoided jargon and acronyms where possible in this book, but some are crucial for clarity. Here I list some scientific terms that might not be familiar to all readers. They are all defined in the text, but you may encounter them further along and need a reminder.

Alleles: different variants of a gene, may be recognized by different phenotypic effect, or by differences in protein or DNA sequences

Allelic exclusion: expression from only one allele of only one gene, often in a multicopy array (e.g., immune genes)

Allopatry: speciation occurring in populations that have become isolated (cf. **Sympatry**: speciation occurring within a population)

Alternative transcripts: RNA transcribed from different start sites, or spliced differentially

AMH, the anti-Müllerian Hormone (also called **MIS, Müllerian Inhibiting Substance**): directs degeneration of Müllerian ducts in male embryos

Amino acids: the chemical subunits of proteins, 20 different types used in nature

Amniote: animal with an amniotic sac during embryogenesis, includes reptiles and mammals

Amplicon: genome region that is present in multiple copies as a result of **gene amplification**

Androgenic, gynogenic: embryo constructed from nuclei, both of which come from the male, or the female parent

Aneuploidy: animals with extra, or deficient chromosomes

Annotation: search of DNA sequence for the hallmarks of genes; open reading frame, signals to denote transcription, splicing, translation

Apoptosis: programmed cell death (note correct pronunciation is apo-ptosis)

Autoregulation: regulation of a gene by its product

Autosomes: all chromosomes other than sex chromosomes

Azoospermic: making few or no sperm

BAC: Bacterial Artificial Chromosomes, large insert DNA clone

Barr body (sex chromatin): condensed inactive X visible at interphase in nucleus of female mammal cell

Bases in DNA: large flat ring molecules made of carbon, nitrogen and oxygen atoms

Base ratio: ratio of A+T or G+C in a stretch of DNA (AT ratio, GC ratio)

Biallelic expression: expression from both alleles of a particular gene in a diploid

Blastocyst: early mammalian embryo, a ball of up to about 100 cells

Brains-and-balls genes: X-borne genes, mutations of which produce mental retardation and gonadal abnormalities

c-banding (<u>c</u>entromere banding): treatment that differentially stains heterochromatin including repetitive sequences at centromere

cDNA, <u>C</u>omplementary <u>DNA</u>: DNA reverse transcribed from messenger RNA, mixture of cDNAs from whole mRNA is a **cDNA library**

Centromere: region of chromosome that maintains attachment of daughter chromosomes after replication and contains kinetochore that attaches to spindle fibres that pull chromosomes apart. Morphologically divides chromosome into short (p) and long (q) arms

Chiasma: (pl. **chiasmata**), the site at which **crossing over** and **recombination** between homologues occurs in meiosis

ChIP-seq (<u>Ch</u>romatin <u>I</u>mmuno<u>p</u>recipitation <u>seq</u>uencing): method for identifying DNA sites that bind particular proteins

Chordate: animals with a notochord, the predecessor of a backbone

Chromatids: daughter chromosomes after chromosome replication

Chromatin: DNA complexed with protein, residing in the cell nucleus. **Euchromatin**: normally stained chromatin. **Heterochromatic**: chromatin staining atypically dark or light

Chromosome: coloured body seen under the microscope, representing a large DNA molecule bound with proteins, a fragment of the DNA genome. Note: a DNA sequence on the computer, no matter how complete, cannot be called a chromosome!

Chromosome conformation capture (3C, HiC): techniques that bind physically close DNA fragments that are sequenced to reveal the spatial relationships of DNA sequences

Chromosome painting: FISH using fluorescence-tagged DNA prepared from isolated chromosomes

Chromosome sorting: physical separation of chromosomes by DNA content

Cis, trans: interaction of genes that are on the same chromosome (in cis), or are on different chromosomes (trans)

Cloning: self-replicating DNA molecules make more of themselves and of the viruses or bacteria they specify, forming a colony of genetically identical individuals (**clone**). If pieces of DNA from, *for example*, human are spliced into these genomes, a colony will contain many copies of this isolated (**cloned**) human sequence

Cohesin: the protein that glues homologous chromosomes together at meiosis

Comparative mapping: mapping orthologous genes in different species

Contig: contiguous stretch of DNA sequence

CRISPR: <u>C</u>lustered <u>R</u>egularly <u>I</u>nterspaced <u>S</u>hort <u>P</u>alindromic <u>R</u>epeats, used to cut out and replace a DNA sequence by another sequence

Deletion: loss of a gene or chromosome region. Lining up deletions produces a **deletion map**

Dimer (polymer): protein with two (or more) subunits

Diploid: cells with two genome copies (e.g., somatic cells of vertebrates)

Distal: further from the centromere

DSD: <u>D</u>isorders (or <u>D</u>ifferences) of <u>S</u>exual <u>D</u>evelopment: genetic conditions which cause atypical sex development

DMR: <u>D</u>ifferentially <u>M</u>ethylated <u>R</u>egions that regulate expression of imprinted genes

DNA: <u>D</u>esoxyribose <u>N</u>ucleic <u>A</u>cid, molecule consisting of two polynucleotide chains wound in a double helix; the stuff of genes

DNA methylation: covalent attachment of a methyl group, usually to the ring of a cytosine base

DNAse: enzyme that breaks down DNA. **DNAse sensitivity** is higher in chromatin that is open and active

Dominant: an allele whose phenotypic effect overrides another (**recessive**) allele

Dosage compensation: mechanism to equalize expression of genes on sex chromosomes in male and female, and balance with autosomes

Double strand breaks: breakage of both DNA strands at same or nearby sites

Dpc: D̲ays P̲ost C̲oitum: measurement of time after fertilization of the egg
Enhancer: DNA sequence that acts on gene promoter(s) to upregulate transcription
Enzyme: protein that catalyzes a biochemical reaction
Epigenetic change, epimutations: somatically heritable changes, not in DNA sequence, but in factors that regulate gene expression
Exome: DNA that represents all the exons in the genome
Exon: protein-coding region of a gene
Eukaryotes: Organisms with a nucleus
Eutheria: placental mammals
Fast X: theory that selection for male-advantage functions encoded by the X leads to faster evolution of genes on the X chromosome
FISH, F̲luorescence I̲n S̲itu H̲ybridisation: hybridization of DNA or RNA sequences trapped on a substrate (e.g., microscope slide) with probe sequences specific for a particular gene. **Fibre-FISH**: microscopic detection of hybridization onto pulled out DNA fibres
Flow cytometry: measurement of chromosomes sorted by DNA content (**chromosome sorting**)
Follicles: cyst-like structures within which oocytes develop
Four core mouse model: comparisons between XX females, XX males, XY females, XY males can ascribe phenotypic differences to genes/chromosomes, or to sex hormones
Frameshift mutations: mutation (e.g., deletion or insertion) that disrupts the reading frame of base triplets and produces nonsense protein, often prematurely terminated
Gb: gigabase, a billion bases (or base pairs), cf. **Mb** (**Megablase**, a million bases) and **kb** (**kilobase**, a thousand bases)
G-banding (G̲iemsa banding): procedure that stains mammalian chromosomes into reproducible patterns of light and dark bands
Gamete: haploid germ cells, either sperm (spermatozoa) or eggs (ova)
Gene conversion: two homologous DNA molecules carrying allelic differences that undergo breakage and repair that results in copying one or other allele
Genetic drift: genome changes that are not selected for or against (neutral), so variation in frequency occurs by chance
Genital ridge: ridge of cells that constitute the bipotential (undifferentiated) gonad
Genitalia: an animal's external organs of reproduction
Genome: all the DNA that specifies an organism, including genes, regulators, junk
Genotype: alleles at a particular locus or loci
Germ cells: cells that will develop into sperm or eggs
GO (G̲enetic O̲ntgeny) terms: standardized representation of properties of gene product, includes molecular function, location, biological process
Gonad: organ in which germ cells are made; the **testis** makes sperm and the **ovary** makes eggs. **Bipotential gonad** refers to the early developmental stage at which the gonad is undifferentiated
Gonadal dysgenesis: atypical development of the reproductive system that leads to infertility and atypical development of secondary sexual differentiation
GSD, G̲enetic S̲ex D̲etermination: sex determination by genes on sex chromosomes
Haldane's Rule states that in an interspecific hybrid, if one sex is infertile or inviable, it is always the heterogametic sex
Haploid: cells with a single genome (e.g., mammal sperm and eggs, adults of many lower eukaryotes)
Haploinsufficiency: mutant phenotype resulting from loss or mutation of one allele, denoting dosage sensitive gene
Helicase: enzyme that unwinds DNA helices

Hermaphrodite: organism with both male and female sex organs

Heterochromatin: chromosome regions that stain differentially because of their atypical base composition. **Facultative heterochromatin:** chromosome regions containing silenced genes that stains differentially only in some tissues or some environments

Heterogametic sex produces two kinds of gamete (e.g., XY male mammals produce sperm with X or Y)

Histones: small basic proteins that associate with DNA to produce the nucleosome structure of chromatin

Histone modifications: post-translational covalent attachment of chemical group such as acetyl, methyl, to the protruding tail of one of the core histones (H2, H3, H4)

Hitchhiking: process by which a Y chromosome carrying a beneficial new mutation is selected along with deleterious mutations of other genes, which hitchhike to a high frequency

HMG box: stretch of 79 amino acids that constitute a homeobox region shared by High Mobility Group proteins

Homoeothermic: animals that maintain a uniform body temperature (e.g., mammals, birds)

Homogametic sex that produces only one kind of gamete (e.g. XX produces only X gametes).

Homologues: two copies of a particular chromosome, one from each parent, that pair and recombine at meiosis

Homology: similarity of genes or chromosomes at the level of gene sequence. **Orthology:** genes (or chromosomes) in different species that derived from the same ancestral gene (chromosome). **Paralogy:** genes in the same species that diverged from a common ancestor (e.g., specialized genes on X and Y chromosome, sometimes called **gametologues**)

Homozygote, heterozygote: diploid organisms may have two copies of the same allele (**homozygous**) or one each of different alleles (**heterozygous**)

Hormones: chemical messengers made in one tissue with effects on other tissues of the body. **Sex hormones** comprise **polypeptide hormones** (made of proteins) and **steroid hormones** (with four rings made of carbon) like androgens and oestrogens

HY antigen, HYA: detected by immunizing female mice with cells from male mice of the same strain, and using antibody to detect male-specific proteins

Hybrid: progeny of parents that differ genetically; **species hybrids**: progeny of male from one species and female from another

ICSI: Intracytoplasmic Sperm Injection, a method of assisted reproduction

Imprinting: difference of activity of the maternal and paternally derived alleles of a gene, imposed by an **imprint** that is retained through somatic growth, occasionally passed to offspring

Imprinting Control Element (ICE): controls transcription of imprinted genes

Introgression: incorporation of allele(s) from a different species into a population

Intron: non-coding sequence within a gene that is excised ("**spliced**") from messenger RNA

Karyotype: chromosomes of a cell cut out and pairs arranged in order of size

Kinetochore: part of centromere that attaches to spindle fibres that separate daughter chromosomes at cell division

Kilobase (kb): 1,000 bases (base pairs)

Knockout: targeted inactivation or removal of a particular gene or sequence

Leydig cells: steroid-producing cells of the testis

LINE elements: Long Interspersed Nuclear Elements are a group of repetitive retrotransposed elements widespread through eukaryote genomes

Linkage: relationship between genes that are close together (**linked**) on the same chromosome so alleles tend to be inherited together. Recombination between markers is measured in centimorgans (1 **cM** = 1% recombination), and a **linkage map** constructed

Locus: gene or gene group responsible for a phenotype

Longread sequencing: sequencing of isolated single, long DNA molecules

Macrochromosomes and **Microchromosomes:** large and tiny chromosomes in reptiles, birds

Maternal, paternal: allele coming from the mother or father

Marsupials, Metatheria, therian mammals that diverged from eutherians 148 Mya

Megabase Mb: a million bases (base pairs)

Meiosis: reduction division for making haploid gametes from diploid cells

Metagenomics: DNA sequencing of an assemblage of organisms

Methylome: sequence of the genome including positions of methylcytosine

MHM, Male Hyper Methylated region on the bird Z, thought to be involved in dosage compensation

Microsatellite: different numbers of tandem repeats generating fragments of different sizes that can be detected by PCR or sequencing

Mitosis: division of somatic cells involving chromosome duplication and segregation

Monoallelic expression: transcription of only one of the two alleles at a locus in a diploid cell

Monotremes: egg-laying mammals confined to Australasia, prototherian mammals such as **platypus and echidna**, that diverged from therians 166 Mya

Monophyletic: derived from the same common ancestor

Mosaic individuals have a mixture of cells with different genetic characteristics; for instance XX and XO cells

mRNA, Messenger RNA: RNA copy of a gene, from which introns have been spliced out

MSCI: Meiotic Sex Chromosome Inactivation of the mammalian X and Y chromosomes during male meiosis in therian mammals

MSUC: Meiotic Silencing of Unpaired Chromosomes

MSY Male-Specific region of the Y chromosome

Müller's ratchet: the process of Y degradation in which the class Y chromosomes with 0, 1, 2... mutations are progressively lost from a population by genetic drift

Mutation: change in the DNA sequence of a gene, includes base substitution, deletion, insertion

Mya: Millions of years ago

N-terminal, c-terminal: the start and the end of the protein. The first to be translated has a free amine group (NH_2), the last a free carboxyl group (COOH)

ncRNA: noncoding RNA that is transcribed from DNA but not translated into protein. There are several types including **long ncRNA, micro RNA, piRNA**

Northern blots: separation of mRNA molecules by size on a gel, blotted to a nylon substrate, hybridized with radioactive probe and autoradiographed to reveal a particular mRNA

Nucleolus: accumulation of ribosomal RNA and protein forming a dark body in the nucleus

Nucleolar Organiser Region (NOR): position on a chromosome of genes that are transcribed into ribosomal RNA

Nucleosome: components of chromatin, structures composed of DNA wound around a core of histone proteins (H2A, H2B, H3, H4)

Ohno's hypothesis states that the first step of the evolution of X inactivation was the upregulation of X-borne genes in males to maintain parity with the expression of interacting autosomal genes. This resulted in overexpression of X genes in females, countered by X inactivation

Ohno's Law states that the gene content of mammal X chromosome is conserved because disruptions to X chromosome inactivation would be fatal

Oogonia, oocyte, ovum: the stages of development of a mature egg from primordial germ cells

ORF, <u>O</u>pen <u>R</u>eading <u>F</u>rame: a region without stop codons that could be translated into protein

Outgroup: species that are more distantly related that can be used as a reference when comparing two more closely related species

Palindrome: sequence that reads the same forward as backward

Parental: allele coming from the mother (**maternal**) or father (**paternal**)

Parthenogenesis: asexual reproduction, in which a female makes diploid eggs without sperm

Pathogen: an agent that causes illness

Phenotype: the visible or assayable characteristics of an individual; for instance eye colour, disease state

Ploidy: number of copies of the genome. **Haploid** cells (including sperm and eggs) have a single copy. **Diploid** cells (as in human somatic cells) have two copies. Some cells, or some organisms (often plants) may have three (triploid) or four (tetraploid), or even higher numbers of genomes (**polyploid**)

Pluripotent, totipotent: cells that retain the ability to differentiate into many, or all, cell types of the body

Poly A addition: addition of a string of adenine residues to messenger RNA molecules, **polyadenylation** is one feature of a gene

Polygenic: controlled by two or more genes acting together

Polymerase: enzyme that adds chemical units (bases) to a growing chain of DNA (**DNA polymerase**) or RNA (**RNA polymerase**)

Polymerase chain reaction (PCR): use of pairs of short DNA 'primers' to synthesize many copies of the DNA between them

Polymorphism: normal variant of a gene that is reasonably frequent in a population (**polymorphic**)

Positional cloning: identification and isolation of a gene by mapping its position on a chromosome

Primordial germ cells (PGC): cells that move into the developing gonad and become eggs or sperm

Promoter: sequence upstream of a gene that controls its transcription

Prosimian: branch of primates including lemurs, tarsiers, distantly related to human

Proteome: catalogue of all the proteins in a cell or tissue

Proto-sex chromosomes: autosome pair that differentiates into pair of sex chromosomes

Prototheria: monotreme mammals

Proximal: toward the centromere

Pseudoautosomal region (PAR): pairing region shared by X and Y, or Z and W. PAR genes show an autosomal pattern of inheritance. The boundary between the PAR and the X- and Y- specific regions is called the **pseudoautosomal boundary**

Pseudogene: inactivated gene whose DNA sequence is no longer translated into functional protein

Radiation hybrids: hybridization of irradiated human somatic cells with mouse cells and correlation of the presence/absence of human genes to make **radiation hybrid maps**

Recessive: an allele whose phenotypic effect is overridden by another (**dominant**) allele

Recombinant DNA: DNA from different sources that has been chemically spliced together; often human DNA inserted into viral or bacterial genomes

Recombination: breakage and rejoining of DNA molecules to make new combinations of alleles

Retinoic acid: metabolite of vitamin A, interacts with *STRA8* to control entry of germ cells into meiosis

Reverse transcription: copying of RNA back into DNA to make cDNA

RNA FISH: Fluorescence in situ hybridization of labelled probe to primary RNA transcript trapped at its chromosomal site of origin

RFLP, Restriction Fragment Length Polymorphism: restriction fragments are lengths of DNA produced by **restriction enzymes** (called **endonucleases**) that cut DNA at defined sequences that may be different for two alleles. They can be lined up to produce a **restriction map** of a region

RNA: Ribose Nucleic Acid, single-stranded polynucleotide. In eukaryotes, RNA is transcribed from DNA and is involved in gene expression

RNA-seq: technique for assessing the presence, and the quantity of RNA transcripts from every gene in the genome in a particular tissue at a particular developmental stage

Ribosomes: cytoplasmic bodies that translate messenger RNA into protein, composed of **ribosomal RNA (rRNA)** and protein

Secondary structure: DNA sequences that are repeated or inverted can form loops when bases on the same strand pair, forming cross-shaped (**cruciform**) structures

Segmental duplications: large genome regions that are copied, inserted and rearranged

Selective sweep: selection of, e.g., a Y chromosome with a beneficial mutation, displacing other Y haplotypes

Sequence Tagged Site, STS: short stretch of DNA with unique sequence

Sertoli cells: cells of the embryonic gonad that specify testis development

Sex-biased genes: genes that are more active in one sex than the other

Sex chromosomes: a pair of chromosomes that are different in male and female, defined by a gene that determines sex

Sex chromatin body, sex vesicle: X and Y chromosomes in meiosis forms a condensed **(Barr) body**

Sex determination: direction toward male or female development of an embryo

Sexual differentiation: development of the sexual characteristics of an animal

Sex linkage: pattern of inheritance in families of genes on a sex chromosome

Silent sites (synonymous substitutions) at which a base change occurs that doesn't affect the amino acid encoded. Ka/Ks (the ratio of non-synonymous to synonymous substitutions for a gene) gives an idea of whether this gene is under selection

SINE, Short Interspersed Nuclear Elements: non-coding sequences of 100–700 bp, derived from transposable elements and repeated all over the genome

SNP Single Nucleotide Polymorphism: alleles differ by a single base pair in a sequence

Somatic: cells of the body

Somatic cell genetics: genes can be mapped by fusing somatic cells from different species and correlating retained genes and chromosomes

Southern blotting: a procedure invented by Ed Southern to detect and separate out particular DNA sequences from a mixture. DNA fragments are separated by size on a gel, blotted to a nylon substrate and autoradiographed to reveal a particular fragment(s)

Spermatogonia, spermatocytes, spermatids: the stages of sperm formation from primordial germ cells

Spindle: the molecular machine that separates chromosomes at mitosis and meiosis, made of protein fibres that join to centromeres of daughter chromosomes

Stem cells: undifferentiated cells that are pluripotent or totipotent, including Embryonic Stem Cells (**ESC**) Embryonic Germ Cells (**EGC**), as well as Induced Pluripotent Stem Cells (**iPSC**) reprogrammed from differentiated cells

Strata: evolutionary strata on the Y chromosome that ceased to recombine with the X at different times, detected by step differences in the ratio of non-synonymous and synonymous mutations

Stochastic: subject to random variation

Synapsis: pairing of homologous chromosomes at meiosis by means of a **synaptonemal complex** that zippers them together

Sympatry: speciation occurring within a population (cf. **Allopatry**)

Synteny: co-retention of genes anywhere on the same chromosome

T2T: telomere-to-telomere sequence that spans a whole chromosome

TAD, Topologically Associated chromatin Domain, identified by chromatin configuration capture

Tandem repeats: short sequences repeated side-by-side

Telomeres: protein/RNA structures that stabilize ends of chromosomes

Testis cords: structures that form in the embryonic testis and surround germ cells

Testis Determining Factor, TDF (Tdy in mice): male determining factor on the Y chromosome

Transcript: RNA copy of a DNA base sequence, **transcription**

Transcriptome: genome wide analysis of RNAs in a tissue

Transfer RNA (tRNA): small RNA molecules that attach to specific amino acids and recognize the appropriate base triplet in mRNA

Transgenic: insertion of a gene from one organism into the genome of another

Translation: sequence of bases on an mRNA molecular is read in triplets into a sequence of amino acids in a protein product

Transmembrane protein: forms channel in the lipid bilayer that allows passage of molecules between nucleus and cytoplasm, or between cells

Trophoblast: cells of the early embryo that develop into part of the placenta

Translocation: two chromosomes swap regions

TSD, Temperature Sex Determination: determination of male or female development in response to incubation temperature; more generally **ESD, Environmental Sex Determination**

Upstream sequences refers to DNA further from the transcription start site at the 5' end of a gene

Vector: self-replicating DNA molecule (usually from virus or bacteria) that can accommodate and replicate inserted bits of DNA from other sources

Vertebrates: animals with backbones, including mammals, reptiles (including birds), amphibians and fish

Western blot: separates proteins by size on a gel, blotted onto a filter and detected with specific antibody

XCI, X Chromosome Inactivation: genetic silencing of one X in somatic cells of a female mammal

XCR/XAR: X Conserved Region (XCR) maps to the X in all therians and represents the ancient mammal X. X Added Region (XAR) was fused to the X in a eutherian ancestor

XIC, X Inactivation Centre: region of the mammal X chromosome containing gene(s) that control X inactivation

YAC: Yeast Artificial Chromosome, large fragment of DNA inserted into yeast genome

YCR/YAR: Y Conserved Region (YCR) maps to the Y in all therians and represents the ancient mammal Y. Y Added Region (YAR) was fused to the Y in a eutherian ancestor

Zygote: fertilized egg, diploid

Nomenclature – Genes, Proteins and Diseases

Here I explain how genes, chromosomes and chromosome regions (and diseases) are named and list those that are an essential part of the story.

Chromosome Nomenclature

Chromosomes of a cell are photographed, then cut up and lined up and numbered in order of size (karyotype). In a diploid species there are two copies of each chromosome (one from the mother and the other from the father).

The shape of chromosomes at mitosis depends on the position of their centromere; a central location produces an X shape (metacentric), and a position nearer to one end produces a short arm ('p') and a long arm ('q'). Location of the centromere near or at one end produces a hairpin shape ('acrocentric' or 'telocentric'). Chromosomes are arranged with their short ends up.

Position within chromosome arms is relative to the centromere; genes that are nearer are termed 'proximal' and further from the centromere are 'distal'. A uniform nomenclature that refers to g-band patterns relative to the centromere has been developed for human chromosomes, so position is given, for example, as '15q11.8' or '2p2.3'.

Chromosome sequence (in megabases) is read from the top (that is, the terminus of the p arm).

Gene Names

Here I list the genes that are significant in the story of sex chromosome evolution or function. I have not included the many genes that get just a cursory mention, for instance, as markers for mapping.

Generally, gene names are italicized, and the proteins they encode are not. Human genes are written as upper-case italics. This has been adopted for all mammals except rodents, which for historical reasons use sentence case italics. Other vertebrates have their own conventions; reptiles, amphibians and fish have settled on lower-case italics. I will underline letters in the gene name that are used in the gene symbol.

The names given to genes need some explanation. Genes are often named for the phenotype specified by mutant alleles, although the gene's normal function is to supply a product that is missing or inactive in the mutant. For instance, the job of the normal *DMD* allele is to make a protein (dystrophin) that strengthens muscles by binding fibres to an extracellular matrix. Mutant *DMD* alleles make defective, or no, dystrophin and cause <u>D</u>uchenne's <u>m</u>uscular <u>d</u>ystrophy.

Many other genes are named for mutant phenotypes in the organism in which they were first discovered (often the fruit fly *Drosophila melanogaster*). For instance, *SPEN* was named for homology with an RNA-binding protein, mutation of which causes <u>sp</u>lit-<u>en</u>ds of bristles on flies. My personal favourite is the poor *flightless* mouse.

A considerable effort was put into ensuring that X and Y homologous genes had sensible names that distinguished their location, for instance, *ZFX/ZFY*. However, several genes flout this convention; for instance, *SMCX/SMCY* became, unhelpfully, *KDM5C/KDM5D* when its product was discovered to be a lysine (<u>K</u>) <u>dem</u>ethylase.

AMELX/AMELY: <u>Amel</u>ogeninin (tooth enamel) paralogues on the X and Y chromosomes
AR: X-borne gene encodes <u>A</u>ndrogen <u>r</u>eceptor protein
ATRX/ATRY: <u>A</u>lpha-thalassemia/mental retardation, <u>X</u>-linked, with a Y copy in marsupials
AZF: <u>Az</u>ospermia <u>f</u>actors, loci identified on the Y chromosome
β-catenin: blocks testis and promotes ovary differentiation
Bkm: <u>B</u>anded <u>k</u>rait <u>m</u>inor repetitive sequences on the mouse Y chromosome
BMP: <u>B</u>one <u>m</u>orphogenetic <u>p</u>rotein, active in development
CBX2: <u>C</u>hromo<u>box</u> homolog 2, chromatin remodelling factor, involved in developing bipotential gonad
CT: <u>C</u>ancer-<u>t</u>estis genes
CTCF: encoding a protein factor binding to DNA sequence CCCTC
CYP16B1: <u>Cy</u>tochrome <u>P</u> family member encodes an enzyme that breaks down retinoic acid and controls meiosis
CYP19B: <u>Cy</u>tochrome <u>P</u> family member that encodes aromatase, the enzyme that makes oestrogen
DAX1: <u>D</u>osage sensitive sex reversal, <u>a</u>drenal hyperplasia critical region, <u>X</u> chromosome gene <u>1</u>, codes for a nuclear receptor, affects gonad differentiation
DAZ: <u>D</u>eleted in <u>Az</u>oospermia; repeated genes on the human Y associated with male infertility/azoospermia. Has a paralogue *DAZL* on chromosome 4 (the original source of DAZ)
DHH: <u>D</u>esert <u>H</u>edge<u>h</u>og, a growth factor important for testis development. Named for spiny bumps on *Drosophila* cuticle in mutants
DMD: X-borne gene encoding dystrophin, required by muscle cells. Delections cause <u>D</u>uchenne <u>m</u>uscular <u>d</u>ystrophy

***DMRT1*:** Doublesex Mab3-related transcription factor 1, the bird sex-determining gene, related to sexual differentiation genes *doublesex* in *Drosophila* and *mab3* in *Caenorhabditis elegans*

DNMT1,3,4: DNA methyltransferases that add methyl groups to cytosine residues in DNA

***Dsx*:** *doublesex* promotes female development in *Drosophila*

***DXZ4*:** DNA sequence defining the hinge between two large domains on the inactive X chromosome

***EIF1AX/EIF1AY*:** Eukaryote initiating factor, X and Y copies code for translation initiation factor

F8, F9: genes coding for two protein blood clotting factors

FGF9: Fibroblast growth factor 9, important for maintaining *SOX9* transcription

FIRRE: Functional intergenic repeating RNA element, X-borne noncoding gene transcribed from inactive X that affects its conformation

FMR1: Fragile X mental retardation, caused by amplification of triplet repeats

FOXL2: Forkhead box transcription factor L2, maintains ovary function

GATA4: transcription factor that binds the nucleotide sequence GATA, involved in development of bipotential gonad

G6PD: Glucose-6-phosphate dehydrogenase, classic enzyme gene on the X

***GBY*:** locus on the Y that causes gonadoblastoma in dysgenic testis (possibly *TSPY*)

***GLA*:** classic X-borne gene that encodes enzyme alpha-galactosidase

***HEMGN*, hemogen:** involved in development of bipotential gonad in birds

Histone modifying enzymes: see Figure 6.3

Hormones: synthesized by enzymes including **5-α reductase** and **aromatase**

HPRT* (previously *HGPRT): Hypoxanthine guanine phosphoribosyl transferase, deficient in Lesch-Nyhan disease

HYA: Human leukocyte antigen, gene(s) encoding male-specific proteins detected by immunising female mice with male cells

***IRF9*:** Interferon regulatory factor 9), ancestor of fish sex-determining gene *Sdy*

***JARID2*:** member of Jumonji gene family, encodes a component of PRC2 (Polycomb repressive complex)

JMJD1A (alt. ***KDM3A):*** member of the Jumonji gene family, encodes enzyme that removes methylation from histone and affects *SRY* transcription

JMJD3*:** (alt. ***KDM6B) demethylates H3K27 to accomplish epigenetic remodelling

***KDMC5/ KDMD5*:** histone lysine demethylase (H3K4 (originally *SMCX/SMCY*)

MECP2: methyl-CpG binding protein 2, regulates transcription of methylated DNA

Msl, *male-specific lethal* locus in *Drosophila* that specifies male development, and is repressed by ***Sxl*** (sex lethal) in females

PGK: Classic X-borne gene encoding enzyme phosphoglycerate kinase. *PGK1* lies on the X, *PGK2* is an autosomal paralogue

Pluripotency genes: *Oct4, Sox2, Klf4* and *c-Myc*

***RBMX/RBMY*:** RNA binding motif protein on the X and Y. *RBMX* retains its original function in RNA splicing and *RBMY* has evolved into a spermatogenesis gene

RCP: X-borne gene that encodes Red cone pigment. Mutations cause colour blindness

RNF12: Ring finger protein family with RING domain (really interesting new gene) attaches ubiquitin to target proteins. X-borne member binds to *XIST* and mediates effects of differentiation

***roX1, roX2*:** *Drosophila* genes that transcribe ncRNA and affect dosage compensation

RSPO1: R-spondin 1, involved in ovary differentiation

RSX: RNA on the silent X: gene that produces a long noncoding RNA that builds a silencing domain on the inactive X of marsupials

***Sdy*:** Sexually Dimorphic on the Y), sex-determining gene in salmonid fish
***SF1*,** Steroidogenic factor 1: member of the Nuclear Receptor family, so it is also called *NR5A1*
***SHOX1*:** Short stature homeobox-containing gene in the pseudoautosomal region
***SMCHD1*:** structural maintenance of chromosomes containing flexible hinge domain
***SOX* gene family:** *SRY* related, HMG box containing. Includes *SOX3*, the conserved X-borne gene that evolved into *SRY*, and *SOX9*: an autosomal gene encoding protein essential for male sex determination as well as cartilage development
SRY: Sex Region on the Y chromosome, the mammalian sex-determining gene
STAT: signalling protein in fish sex determination pathway
STRA8: Stimulated by retinoic acid, controls entry of germ cells into meiosis
STS: Steroid sulfatase; deficiency causes the disease ichthyosis
TES: Testis enhancer sequence
TESCO: *TES* core element
***Tra*:** Transformer gene promotes female development in *Drosophila*
***Tsix*:** antisense to *Xist* in mouse, transcribed into noncoding RNA
TSPX/ TSPY: Testis-specific Y-encoded protein 1, copies on the X and Y
UBE1X/ UBE1Y: Ubiquitin-like modifier activating enzyme 1, copies on X and Y involved in protein folding and degradation
***USP9X/ USP9Y*:** Universal stress protein, copies on X and Y code for deubiquinitase, mitigates environmental stress
UTX/ UTY: Ubiquitously transcribed tetratricopeptide, X and Y borne genes that code for histone lysine demethylase (H3K27)
WNT4: wingless in *Drosophila*, activates mammalian ovary differentiation
***WT1*,** Wilms' tumour suppressor 1: gene involved in development of bipotential gonad, +KTS and -KTS forms
XACT: X-active specific transcript, human X-borne noncoding gene, competes with *XIST* to coat inactive X
***XGA*:** X-borne gene that encodes surface protein recognized as a blood group
***XIST*:** X-inactive specific transcript; gene that produces a long noncoding RNA that builds a silencing domain on the inactive X of placental mammals
***Xite*:** X-Inactivation Intergenic Transcription Elements mouse noncoding gene downstream of *Xist*, boosts *Tsix*, regulates *Xist*
XLMR: X-linked mental retardation genes
ZFY* (**and X-borne partner ***ZFX): Zinc finger gene on the Y (X). Initially claimed to be the sex-determining gene, *ZFY* may function in spermatogenesis or control of autosomal gene activity

Protein Names

Cohesin: protein that sticks homologous chromosomes together at mitosis and meiosis
DCC: Dosage Compensation Complex: protein complex that regulates structure and transcription of fruit fly X chromosomes
LBR: Lamin B Receptor, transmembrane protein
PRC (PRC1, PRC2): Polycomb repressive complexes, involved in epigenetic silencing
Protein kinases: enzymes that phosphorylate proteins
RNA polymerase II: enzyme that adds units to RNA during transcription from DNA
YY1: Yin-yang tethers *XIST* to its site of transcription

Human Diseases

Androgen Insensitivity Syndrome, AIS: female development of XY embryo with mutation in the *AR* androgen receptor gene

Colour blindness: red-green colour blindness caused by mutation/deletion in pair of RCP and GCP genes on the X

CAH, Congenital adrenal hyperplasia: life-threatening hormone deficiency causing salt loss as well as atypical sexual development

CD, Campomelic dysplasia: caused by deletion/mutation of *SOX9* gene, affects long bone development and sex determination

Cystic fibrosis: autosomal recessive respiratory disease caused by mutation of a gene that encodes an ion channel that regulates mucus in the lungs

Duchenne (Becker) Muscular Dystrophy: Progressive muscle weakness and early death caused by mutation in the X-borne *DMD* gene that encodes dystrophin, usually deletions that shift the reading frame. Becker is a less severe disease caused by in-frame deletions

Haemophilia: blood clotting disease caused by mutation in one of the genes that encode proteins in the blood clotting cascade; *F8* and *F9* lie on the X chromosome

ICF Syndrome: Immune deficiency, Centromere instability and Facial dysmorphology) caused by reactivation of genes on the inactive X caused by mutation of autosomal gene *DMNT* encoding DNA methyltransferase

Ichthyosis: X linked 'fish skin' disease caused by mutation in *STS* gene that encodes steroid sulfatase

Lesch-Nyhan Disease: X linked deficiency of enzyme HPRT causes mental retardation, behaviour and muscle problems

Prader-Willi Disease: (eating disorder, mild mental retardation) and **Angelman Syndrome**: (ataxia and severe mental retardation) are caused by deletions of a region from either the paternal, or the maternal chromosome

Testicular feminization: caused by mutation in the androgen receptor gene *AR*

Rett Syndrome: neurological and developmental disorder caused by mutation of X-borne *MECP2* gene, characterized by loss of language and motor skills

Index

For EU product safety concerns, contact us at Calle de José Abascal, 56–1°,
28003 Madrid, Spain or eugpsr@cambridge.org.